Best wishes

Denn L. Weam

A COURSE OF

LECTURES

IN

Natural Philofophy.

By the late

RICHARD HELSHAM, M. D.

Profeffor of PHYSICK and NATURAL PHILOSOPHY
in the Univerfity of DUBLIN.

PUBLISHED BY

BRYAN ROBINSON, M. D.

The FOURTH EDITION

LONDON:

Printed for J. NOURSE, oppofite *Katherine-Street* in the
Strand, Bookfeller in Ordinary to his MAJESTY.

M.DCC.LXVII.

This reprint of Richard Helsham's *Course of Lectures on Natural Philosophy* was taken almost entirely from a copy of the fourth edition (1767), kindly donated by Trinity College Dublin by Norman McMillan.

Its publication to mark the Millennium also benefited from the Academic Development Fund of the Provost of Trinity College and the award of a visiting fellowship to D A Attis by the National Science Foundation and the Social Science Research Council.

Editors: D Weaire, Trinity College Dublin
                  P Kelly, Trinity College Dublin
                  D A Attis, Princeton University

Published by the Physics Department of Trinity College Dublin, in association with

Institute of Physics Publishing, Bristol, UK

Verlag MIT, Bremen, Germany

ISBN 1 898 706 17 4

Helsham and the Rise of Newtonian Physics

Physician and Physicist

Richard Helsham was born near the town of Kilkenny, probably in 1682.[1] He received his early education at the distinguished Kilkenny College where Jonathan Swift and William Congreve had studied a decade and half before and where George Berkeley was a fellow student. In 1698 Helsham entered Trinity College, where he performed well enough to win a scholarship in 1700 and a bachelor's degree in 1702. Shortly after, in 1704, he was elected a Ffellow of the College. At that time the Laudian statutes of 1637 ostensibly governed the course at Trinity. They laid out a typically scholastic course of reading with an emphasis on Aristotelian logic and metaphysics and the study of Greek and Roman literary models.

By the beginning of the eighteenth century, however, even this conservative university could not ignore the revolutionary scientific developments of the previous century. A letter from a Trinity student in 1703 indicates that by then the course included the ideas of such modern thinkers as Descartes, Gassendi, Malebranche and Locke.[2] Mechanical and experimental philosophy had gradually begun to make their appearance, and Helsham himself would soon become a major force in promoting this trend.

---

[1] For his life see DNB, McMillan, and Kirkpatrick, pp. 79-82.
[2] McDowell and Webb, pp. 31-32.

In the two decades before Helsham went up to Trinity, Dublin's intellectual culture had been transformed. Indicative of the change in atmosphere was the foundation of the Dublin Philosophical Society in 1683.[3] It was modeled on the Royal Society of London (founded in 1660). In fact, many of the original members of the Dublin Society were or became members of the Royal Society and published in its *Transactions*. William Petty, the first president of the Dublin Society, had been an original member of the Royal Society. William Molyneux, generally regarded as the founder of the Dublin Society, also had close ties with the English scientific community. He corresponded with John Locke and published an important work on optics.[4] He also translated some of the great scientific works of Descartes, Galileo and Torricelli. Together with Petty and Molyneux, St. George Ashe was also a major proponent of what was often called "the New Learning." He contributed to the *Transactions* of the Royal Society, established a small observatory at Trinity College Dublin in 1685, and, as Provost of Trinity, helped to introduce science into the course at Trinity. On Molyneux's advice, he ordered all of the bachelors to read Locke's *Essay Concerning Human Understanding*. Ashe was Jonathan Swift's tutor and may have been the first to introduce him to the science he so successfully parodied in *Gulliver's Travels*. Interestingly, it was also Ashe who ordained Berkeley for the Anglican priesthood.

Of course, in addition to growing scientific activity this was also a period of political upheaval. The appointment of Tyrconnell as Viceroy in 1687 forced the Dublin Society to disband as members fled to England and the Jacobite army occupied Trinity College. In 1693, the Society was reestablished, and Ashe allowed members to meet at the Provost's House. The Society once again ceased to meet in 1698, on the death of William Molyneux, but his son, Samuel Molyneux, briefly revived it in 1707. Helsham, Berkeley, and Bryan Robinson (the editor of Helsham's lectures) were all involved with the society during this final period, though scant records make it impossible to determine the full scope of their

---

[3] See especially Hoppen.
[4] See Kelly.

participation. The ultimate collapse of the Dublin Philosophical Society marked the end of the first period of sustained interest in science in Dublin. The next such period would not begin until the creation of the (Royal) Dublin Society in 1731, just before Helsham's death.

The School of Physic and the New Laboratory

One group with both intellectual and professional interests in the new science was that of the physicians. After finishing his MA degree in 1705, Helsham decided to follow a medical career. He earned the degree of doctor of medicine in February of 1709/10 and became a fellow of the Royal College of Physicians in Ireland in October of 1710. Chartered in 1667, the Royal College of Physicians had been created as the medical school for Trinity College. By the time Helsham arrived, however, it had become essentially independent, forcing the authorities at Trinity to make new arrangements for the instruction of students in medicine. To that end, the Board of Trinity decided to establish a new medical school, ordering in June of 1710, "that ground be laid out in the south east corner of the physick garden sufficient for erecting an Elaboratory and an anatomical theatre thereupon."[5] The physic garden, originally the kitchen garden, had been converted to grow medicinal plants. The Board soon voted a sum of one hundred pounds for the new laboratory to serve as the core of a new medical school. In August of 1711 the building, which included a chemical laboratory, lecture room, dissecting room and a museum, was opened. As part of the foundation of the school, Trinity also established lectureships in anatomy, chemistry and botany. Befitting this first attempt by Trinity to institutionalize the teaching of science, the opening ceremony included scientific disquisitions by the new lecturers. The Professor of Physic, Thomas Molyneux, also spoke, and Richard Helsham gave a lecture on natural philosophy. He appears to have lectured regularly on the subject from this point on, although it was to be over a decade before he would receive official recognition for his work.

---

[5] Wormell, p. 21. See also Kirkpatrick, pp. 76-78.

The Circle of Jonathan Swift

Meanwhile, as Helsham was lecturing to medical students on natural philosophy, his own medical career was taking off. He was elected president of the Royal College of Physicians in 1716 (a position he held again in 1725). After Jonathan Swift's return to Dublin as Dean of St. Patrick's in 1714 Helsham became his personal physician. (Swift later described him as "the most eminent Physician of this City and Kingdom."[6]) Another member of Swift's social circle and a close friend of Helsham's was Patrick Delaney, a Fellow of Trinity and the first professor of Oratory and History. Like Helsham, Delaney shared Swift's Tory politics and his opposition not only to the ruling Walpole administration but also to the Whig Provost of Trinity College, Richard Baldwin. Fearful of possible Tory sympathy for the Jacobite cause, Whigs like Baldwin enforced a narrow political orthodoxy. At Trinity students were expelled for tongue-in-cheek toasts to William III or for 'decorating' his statue in College Green. Baldwin ruled Trinity with an iron hand as provost from 1717 to 1758, attracting the ire of Helsham and his Tory friends. McDowell and Webb, the principal historians of Trinity College in recent times, describe Helsham, Delaney and other members of Swift's circle as, "a group of men who had been driven by the narrow and heavy Whig orthodoxy represented by Baldwin into rather futile and cantankerous political opposition, and trivial punning and parlour games."[7] One example of their playful political opposition was *A Long History of a Certain Session of a Certain Parliament, In a Certain Kingdom*, written by Helsham and Delaney and published anonymously in 1714. Helsham also published *An Humble Remonstrance In the Name of All the Lads in All the Schools of Ireland, Where Latin and Greek are Taught; And of the Young Students Now in the University of Dublin, Together With a Protest of All the Senior Fellows in Trinity College, Dublin (Except One) Against the Provost* (Dublin 1727-8). In it, Helsham and a number of the other Fellows (including Delaney) protested against

---

[6] Williams, Vol. IV, p. 360.
[7] McDowell and Webb, p. 38.

Baldwin's autocratic style of governance. The situation grew so bad that Delaney was eventually forced to resign his fellowship after being publicly humiliated by Baldwin.

In addition to their politics, Helsham and Delaney were also known for their congenial hospitality. They built a villa about a mile north of Dublin in Glasnevin where they entertained Dublin's literati. It was here that Mrs. Pendarvis, the future Mrs. Delaney, first met Swift, Delaney and Helsham, whom she described as "a very ingenious entertaining man."[8] The villa, she explains, "was known as Hel-Del-Ville, formed from the initial syllables of the names of the proprietors, to intimate their joint property in the place, but the *first* was soon dropped, as having a *strange association*."[9] Helsham flourished in this atmosphere of intellectual conversation and easy sociability.

Professor of Natural and Experimental Philosophy

While Helsham lectured unofficially in natural philosophy from the opening of the laboratory in 1711, his first official position was the Donegal Lectureship in Mathematics, which he held from 1722/3 to 1730. Just before his appointment, the Board of the College voted to spend one hundred pounds on instruments for the teaching of natural philosophy, some of which would presumably have been for Helsham's use. In 1724, after thirteen years of teaching his course of experimental philosophy without payment, the College of Physicians voted to reward him with a gold plate. At the same time, Trinity appointed him as the first holder of the new Erasmus Smith Professorship of Natural and Experimental Philosophy, the College's first professorship in the sciences.[10] With his new position, Helsham could continue to deliver the lectures on natural philosophy as he had for over a decade. He, like many other Trinity dons, seems to have had little interest in publishing the results of his work. In fact, McDowell and Webb were unable to find any publications by current fellows of

---

[8] Llanover, Vol. I, p. 396.

[9] Llanover, Vol. I, p. 398.

[10] See Holland.

the college during the period 1722 to 1753. Thus began Trinity's reputation as the 'Silent Sister' of Oxford and Cambridge.

Though he resigned his fellowship in 1729/30, upon his marriage to Jane Putland, Helsham served as Regius Professor of Physic in 1729/30. He seems to have been active outside the university as well, putting his knowledge to use for the good of the city. In 1737 Dublin Corporation granted Helsham the freedom of the city for "his readiness to assist this citty with respect to the being better supplied with pipewater."[11]

Even in death, Helsham played the man of science, leaving orders that his body be dissected. The results of the postmortem were reported in the *Gentlemen's Magazine*:

> It was imagin'd that his disorder proceeded from a twisting of the guts, and he took quicksilver, which proved ineffectual. He desired that his body might be opened for the benefit of mankind, which being done there was found in one of his guts an excresence of three pieces of Flesh, the smallest as large as a hen's egg, and resembling the Flesh of the liver.[12]

His will also included explicit and bizarre instructions for the treatment of his corpse:

> As to my funeral it is my will (and I do adjure my executor not to fail in the execution of it) that before my coffin be nailed up my head be severed from my body and that my corps be carried to the place of burial by the light of one taper only at the dead of night without Herse or Pomp attended by my Domesticks only.[13]

---

[11] Kirkpatrick, p. 80.

[12] Quoted in Kirkpatrick, p. 82.

[13] Quoted in Kirkpatrick, p. 82.

Helsham also left something of much greater value for science, the notes to the lectures on natural philosophy he had been delivering for over twenty-five years.

A Course of Lectures on Natural Philosophy

After Helsham's death, his pupil Bryan Robinson (who would later succeed him as Professor of Physic at Trinity) took up the task of editing Helsham's lectures on natural philosophy for publication. The *Course* would eventually go through eight editions in Dublin, London and even Philadelphia, the last as late as 1834 The publication history is as follows:

> *A Course of Lectures in Natural Philosophy By the late Richard Helsham, M.D., Professor of Physick and Natural Philosophy in the University of Dublin,* Published by Bryan Robinson, M.D. (Dublin: Printed by R. Reilly at the University Press And Sold by G. Risk, G. Ewing, and W. Smith, Booksellers in Dame-street, 1739)

> *A Course.........* (London: Printed for John Nourse at the Lamb without Temple Bar, 1739)

> *A Course.........* Second Edition (London: Printed for J. Nourse, at the Lamb, without Temple-Bar, 1743)

> *A Course.........* Third Edition (London: J. Nourse, 1755)

> *A Course.........* Fourth Edition (London: Printed for J. Nourse opposite Katherine-Street in the Strand, Bookseller in Ordinary to his Majesty, 1767)

> *A Course.........* Fifth Edition (London: Printed for J. Nourse, in the Strand, Bookseller to his Majesty, 1777)

> *A Course.........*Sixth Edition (Dublin: Printed by William Sleater and William M'Kenzie, 1793)

A Course.......... Seventh Edition (Philadelphia: P. Byrne, 1802)

Select Parts from Helsham's Lectures on Natural Philosophy (Dublin: Printed by Graisberry and Campbell, Printers to the University, 1818)

Select Parts from Helsham's Lectures on Natural Philosophy (Dublin: R. Millikan, 1827)

Select Parts from Helsham's Lectures on Natural Philosophy (Dublin: John S. Folds for Richard Millikan, 1834)

All of the editions are identical in content, except for those under the title *Select Parts*, which reprint only lectures VI, VII, VIII, X, XII, XIII, XV, XVI, XVII. All but the first Dublin edition include five appendices with problems and solutions.

At first, the continuing popularity of Helsham's *Course* might appear puzzling. How did a textbook in a field of rapidly expanding knowledge remain in use for over a century? A closer look at the text itself and the context in which it was read and used provides a number of clues to its success.

First of all, Helsham's book was one of the early attempts to explain Newton's science in a simple fashion. Newton's *Philosophiae naturalis principia mathematica* (1687) and his *Opticks* (1704) marked an epoch in the history of science.[14] They not only solved the problems of the motion of the planets, falling bodies, the tides and the relation of colors to white light, they also introduced simple mathematical laws and methodological precepts that promised to lead to even more discoveries. As Newtonian ideas came to dominate eighteenth-century science and to affect literature, philosophy and religion, enlightened readers everywhere wanted to understand what it was all about.

Newton's own works, however, were written for trained mathematicians. So, beginning in the 1720s and 1730s popular

---

[14] See especially Schaffer and Dobbs and Jacob.

accounts of Newtonian science began to appear.[15] Science writing like other genres benefited from the rapidly expanding reading public in the early eighteenth century. One of the earliest and most popular was William Jacob 'sGravesande's *Mathematical Elements of Natural Philosophy confirm'd by Experiments; or an Introduction to Sir Isaac Newton's Philosophy* (1720). In 1728 Henry Pemberton published *A View of Sir Isaac Newton's Philosophy*, on whose subscription list Helsham's name appeared. Meanwhile, Francesco Algarotti's *Newtonianism for the Ladies* (1737) illustrated the breadth of the audience for Newtonianism, while Voltaire and the Marquise du Chatelet made its philosophical implications clear in their *Elements of Newton's Philosophy* (1738).

Helsham's book was part of this popular trend, and on the very first page he describes natural philosophy as "entertaining and delightful." Unlike many other early Newtonian texts, however, Helsham's was aimed particularly at university students rather than at a general educated audience. In fact, Helsham's *Course* was one of the first books to be published by the new Dublin University Press whose printing house had been completed in 1734. The new science had only recently been introduced to the university, and students were far from qualified to read Newton's original work. Helsham glossed over Newton's complicated mathematical proofs, added simple experiments on nearly every page, and organized the diversity of current physical knowledge into a coherent whole. Newton's own works, while full of brilliant proofs and useful suggestions, never presented a systematic account of the entire range of natural phenomena. Helsham's *Course*, on the other hand, was written in a style that reassured students that all important questions had been answered and that modern science rested on secure foundations. As the purpose of science education in the eighteenth-century university was to appreciate examples of God's workmanship in the natural world, Helsham's text was a perfect element in a liberal education designed primarily for the training of Anglican clergymen.

---

[15] Knight, pp 63-79.

Structure of the Work

The *Course of Lectures* begins with a preface by Bryan Robinson laying out Newton's precepts of scientific reasoning, drawn in direct quotation from both Newton's *Opticks* and the *Principia*. Newton believed that the greatest obstacle to natural knowledge from Aristotle to Descartes had been the use of hypotheses or undemonstrated assumptions about the physical world. Instead he advocated the use of induction, moving gradually from experiments and observations to general conclusions. These conclusions could never be demonstrated with certainty. Often, as was the case with gravity, they could not even be explained. However, by generalizing from numerous experiments, science, he believed, could establish a secure body of knowledge about the physical world.

Helsham elaborates on this method in his first lecture, attributing the tremendous progress made recently in science to the emphasis on experiment rather than hypothesis. His entire text illustrates this point by its ubiquitous experimental examples and its attempt to avoid theoretical explanations for such phenomena as gravity. Helsham is generally content to demonstrate, classify and group experimental phenomena rather than to explain them by a more general theory. His goal is to collect "the general powers and laws of nature" from experiments and observations.[16]

However, this does not mean that Helsham's science is entirely without assumptions. A major claim, in common with that of Newton, is that there exist "certain principles, forces, or powers, wherewith all parts of matter of what kind soever, so far as experience reaches, seem to be endued; and whereby they act upon one another for producing a great part of the phaenomena of nature" [p. 2]. Unlike Descartes, who believed that matter is entirely passive and homogeneous and only interacts through collisions, Helsham believes with Newton that matter has certain powers such as attraction, whose cause may be unknown but whose existence can be demonstrated experimentally. In the *Opticks* Newton speculated that God created solid massive particles and

---

[16] See Schofield, pp. 32-34.

endowed them with certain powers that allow them to act at a distance on each other. These forces, Newton claimed, produce the familiar phenomena of nature, such as the refraction and reflection of light, gravity, magnetism, electricity, chemical composition and decomposition, crystallization, capillarity, evaporation, fermentation and elasticity.

Newton and Helsham saw physics as the science of particles and forces. Not everyone was convinced by the Newtonian position, however. Many objected that Newton's forces were 'occult' or mysterious. And in fact, Newton toyed with the idea of a subtle aether that would serve to transmit forces between particles.

Helsham, however, was untroubled by such questions. He simply states as a fact that nature can be divided into active and passive principles, with attraction and repulsion active principles and inertia passive. In addition to attractive forces such as electricity, magnetism and gravity, Helsham explains that there are also repulsive forces such as electricity and magnetism, the elasticity of the air and the expulsion of particles of light by luminous bodies. All the forces are demonstrated by experimental examples, and some, like gravity, are also described mathematically. After introducing these forces, Helsham moves on to mechanics, and here his text will look very familiar to any modern student of physics. Topics include the composition of motion, the laws of collisions, simple machines such as the balance, lever and pulley, friction, motion on an inclined plane and projectile motion. Helsham's simple explanations with a minimum of mathematics remained useful for beginning students even after more advanced students began to be trained in the methods of the calculus.

Mechanics is followed by hydrostatics, particularly the examination of the flux of water through pipes. This was no doubt related to Helsham's work on the Dublin water supply as well as Bryan Robinson's detailed mathematical investigation of flow (see below). Helsham's analysis of pneumatics (also utilized by Robinson) treats air as a fluid composed of mutually repulsive particles. This allows Helsham to derive the velocity of sound in air, following Newton, and enabled Robinson to present an elaborate mathematical theory of the aether, which he believed to be similarly composed of small repulsive particles. Helsham, however, makes no mention of the aether here, and keeps the mathematics to a minimum.

The rest of the *Course of Lectures* is devoted to light. Helsham follows Newton in presenting light as "a most subtile fluid, consisting of particles exceedingly small, but of different magnitudes... which are thrown off from luminous bodies by the vibrating motions of their parts...". [p. 287] While Helsham is particularly impressed by Newton's explanation of the spectrum, he also provides a detailed description of the anatomy of the human eye, referring both to Bryan Robinson's anatomical investigations and George Berkeley's work on vision. Interestingly, while both Berkeley and Robinson were engaged in philosophical as well as physical questions about vision and perception, Helsham ignores such difficulties, presenting the eye simply as a wonderful, but fully intelligible, mechanism. While Berkeley saw certain optical phenomena as evidence of the failure of geometrical optics and offered a radical critique of Locke's theory of perception, Helsham simply presents the phenomena as an untroubling case of mental confusion. This is true throughout the *Lectures;* phenomena and theories that were commonly seen as contentious or deeply problematic are presented by Helsham as obvious and in no need of elaborate analysis. The experiments, he seems to say, speak for themselves. This non-polemical style, common to later textbooks but not as common during this period, may be one of the reasons for the great success of the book and its appeal to both beginning students and teachers anxious to present Newtonian science as the bulwark of certainty and rationality. Bryan Robinson's work, on the other hand, exemplifies a more contentious style of natural philosophy.

Bryan Robinson

Bryan Robinson was not only Helsham's colleague at Trinity College and the editor of his lectures but also a well known physician and natural philosopher in his own right. Born in Dublin,

Robinson received the degree of bachelor of medicine from Trinity College Dublin in 1709 and his MD in 1711.[17] He would later serve as president of the College of Physicians and was also a member of the Royal College of Surgeons. Robinson lectured in anatomy at Trinity College briefly in 1716 and was both a governor and a physician at the new Steeven's Hospital when it opened in 1733. In 1745 he was appointed public Professor of Physic at Trinity College Dublin, and he appears to have had a thriving practice in Dublin.

Ironically, Robinson's success lay in precisely those areas that Helsham's text avoided, namely the creation of elaborate theories to explain the operations of physical and biological forces and the development of highly mathematical treatments of a wide range of phenomena. Robinson's primary scientific goal was to apply Newton's mathematical principles to medical problems, and his most important contribution was the revival of Newton's theory of the aether.[18]

Newton had discussed the aether in the General Scholium of the 1713 edition of the *Principia* and in certain queries in the 1717 edition of the *Opticks*, presenting it as a subtle spirit pervading all space and all bodies. The aether allowed him to explain action at a distance in a less mysterious way, providing a material cause for gravity, electrical attraction and repulsion, together with other forces. Newton, however, never made his aether theory mathematical, nor did he stand firmly behind it, and most early Newtonians abandoned the theory, preferring to focus on forces acting a distance between particles in empty space.

Robinson became interested in the aether, he claims, because of Newton's speculation that the vibration of the aether along the nerves causes sensation. His first major presentation of his aether theory occurs in his 1732 *Treatise on the Animal Oeconomy*, where he uses it to explain muscular motion. The text includes eighty pages of hydrodynamical computations backed up by experiments on the motion of liquids in pipes of various sizes and shapes. All of this is designed to offer a mechanical theory of

---

[17] See DNB, Kirkpatrick, pp. 109-112.
[18] See Schofield, pp. 106-114 and Thackray, pp.135-41.

the animal body in which fluids like aether and blood flow through nerves and veins just as water flows through a pipe. In 1743 he published a *Dissertation on the Aether of Sir Isaac Newton*, and in 1745 he published *Sir Isaac Newton's Account of the Aether*. This latter work was primarily a reprinting of Newton's 1678/79 letter to Robert Boyle on the aether, recently discovered and published for the first time in Thomas Birch's 1744 *History of the Royal Society*. New developments in chemistry, electricity, physiology and theology led to a renewed interest in aether theories in the 1740s, but Robinson stood out by his success in transforming Newton's cautious hints into a fully developed mathematical "theory of everything" that he believed could explain all physical, chemical and physiological phenomena.

Interestingly, Robinson's work was only the beginning of aether theories in Dublin. With the development of the wave theory of light in the early nineteenth century, many of Dublin's finest mathematicians and physicists became involved in aether theories, including William Rowan Hamilton, Humphrey Lloyd, James McCullagh and George Francis Fitzgerald.[19] There is a distant echo of that work even in the most recent work of the Department of Physics[20]

The Context of Newtonianism

In contrast to Robinson's radical theories, Helsham's matter of fact representation of Newton's science was a perfect fit for a period in which Newtonianism was increasingly adopted by those seeking a source of stability in politics and religion. The straightforward style of Helsham's text disguises the tumultuous political and religious debates surrounding the new science.

As Dobbs and Jacob have argued, many of the earliest supporters and popularisers of Newtonian science were latitudinarian Whigs anxious to find ways to justify the new balance in the English Church and constitution following the

---

[19] See Cantor and McConnell.
[20] See Weaire

Glorious Revolution. Perhaps the best example of this is that of the Boyle lectures, endowed by Robert Boyle on his death in 1691, "to prove the truth of the Christian religion against infidels." Both Newton and Boyle had been extremely concerned with the religious implications of their science and took care to avoid the traditionally heretical implications of materialism and mechanism. Newton saw gravity as the direct manifestation of God's presence and power in the world. Boyle lecturers like Samuel Clarke and Richard Bentley, according to Jacob, were "dedicated to showing how the Newtonian system demonstrated the wisdom and providence of God, justified order and stability in polity, and supported the contemporary balance in the constitutional settlement."[21] They sought to ground order, hierarchy and human authority in the natural order of the universe, rather than the divine right of kings or revelation. These broad church Whigs found that Newtonianism could be used to attack materialist republicans like John Toland as well as obscurantist High Church Tories more interested in faith than reason. What John Gascoigne has called the "Holy Alliance" between Anglicanism and Newtonian science served to promote the latter in the universities, particularly Cambridge, where future Anglican divines were trained to utilize Newtonian science in their sermons.[22]

High Church Tories like John Arbuthnott, Jonathan Swift and George Berkeley attacked this new alliance of Newtonianism, Whiggism and latitudinarianism.[23] While Arbuthnott and Swift satirized the pretensions of natural theology and inappropriate applications of science to religion and morality, Berkeley went for the jugular. He attacked the mathematical foundations of Newton's work, questioned the possibility of mathematical optics and criticized the Lockean epistemology that grounded the standard interpretation of Newton's science. The arrogance and deistical tendencies of the major Newtonians antagonised these men. Newton and his followers Samuel Clarke and William Whiston, for example, were all anti-Trinitarians, and High Church Anglicans

---

[21] Dobbs and Jacob, p. 57.

[22] See Gascoigne.

[23] See Olson.

tended to see Newtonian natural theology as an attempt to supplant revealed religion. Berkeley's idealist philosophy and his aetherial medical theory (in which tar water is the vehicle for aether) were both designed to reinforce the importance of spirit in a world increasingly populated by materialists.[24] Materialist theories of soul and mind (like Robinson's) left little room for God to play a role in the universe and implied that there were no questions that could be answered by faith.

We know little about Richard Helsham's personal opinions on these matters, and perhaps that is how he wanted it. In an intellectual world divided by political, religious and scientific questions, Helsham looked for the common ground. By the avoidance of hypotheses and the focus on experimentation, he sought to convince students that Newtonian science really did offer a way out of such divisive debates. The success of his textbook illustrates the continuing hope that through the study of orbiting planets, inclined planes and telescopes, man can come to know something with certainty.

D.A. Attis, P.Kelly, D. Weaire
Trinity College Dublin, 1999

---

[24] See Benjamin.

References

M. Benjamin, "Medicine, Morality and the Politics of Berkeley's Tar-Water," pp. 165-93 in A. Cunningham and R. French (eds.), *The Medical Enlightenment of the Eighteenth Century* (Cambridge: Cambridge University Press, 1990)

G.N. Cantor, *Optics After Newton: Theories of Light in Britain and Ireland, 1704-1840* (Manchester: Manchester University Press, 1983)

B. J. Teeter Dobbs and M. C. Jacob, *Newton and the Culture of Newtonianism* (Atlantic Highlands, NJ: Humanities Press International, 1995)

J. Gascoigne, *Cambridge in the Age of Enlightenment: Science, Religion and Politics from the Restoration to the French Revolution* (Cambridge: Cambridge University Press, 1989)

K.T. Hoppen, *The Common Scientist in the Seventeenth Century: A Study of the Dublin Philosophical Society, 1683-1708* (London: Routledge & Kegan Paul, 1970)

P. Kelly, "Locke and Molyneux: The Anatomy of a Friendship," *Hermathena* (1979) 126: 38-54.

T.P.C. Kirkpatrick, *History of the Medical Teaching in Trinity College Dublin and of the School of Physic in Ireland* (Dublin: Hanna and Neale, 1912)

D. Knight, *Natural Science Books in English 1600-1900* (London: Portman Books, 1972)

Lady Llanover (ed.), *The Autobiography and Correspondence of Mary Granville, Mrs. Delany* (London, 1861)

A.J. McConnell, "The Dublin Mathematical School of the First Half of the Nineteenth Century," *Proceedings of the Royal Irish Academy* (1944) 50A: 75-88.

R.B. McDowell and D.A. Webb, *Trinity College Dublin 1592-1952: An Academic History* (Cambridge: Cambridge University Press, 1982)

N. D. McMillan, "Richard Helsham M.D. (1683-1737), Medical Man, Virtuoso and Educationalist: Author of the First Purpose-Written Student Textbook on Natural Philosophy in the Vernacular," *Science (Journal of the Irish Science Teachers Association)* (1988) 24: 13-18.

R.G. Olson, "Tory-High Church Opposition to Science and Scientism in the Eighteenth Century: The Works of John Arbuthnott, Jonathan Swift, and Samuel Johnson," in J. G. Burke (ed.) *The Uses of Science in the Age of Newton* (Berkeley: University of California Press, 1983) pp. 171-204.

S. Schaffer, "Newtonianism," pp. 610-26 in R.C. Olby, G.N. Cantor, J.R.R. Christie and M.J.S. Hodge (eds.) *Companion to the History of Modern Science* (London: Routledge, 1990)

R. Schofield, *Mechanism and Materialism: British Natural Philosophy in an Age of Reason* (Princeton: Princeton University Press, 1970)

A. Thackray, *Atoms and Powers: An Essay on Newtonian Matter-Theory and the Development of Chemistry* (Cambridge: Harvard University Press, 1970)

C.H. Holland (ed.), *Trinity College Dublin & the Idea of a University* (Dublin: Trinity College Dublin Press, 1991)

D. Weaire (ed.), *The Kelvin Problem* (London: Taylor & Francis 1997)

H.Williams (ed.), *The Correspondence of Jonathan Swift* (Oxford: Clarendon, 1965)

D.E.W. Wormell, "Latin Verses By W. Thompson Spoken at the Opening in 1711 of the First Scientific Laboratory in Trinity College Dublin," *Hermathena* (1962) 96: 21-30.

THE

PREFACE.

THAT the reader may be duly prepared for the perufal of the following Treatife, it will be neceffary that he firft acquaint himfelf with the genuine Method and Rules of Philofophizing, as they have been delivered by Sir ISAAC NEWTON.

His Method of Philofophizing is thus laid down in his *Opticks* *.

"As in mathematicks, fo in natural philo-
"fophy, the inveftigation of difficult things
"by way of *Analyfis*, ought ever to precede
"the method of compofition. This *Analyfis*
"confifts in making experiments and obfer-
"vations, and in drawing general conclufions
"from them by induction, and admitting of
"no objections againft the conclufions, but
"fuch as are taken from experiments or other
"certain truths. And although the arguing
"from experiments and obfervations by in-
"duction be no demonftration of general
"conclufions; yet it is the beft way of ar-
"guing which the nature of things admits
"of, and may be looked upon as fo much
"the ftronger, by how much the induction

* Opt. p. 380.

"is

" is more general.　And if no exception oc-
" cur from *Phænomena,* the conclusion may
" be pronounced generally.　But if at any
" time afterwards, any exceptions shall occur
" from experiments, it may then be pro-
" nounced with such exceptions as shall occur.
" By this way of *Analysis,* we may proceed
" from compounds to ingredients, and from
" motions to the forces producing them;
" and in general from effects to their causes,
" and from particular causes to more general
" ones, till the argument ends in the most
" general.　This is the method of *Analysis:*
" And the *Synthesis* consists in assuming the
" causes discovered, and established as prin-
" ciples, and by them explaining the *Phæ-*
" *nomena* proceeding from them, and proving
" the explanations."

His Rules of Philosophizing, delivered in
his *Principles* *, are these four.

R U L E   I.

" *More causes of natural things are not to be*
" *admitted, than are both true and sufficient*
" *for explaining their phænomena.*

" Thus Philosophers say; nature does no-
" thing in vain, and in vain that is done by
" more causes, which can be done by fewer.
" For nature is simple, and delights not in
" superfluous causes of things."

* Philof. Natur. Princip. Mathemat. p. 387.

R U L E

RULE II.

" *Of natural effects therefore of the fame*
" *kind the fame caufes are to be affigned, as far*
" *as it can be done.*

" As of refpiration in a man and in a
" beaft; of the defcent of ftones in *Europe*
" and in *America*; of light in a culinary fire
" and in the fun; of the reflexion of light in
" the earth and in the planets."

RULE III.

" *The qualities of bodies which cannot be*
" *increafed and diminifhed, and which agree*
" *to all bodies in which experiments can be*
" *made, are to be reckoned as qualities of all*
" *bodies whatfoever.*

" For the qualities of bodies are not known
" but by experiments; and therefore, as ma-
" ny are to be reckoned general as generally
" agree with experiments, and thofe which
" cannot be diminifhed cannot be taken
" away. Certainly dreams are not to be de-
" vifed at pleafure contrary to the tenor of
" experiments; nor muft we depart from the
" analogy of nature, fince fhe is wont to be
" fimple, and always confonant to herfelf.
" The extenfion of bodies is not known but
" by the fenfes, nor is it perceived in all bo-
" dies: but becaufe it agrees to all bodies
" which are perceiveable, it is affirmed of all
A 3 " what

" whatſoever. We experience many bodies
" to be hard. But the hardneſs of the whole
" ariſes from the hardneſs of the parts, and
" thence with good reaſon we conclude the
" undivided parts not only of thoſe bodies
" which are perceived, but alſo of all others,
" to be hard. We gather all bodies to be
" impenetrable, not by reaſon, but by ſenſe.
" We find the bodies we handle to be im-
" penetrable, and thence conclude impene-
" trability to be a property of all bodies what-
" ſoever. That all bodies are moveable, and
" by certain forces (which I call *vires iner-*
" *tiæ*) perſevere in motion or reſt, we gather
" from theſe ſame properties in bodies which
" are ſeen. Extenſion, hardneſs, impenetra-
" bility, mobility, and *vis inertiæ* of the
" whole, ariſe from the extenſion, hardneſs,
" impenetrability, mobility, and *vis inertiæ*
" of the parts; and thence we conclude that
" all the leaſt parts of all bodies are extend-
" ed, and hard, and impenetrable, and move-
" able, and endued with *vires inertiæ*. And
" this is the foundation of all Philoſophy.
" Farther we know from the *Phænomena*, that
" the parts of bodies which are divided
" and mutually contiguous to one another,
" may be ſeparated from one another; and it
" is certain from mathematicks, that the un-
" divided parts may by reaſon be diſtin-
                                    " guiſhed

" guifhed into lefs parts. But whether thofe
" parts diftinct, and not yet divided, can
" by the powers of nature be divided and
" feparated from one another, is uncertain.
" But if it fhould appear, even by one fingle
" experiment, that by breaking a hard and
" folid body, any undivided particle fuffered
" a divifion; we might conclude by the
" force of this Rule, that not only the di-
" vided parts were feparable, but that the un-
" divided parts might be divided *in infinitum.*

Laftly, If it be univerfally evident by
" experiments and aftronomical obfervations,
" that all bodies round the earth gravitate
" towards the earth, and that in proportion
" to the quantity of matter in each, and that
" the moon gravitates towards the earth in
" proportion to its quantity of matter, and in
" like manner our fea gravitates towards the
" moon, and that all the planets mutually
" gravitate towards one another, and that
" there is a fimilar gravity of comets to-
" wards the fun; we muft pronounce by
" this Rule, that all bodies gravitate mutu-
" ally towards one another. For the argu-
" ment from the *Phænomena* will be ftronger
" for an univerfal gravity, than for the im-
" penetrability of bodies, concerning which
" in the heavenly bodies we have no experi-
" ment, no obfervation at all."

A 4   RULE

RULE IV.

" *In experimental Philosophy propositions*
" *collected from the* Phænomena *by induction,*
" *are to be deemed, notwithstanding contrary*
" Hypotheses, *either accurately or very nearly*
" *true, till other* Phænomena *occur, by which*
" *they may be rendered either more accurate*
" *or liable to exceptions.*

" This ought to be done, left arguments
" of induction fhould be deftroyed by *Hy-*
" *potheses.*"

This Method and thefe Rules, have been
carefully obferved by our Author in thefe
LECTURES; which, from the clearnefs and
diffufivenefs of the ftile, and the eafy and
juft manner of reafoning, are, in my opinion, better fitted for the inftruction of youth,
than any thing which I have feen on this
fubject.

I have added a few *Problems* by way of
APPENDIX.

THE

THE
CONTENTS.

LECTURE I.

LECTURE XV.

LECTURE I.

OF ATTRACTION.

AS natural philofophy is a fcience in its own nature entertaining and delightful, and withal conducive in many inftances to the eafe and convenience of life; it is not to be wondered that there have been men in all ages who have laid themfelves out in the improvement and cultivation of it. But it is a matter of no fmall furprize to think how inconfiderable a progrefs the knowledge of nature had made in former ages, when compared with the vaft improvements it has received from the numberlefs difcoveries of later times; infomuch, that fome of the branches of natural philofophy, which at this day is almoft compleat in all its parts, were utterly unknown before the laft century. If we look into the reafon of this, we fhall find it to be chiefly owing to the wrong meafures that were taken by philofophers of former ages in their purfuits after natural knowledge : for they difregarding experiments, the only fure foundation whereon to build a rational philofophy, bufied themfelves in framing hypothefes, for the folution of natural appearances, which as they were creatures of the brain, without any foundation in nature, were generally fpeaking fo lame and defective, as in many cafes not to anfwer thofe very phænomena for whofe fakes they had been contrived. Whereas the philofophers of later times, laying afide thofe falfe lights, as being of no other ufe than to mifguide the underftanding

in

LECT.
I.
in its searches into nature, betook themselves to experiments and observations ; and from thence collected the general powers and laws of nature ; which with a proper application, and the assistance of mathematical learning, enabled them to account for most of the properties and operations of bodies ; and to solve many difficulties in the natural appearances, which were utterly inexplicable on the foot of hypotheses. By this means has natural philosophy within the compass of one century been brought out of the greatest darkness and obscurity into the clearest light ; and this has been chiefly owing to the unparalleled abilities, and indefatigable industry of that great and accurate philosopher Sir ISAAC NEWTON ; who to his great honour has in his principles of natural philosophy, and his incomparable treatise of light and colours, cleared more difficulties, and disclosed more and more important truths relating to nature, than are to be met with in the voluminous writings of all that went before him. To illustrate some of these truths by experiments is the design of this course, which consists of four parts. In the first are considered solid bodies and their properties. In the second water and watery fluids. In the third the elastic fluid of air. And in the last the subtile fluid of light. But before I proceed to these particulars, it will be necessary to say something concerning certain principles, forces, or powers, wherewith all parts of matter of what kind soever, so far as experience reaches, seem to be endued ; and whereby they act upon one another for producing a great part of the phænomena of nature.

Such is first that power whereby the minute particles of matter do in some circumstances tend towards one another, which is commonly called attraction ; the cause whereof is in a great measure unknown, tho' the thing itself is manifest from experiments. For if two polished plates of brass be

Exp. 1.

4

laid

laid one upon another, having their contiguous sides
smeared with oil, they will cohere in *vacuo*, and
with such firmness, that when they are suspended,
the force of gravity in the lower plate will not
suffice to separate and pull them asunder.

That the cohesion of these plates is to be attributed to the mutual attractions of their contiguous
parts, cannot I think admit of a doubt, since the
pressure of the outward air on their external surfaces, (to whose force this effect might otherwise
have been attributed) is in this case taken off.

The use of the oil is to fill up the minute cavities in the surfaces, and by so doing to prevent the
lodgment of air between the plates; which upon
the removal of the outward air would expand itself
by reason of its elasticity, and thereby force the
plates asunder.

The forementioned attraction is in like manner
collected from the following experiments.

If two plane polished plates of glass be laid toge- Exp. 2.
ther, so that their sides be parallel, and at a very
small distance from one another; and their lower
edges be dipped in water; the water will rise up
between them, and the less the distance of the
glasses is, the greater will the height be to which
the water rises. If the distance be about the hundredth part of an inch, it will rise to the height of
about an inch; and if the distance be greater or less
in any proportion, the height will be reciprocally
proportional to the distance very nearly.

The reason why the water ascends between the
plates is, that those parts of the surfaces of the
glasses which lie next above the surface of the water, and are contiguous thereto, attract the water,
and by that means cause it to ascend; and this
ascent continues till the weight of the elevated
water becomes equal to the force of the attracting
surfaces, and then the motion ceases, the water
tending as much downward by the force of its own
gravity,

L e c t.
I.

gravity, as it doth upward by the attraction of the glaſſes.

The reaſon why the water riſes to heights which are inverſly as the diſtances of the glaſſes, is this: the abſolute attractive force of the glaſſes, whereby the water is raiſed, continues unvaried whatever be the diſtance of the glaſſes; for the height and length of the glaſs ſurfaces, whoſe attractions influence the aſcent of the water, are always the ſame, and conſequently the attractive force muſt be ſo too; and for that reaſon will conſtantly ſupport the ſame weight of water; but the quantity and conſequently the weight of the elevated water will always be the ſame, if its height be reciprocally as its baſe, that is, in this caſe, as the diſtance of the plates; for the length of the baſe being equal to the length of the plates, it continues unvaried; and therefore the baſe will ever be as its breadth, that is, as the interval between the plates.

Exp. 3.

If the glaſs plates inſtead of being ſet parallel to one another, be made to meet at one of their ends, and kept at a little diſtance at the other; and their lower edges be then dipped in water, ſpirit of wine, or any other convenient liquor; the inward ſides of the plates being firſt moiſtened with a clean cloth dipped in the liquor; the liquor will riſe between the plates; and the upper ſurface of the elevated liquor will form a curve, the heights of whoſe ſeveral points above the ſurface of the ſtagnating liquor will be to one another reciprocally as their perpendicular diſtances from the concourſe of

Fig. 1.

the plates. For the illuſtration of which, let AE be the ſurface of the ſtagnating liquor wherein the lower edges of the plates are immerſed, A H the concourſe of the plates, and F, G, I, K, L the curve formed by the ſurface of the elevated liquor; from any points in the curve as G, I, K, L taken at pleaſure, let fall the right lines G B, I C, K D, L E perpendicular to A E, and thoſe lines will expreſs the

heights

heights of the refpective points of the curve above the furface of the ftagnant liquor; whilft A B, A C, A D, A E denote the perpendicular diftances of the fame points from the concourfe of the glaffes; now thefe heights and diftances are to one another in a reciprocal proportion: for if we fuppofe the lines G B, I C, K D, L E to be fo many pillars of liquor confifting of four fides, two of which are terminated by the plates, and the other two by the contiguous liquor; and if thofe fides which lie next the plates be of an equal but exceedingly fmall breadth in all the pillars, then will the attracting furfaces of the plates which fupport thofe pillars be likewife equal, and confequently the quantities fupported, that is, the pillars muft be fo too. But in order to have them equal, their heights muft be reciprocally proportional to their bafes; which bafes inafmuch as they are fuppofed to be equally broad muft be as their lengths, that is, as the intervals between the glaffes in thofe parts where the pillars are taken; and therefore the heights of the pillars muft be reciprocally as the intervals between the plates; but from the nature of fimilar triangles the intervals between the glaffes at different diftances from the concourfe are to one another directly as thofe diftances; whence it follows, that the heights of the pillars are to one another reciprocally as their refpective diftances from the concourfe of the plates; that is, if G B be double of I C, then is A C double of A B.

From what has been faid it is plain that the curve formed by the upper furface of the elevated liquor muft be an hyperbola; for from the nature of the hyperbola the external ordinates are reciprocally as the abfciffæ; wherefore if A B, A C, A D, A E be taken for the abfciffæ; then will B G, C I, D K, E L, be the refpective ordinates; and confequently the curve which paffes through the points G, I, K, L is an hyperbola.

As

As water or any other proper fluid afcends between polifhed plates of glafs by the force of their attractions; fo does it likewife in flender pipes of glafs open at both ends; for if fuch tubes be dipped at one end into water, fpirit of wine, or any other convenient fluid, the liquor will rife within the pipes to a confiderable height, and this experiment (as alfo thofe before made) fucceeds in the very fame manner in *vacuo*, as in the open air, for the liquor conftantly afcends to the fame height in both.

That the afcent of liquor in thefe fmall tubes, as alfo between polifhed plates of glafs, is to be attributed to fome power in the glafs ftrongly acting on the liquor, and not to the preffure either of the ftagnating liquor or incumbent atmofphere, is evident from this confideration; that as much of the liquor remains fufpended in the pipes, and between the plates, when they are lifted out of the ftagnating fluid, either in *vacuo* or the open air, as was elevated above the furface of the fluid, while they were immerfed therein: and therefore whatever caufe concurred to the elevating of the liquor while the plates and pipes were therein immerfed, and expofed to the air; the fame contributes as powerfully to keep it up, when the ftagnating liquor is removed, and the preffure of the atmofphere taken off, and confequently muft be fome power inherent in the glafs.

The heights to which the liquor rifes in flender pipes, are to one another reciprocally as the diameters. For the power which raifes the liquor in a flender pipe, being the attractive force of that part only of the internal concave furface which lies next above the liquor, and conftitutes a ring of an indefinitely fmall height, which height is ever the fame whatever be the diameter of the ring, becaufe the diftance to which the attractive force of glafs reaches is unvaried; and the attractive force of fuch an annular furface being as the number of attracting parts whereof it is compofed, that is, as the furface,

which

which becaufe its height is given is as the periphery, that is, as the diameter, the attractive force of the pipe muft be as the diameter. Wherefore if in comparing the forces of two fuch pipes we make F to denote the attractive force of the larger, and f the attractive force of the fmaller, and alfo D and d to denote their diameters; we fhall have this analogy, *viz.* F : f : : D : d, that is, the force of the larger pipe is to that of the fmaller as the diameter of the larger to the diameter of the fmaller: but thefe forces are likewife to one another in the fame ratio with the quantities of liquor which they keep fufpended, for they continue to elevate the liquor till fuch time as the weights, and confequently the quantities of liquor drawn up, become a balance to the attracting forces. Wherefore if H be put for the height of the liquor in the pipe, whofe diameter is D, and h for its height in the pipe whofe diameter is d; then will H multiplied into the fquare of D be as the quantity of liquor in the larger pipe; and h multiplied into the fquare of d as the quantity of liquor in the fmaller pipe; whence we have this fecond analogy $F : f : : H \times D^2 : h \times d^2$; and by fubftituting D and d in the room of F and f, to which they are proportional, as appears from the firft analogy, we fhall have $D : d : : HD^2 : hd^2$; and then multiplying extreams and means, and throwing off fimilar quantities, we fhall have $HD = hd$, and by refolving this equation into an analogy, we fhall have H : h : : d : D, that is, the height to which the liquor rifes in the larger pipe is to the height to which it rifes in the fmaller, as the diameter of the fmaller pipe to that of the larger; fo that the heights of the liquor are reciprocally proportional to the diameters of the pipes.

By virtue of this attractive force, wherewith fmall pipes are endued, plants receive nourifhment from the earth; the flender tubes whereof their roots are compofed, fucking in various juices according to

B   their

their different natures and conſtitutions. From the
ſame attractive force it is that ſponges take in water; and that water aſcends in loaf ſugar, when
any part of it is dipped therein; thoſe parts of the
ſugar which lie next above the water attracting,
and thereby raiſing the ſame. And here it muſt
be obſerved that the water riſes by the action of
thoſe particles alone which are contiguous to, and
lie next above the ſurface of the elevated water;
thoſe particles which are at any the leaſt ſenſible
diſtance above the water being too far removed to
influence the water by their attractions: and what
has been thus obſerved of ſugar, is likewiſe true of
poliſhed plates, ſlender pipes, and every other attracting body, by virtue of whoſe attractions fluids
are raiſed. For if thoſe parts of attracting ſurfaces which are at any ſenſible diſtance above the
ſurface of the fluid, do in any meaſure contribute
to the aſcent; it is evident that the fluid *cæteris pa-*
ribus muſt riſe to a greater height when the attracting ſurfaces are continued to a conſiderable height
above the elevated fluid, than when they terminate
at a very little diſtance above the ſame. But the
Exp. 5. contrary appears from experiment. For if two poliſhed plates of glaſs ſet parallel to one another at
the diſtance of about the hundredth part of an inch,
be immerſed in water ſo far that only an inch and
one tenth be ſuffered to remain above the water,
the water will riſe up between them to the height
of about an inch; and if the ſurface of the ſtagnating water be then depreſſed by drawing off
ſome of the water, the elevated water will likewiſe
deſcend between the plates ſo as ſtill to preſerve the
height of about an inch and no more.

Exp. 6.   If a poliſhed plate of glaſs be laid parallel to the
horizon, and another plate of the ſame kind be laid
thereon, ſo as that they may touch at one of their
ends, and be kept at a very ſmall diſtance at the
other: being firſt moiſtened on their inward ſides
with

with a clean cloth or feather dipped in oil of o-
ranges; and if a drop of the oil be placed between the plates at that end where they are at fome
diftance from each other, fo as that it may be touched by both the plates, it will begin to move towards the concourfe of the glaffes, and will continue
to go on with an accelerated motion till it arrives
at the concourfe. And if during the motion of the
drop, that end of the glaffes where they meet,
and towards which the drop moves, be lifted up, the
drop will neverthelefs continue its motion, and of
confequence muft be attracted; but as the end of
the glaffes is raifed higher and higher, the drop will
afcend more and more flowly, till at laft upon a
certain elevation of the plates the motion ceafes, the
gravity of the drop, wherewith it tends downward,
becoming equal to the attractive force which draws
it upward; as appears from this, that upon giving
the plates the leaft degree of elevation beyond what
is neceffary to ftop the drop, it ftraightway begins
to defcend, its gravity in that cafe overcoming the
attraction.

By the help of this phænomenon may the force
be determined, wherewith the drop is attracted at
all diftances from the concourfe of the glaffes. For
that part of a body's gravity whereby it is carried
down an inclined plane, is to its abfolute weight,
as the fine of the angle of the plane's elevation, to
the Radius, or as the perpendicular height of the
plane to the length thereof; and therefore may be
denoted by the perpendicular height applied to the
length; and where the length of the plane is given,
that force will be every where as the fines of the
angles of elevation, or the perpendicular altitudes
of the plane; as fhall be made appear when I come
to treat of the defcent of bodies on inclined planes.
If therefore the fines of fuch elevations of the plates
as are neceffary to ftop the motion of the drop, be
taken at two different diftances of the drop from

the

the concourſe of the plates ; thoſe ſines will denote the reſpective gravities of the drop, and conſequently the attractive forces, wherewith the plates
act upon the drop at each of thoſe diſtances. Thus
for inſtance, if the diſtances of the drop from
the concourſe of the glaſſes be as one and two ;
and the ſine of the elevation neceſſary to ſtop the
motion of the drop when at the ſmaller diſtance be
as four, and when at the greater diſtance as one ; the
gravity of the drop, wherewith it endeavours to deſcend at the forementioned diſtances of one and two,
will be as four and one. For the illuſtration of
Fig. 2. which, let AB and AC repreſent the plates at different elevations ; F and G the places where the
drop ſtands upon thoſe elevations ; then will BD
and CE denote the forces of gravity wherewith
the drop endeavours to deſcend along the plates in
the points F and G, which forces are equal to the
attractions of the glaſſes in thoſe points ; and if
BF and CG the diſtances of the drop from the
concourſe of the plates be as one and two, and
BD and CE as four and one ; then is the attractive power wherewith the glaſſes act upon the
drop at F, to the force wherewith they act upon
it at G, as four to one, that is, reciprocally as the
ſquares of the diſtances of the drop from the concourſe of the glaſſes ; and this is nearly the caſe, as
Exp. 6. will appear from the experiment.

Tho' the drop be attracted by forces that are in
the reciprocal duplicate ratio of the diſtances of
the drop from the concourſe of the glaſſes ; yet are
the attractions within the ſame quantities of attracting ſurface in the reciprocal ſimple ratio only of
thoſe diſtances : for as the drop moves towards the
concourſe of the glaſſes, it muſt ſpread and touch
each glaſs in a larger ſurface ; and this ſpreading is
always proportional to the leſſening of the interval
between the glaſſes ; and of conſequence from the
nature of ſimilar triangles, it is likewiſe proportional

to

to the diminution of the diſtance from the concourſe. So that the force which acts upon the drop is increaſed as the drop approaches the concourſe in the ſimple reciprocal ratio of the diſtance, on account of the inlargement of the attracting ſurface in that proportion; and therefore in a given quantity of attracting ſurface the force muſt be in the reciprocal ſimple ratio of the diſtance from the concourſe; that is to ſay, any given portion of the glaſs ſurfaces taken at the diſtance of one inch from their concourſe muſt act with twice the force that it does at the diſtance of two inches, and with thrice the force that it does at the diſtance of three inches, and ſo on. Hence it will be found that the attractive force of one and the ſame ſlender pipe of a conical figure is given; or in other words, that the attractive force wherewith a conical pipe is indued at any one diſtance from the vertex of the cone, is equal to the attractive force of the ſame at any other diſtance from the vertex; ſo that the attractive force of a conical pipe is in every part equal throughout the whole length of the pipe; and may be expreſſed by the diameter of a circular ſection of the pipe taken at any diſtance from the vertex, applied to that diſtance. For the attraction in any part of ſuch a pipe, is as the quantity of attracting ſurface in that part multiplied into the abſolute force; but the quantity of attracting ſurface in any part is as the diameter of that part, and the abſolute force is reciprocally as the diſtance from the vertex; wherefore if A be put to denote the diſtance of any part from the vertex and D the diameter, $\frac{D}{A}$ will expreſs the attraction of that part; but from the nature of ſimilar triangles the diameters of the circular ſections of a cone taken at different diſtances from the vertex are to one another as the diſtances, conſequently $\frac{D}{A}$ is a ſtanding quantity.

Wherefore

Wherefore since the attractive force in every part of a conical tube is denoted by a quantity which is invariable, it follows that the force is so too; so that in this respect conical pipes do not differ from those of a cylindrical form; but herein lies the difference, that in very slender pipes where the diameters are equal, the attractions of such as are conical do far surpass the attractions of those which are cylindrical. And indeed so exceeding great does this attractive force become with respect to the quantity of attracting surface in that part of a conical pipe, where the diameter is but one part of an inch divided into ten millions (if such minuteness may be supposed) that if the attraction of a cylindrical tube, whose diameter is an inch, were as great with respect to its quantity of attracting surface, it would be able to support a column of water an inch in diameter and upwards of three miles in height. For let us suppose a conical tube whose base is an inch in diameter to be continued till the diameter is so far diminished as to equal only one part of an inch divided into ten millions; it is evident from what was just now said, that the whole attractive force of such a pipe, where its diameter is an inch, is equal to the whole attractive force of the same, where the diameter is but the ten millioneth part of an inch; consequently if a portion of the larger attracting surface be taken equal to the smaller attracting surface, the force of that will be to the force of this, as the force of the smaller surface divided by the number of parts in the larger surface, to the force of the smaller surface, that is, as one divided by ten millions to one. If therefore a conical, or indeed a cylindrical tube an inch in diameter (for where the diameter is so large there is scarcely any difference) was indued with an attractive force as great in proportion to its quantity of attracting surface, as is a conical tube of the ten millioneth part of an inch in diameter, its force would be ten millions of times

greater

greater than it is, and of confequence would raife the water ten millions of times higher than it doth at prefent: but it has been found by experience that in a cylindrical tube of an inch in diameter, the water will rife to the height of about the fiftieth part of an inch, and therefore if the force by which it rifes was augmented in the forementioned proportion, it muft rife to the height of two hundred thoufand inches, which being divided by fixty three thoufand three hundred and fixty, the number of inches in a mile, gives three and a little more in the quotient.

The quantities of liquor fupported by the attractions of flender conical pipes are to one another, as the diameters of the little circular furfaces of the elevated liquor, applied to the refpective diftances of the fame circular furfaces from the vertices of the feveral cones whereof the pipes are portions. For it has been proved that the attractive forces of conical pipes are as thofe quantities; and therefore the weights which they fupport muft be fo too. Hence it follows that the lefs the proportion is, which the diftance of the elevated liquor's furface from the vertex of the cone bears to the diameter of the fame furface, or which amounts to the fame thing, the fafter the fides of the pipe converge, the ftronger is its attractive force, and the greater the quantity of liquor which is fupported.

The firm union and ftrong cohefion of the particles of folid bodies feems to arife from this force, wherewith they mutually attract each other; which as it appears to be exceeding ftrong in the immediate contact of the particles, fo is it found by experience to reach but a very little way beyond the fame with any fenfible effect. At very fmall diftances indeed it is fufficient to raife up liquors, as alfo to produce the many odd and furprizing appearances which are to be met with in chymical operations, and which without the affiftance of this and

fome

some other principles, which I fhall hereafter have occafion to mention, are utterly inexplicable. For want of a due knowledge of thefe powers chymifts have fallen into grofs miftakes and abfurdities in their reafonings. Thus for inftance, fome who were unacquainted with the principle of attraction, have attempted to give a reafon for the floating of the minute particles of folid bodies in menftruums fpecifically lighter than themfelves; by faying that there is an inteftine motion in the parts of the menftruums, by virtue whereof the particles of the folid bodies are driven perpetually from place to place, and by that means are kept from falling : not confidering that Sir ISAAC NEWTON has demonftrated in the nineteenth propofition of the fecond book of his principles, that fluids have not naturally any inteftine motion; but that fetting afide all external caufes of motion, the particles of fluids are as perfectly at reft as thofe of folid bodies. There is indeed during the time of the folution a confiderable motion, but as this is occafioned by the mutual attraction between the menftruum and the body, by means of which attraction the parts of the fluid are driven with great force between the parts of the folid, fo as to loofen and divide them one from another; as foon as the folution is over the motion ceafes, and all the parts are at reft again, and the particles of the diffolved body are kept fufpended by their clofe adhefion to the parts of the menftruum, and not by any imaginary motion, wherewith they are toffed to and fro in the manner of a fhuttle-cock ; and in truth, could fuch an inteftine motion be allowed, as it muft be made in all manner of directions, it would be as apt, nay more apt confidering the confpiring gravity of the particles, to precipitate and caft them down, than to raife and keep them up.

Were it not befide my prefent purpofe, I could produce many more inftances of falfe reafonings in the Writings of chymifts, occafioned by their ignorance of the true principles of nature; but as chymiftry is at prefent out of my province, I fhall reft contented with the fingle inftance which I have given.

LECTURE II.

Of Attraction.

HAVING in my former Lecture proved from experiments, that there is a power in nature whereby the parts of matter, which are brought fo near as to touch, do in fome circumftances mutually attract each other: I fhall now treat of fuch kinds of attraction as extend themfelves to confiderable diftances beyond the point of contact, and on that account affect the mind more ftrongly, fo as to convince it more fully of the reality of fuch a principle. Of this kind is, Firft, that attraction which obtains between glafs and glafs. Secondly, that of electricity. Thirdly, the attraction of magnetifm. And laftly, that of gravity; of all which in their order.

And firft, if a glafs bubble be fet to float on water contained in a glafs veffel, at a fmall diftance from the fide of the veffel, it will from a ftate of reft begin to move towards the fide of the veffel; and its motion will be continually accelerated, fo as to make it upon its arrival at the fide of the veffel to ftrike the fame with fome force.

Perhaps it may be thought that the motion of the bubble arifes from fome declivity in the water towards the fides of the veffel: but whoever obferves the furface of the water will find, that it rifes all about the fides of the glafs, fo as to become of a concave figure, and for that reafon may retard, but can by no means promote the motion of the bubble; and this rifing of the liquor about the fides of the veffel is to be attributed to the fame caufe with the motion of the bubble, namely, the attraction of the glafs.

The

The acceleration obfervable in the bubble's motion arifes from two caufes; the firft is, the continued and uninterrupted action of the attractive force
of the glafs; for if we fuppofe the time of the
bubble's motion to be divided into a number of equal
parts, as for inftance ten; and if the attraction of
the glafs be fuppofed to make equal impreffions on
the bubble in each of thofe parts of time, it is plain
that whatever be the motion which is excited in the
bubble by the impreffion of attraction in the firft
portion of time, the fame will be doubled in the
fecond, tripled in the third, and fo on continually
thro' the feveral portions of time; for the motion
produced in the firft portion of time is not loft, and
therefore by the addition of as much more in the
fecond portion of time it becomes double, and in the
third triple, and fo on.   Now if inftead of ten parts
we fuppofe the time of the motion to be divided
into numberlefs parts indefinitely fmall, in each of
which the attraction of the glafs makes equal impreffions on the bubble, as before; the motion will
be continually accelerated, tho' the attractive force
of the glafs fhould continue the fame at all diftances
of the bubble; but the attractive force acts more
ftrongly the nearer the bubble approaches, on which
account the motion is more and more accelerated
the nearer the bubble comes to the glafs.

By electrical attraction, I mean that kind of attraction which is excited in bodies when their parts
are heated by friction, and which doth not difcover
itfelf by any fenfible effect when the bodies are
cold.   Of this fort are the attractive forces, which
amber, rofin, fealing-wax, and indeed moft ful-
phurous fubftances when heated by rubbing, have
been found to exert towards chaff, feathers, leaf-gold,
lamp-black, and many other light fubftances.   But
as the attraction of thefe bodies have fallen within
the notice of vulgar eyes, I think it needlefs to make
any experiment for the proof thereof; but choofe
rather

rather to lay before you some experiments which plainly shew this power to obtain in glass, and that to a very notable degree, tho' it has not till of late been commonly observed. And first,

If a cylindrical tube of flint glass be rubbed briskly with brown paper, or woollen-cloth till it acquires some degree of heat, and be then held near to small pieces of gold or brass leaf; they will begin to move, and some of them will fly towards the tube with great swiftness, and fix themselves upon it so as to adhere thereto, being acted upon by the attractive force of the glass : whilst others during their ascent towards the tube, will before they can reach the same, be driven backward with great violence, as will likewise some of those which touch the glass, being actuated by another force very different from that of attraction, which I shall endeavour to explain to you hereafter. The hotter the tube is made by rubbing, the farther doth its power reach, so as in some cases to act upon the leaf at the distance of a foot or more.

This electrical attraction of glass doth in like manner appear from the following experiments.

If over a globe of glass fixed on an axis, whose Exp. 2. position is horizontal, a parcel of woollen threads be suspended from a semicircular wire, so as that their lower ends may be distant an inch or a little more from the globe, they will suitably to the nature of all heavy bodies, hang down perpendicular to the horizon, and parallel to each other ; if then the globe be moved pretty briskly round its axis, the threads will immediately change their position, so as to have their ends bent a little upward, pointing that way towards which the motion tends ; the rotatory motion of the globe being communicated to the circumambient air wherein the threads hang, and by means thereof in some measure to the threads themselves. Let then an hand be applied to the lower part of the globe, so as to rub the same, and

as

Lect. as foon as it grows warm from the friction, the
II. threads which were before crooked will dart themfelves out into fo many ftrait lines, all pointing towards the center of the globe; but as foon as the
attrition ceafes, and the globe cools, they quit this
direction, and return to their former pofition;
whence it evidently appears that they are attracted
by the glafs, fince they are made to point towards
its center, notwithftanding the contrary directions
that were given them by the motion of the air and
the force of gravity. In this and the two following
experiments there is one remarkable circumftance,
which tho' it does not concern the matter in hand,
yet becaufe I fhall have occafion to have recourfe
to it hereafter, I fhall to prevent the repetition of
experiments take notice of it here. And it is this;
if while the threads are extended and acted upon by
the attraction of the globe, a finger be moved towards the extremity of any of them, they will immediately recede and fly from the touch, and this
they will do upon every approach of the finger.

Exp. 3.   If the axis of the globe inftead of being parallel
to the horizon be placed perpendicular thereto, and
the femicircular wire which fupports the threads be
in the plane of a circle parallel to the horizon, the
threads muft by reafon of their gravity hang down
in lines parallel to the axis of the globe, yet as foon
as the motion and attrition are given to the globe
as before, the threads will begin to raife and extend themfelves towards the center of the globe,
and appear like fo many rays converging towards
that center in a plane parallel to the horizon: fo
that in this cafe the attractive force of the glafs does
not only draw the threads out of the parallel pofition they have to each other, but likewife raifes them
up in a pofition parallel to the horizon, notwithftanding the force of gravity which is conftantly
acting upon them to carry them down.

If

If the threads inftead of being placed without the
globe, be fixed to the axis at the center, and be of
fuch a length as to reach within about an inch of
the furface ; when the globe is turned round, they
will bend backward contrary to the direction of the
motion ; becaufe the included air, tho' it does in
fome meafure partake of the rotation of the globe,
yet doth it not move with equal fwiftnefs, and for
that reafon muft refift the rotation of the threads
and bend them backward. When the threads are
in this ftate, if the attraction of the glafs be excited by attrition as in the two laft experiments,
they will ftraightway extend themfelves towards the
concave furface of the globe conftituting as it were
fo many rays iffuing from the center, and diverging
from one another in a regular manner.

The reafon why the threads in all thefe experiments are ftretched into lines tending either to or
from the center of the globe, feems to be this.
Whatever be the force wherewith the globe acts on
the threads, the direction of it muft be perpendicular to the furface of the globe; confequently in
the fame direction muft the threads move ; but from
the nature of the globe thofe and thofe lines only are
perpendicular to its furface, which either iffue from
or tend towards the central point.

Having faid thus much concerning electrical attraction I now proceed to that of magnetifm. Many
and furprifing are the properties both of the loadftone and magnetical needle, which however I fhall
not here confider ; my intent at prefent being only
to fhew from experiment the law of magnetical attraction ; or in other words, to fhew in what proportion the attractive power of the loadftone varies
according to the different diftances of the iron which
it attracts. And in order to this, let a loadftone
be fufpended at one end of a balance, and counterpoifed by weights at the other ; let a flat piece of
iron be placed beneath it at the diftance of four
tenth

tenth parts of an inch, the ftone will immediately defcend, and adhere to the iron : let the ftone again be removed to the fame diftance, and a weight of four grains and four tenth parts of a grain be thrown into the fcale at the other end of the balance ; this weight will be an exact counterbalance to the attractive force, and prevent the defcent of the ftone ; but if any part of the weight be taken out, the attraction will prevail, and carry the ftone down. If the ftone be placed at half the former diftance, that is to fay, at the diftance of two tenth parts of an inch above the iron, the weight neceffary to hinder its defcent will be about feventeen grains and an half, that is four times as much as before. Confequently, the attractive force of the ftone at the fingle diftance from the iron, is to the fame at the double diftance as four to one, that is reciprocally as the fquares of the diftances.

Perhaps it may be objected that Sir Isaac Newton (to whofe judgment in natural affairs the utmoft regard is due) has faid that the power of the loadftone decreafes nearly in the triplicate ratio of the increafe of the diftance. But whoever confiders his words in the fifth corollary of the fixth propofition of the third book of his principles, where he mentions this law, will find that he fpeaks of it with diffidence, as a thing which he rather gueffed at from fome rude obfervations, than collected from accurate experiments, for his words are, *Et in receffu a magnete decrefcit in ratione diftantiæ non duplicatâ, fed fere triplicatâ, quantum ex craffis quibufdam obfervationibus animadvertere potui.* So that notwithftanding this objection I fhall ftill venture to affirm the law of magnetical attraction to be fuch as makes it act with forces which are in the reciprocal duplicate ratio of the diftance. Becaufe this law is deduced from an experiment made with fufficient exactnefs, and which does not feem liable to any exception.

Tho'

Tho' the principle of gravity, which comes next to be treated of, be diffufed throughout the folar fyftem, and may probably be extended fo far as to reach the other fyftems of the univerfe ; yet fhall I confider it at prefent with refpect only to the globe of earth, which we inhabit; the parts whereof would by reafon of the diurnal rotation be apt to fly afunder, were they not kept together by the influence of this principle ; whereby likewife all bodies on or near the furface of the earth are made to tend towards its center. This power at equal diftances from the center of the earth is always proportional to the quantity of matter in the body whereon it acts ; for all bodies, the light as well as heavy, being let fall from the fame height defcend with equal fwiftnefs, provided they meet with no refiftance from the air, as will appear from the following experiment. Let a piece of gold and a fea- Exp. 7. ther be let fall from the top of an exhaufted receiver at the fame inftant of time, and they will both arrive at the bottom at the fame time very nearly.

The reafon why the feather doth not reach the bottom quite fo foon as the gold, is, that the receiver cannot be perfectly exhaufted, and therefore the fmall portion of air which remains within, though very much rarified, gives fome fmall refiftance to the defcending bodies, which fuitably to the nature of all refiftance muft retard the lighter body more than the heavier, and confequently caufe fome little difference in the times of the defcent, which otherwife would be exactly equal. This then being the cafe, it evidently follows, that the forces of gravity, whereby bodies defcend, muft at equal diftances from the center be as the quantities of matter in the defcending bodies ; for if a certain force of gravity be requifite to carry down a certain quantity of matter with a certain fwiftnefs, then is double the force neceffary to carry down a double quantity of matter with the

4 fame

same fwiftnefs; and triple the force to carry down a triple quantity, and fo in proportion whatever be the quantity of matter: fo that the weights of bodies at equal diftances from the center of the earth are always proportional to the quantities of matter which they contain; and therefore the quantity of matter in any body may be meafured by its weight.

The gravity of a body at any place beneath the furface of the earth has been proved by Sir ISAAC NEWTON, to be directly as the diftance from the center; that is, fuppofing the earth's radius to be four thoufand miles, a body which on the furface of the earth weighs a pound, will within the earth at the diftance of two thoufand miles from the center weigh only half a pound, at the diftance of one thoufand miles only a quarter, and fo on till at the center it lofes all its gravity.

It has been likewife proved that the force of gravity on the furface of the earth, and at all diftances beyond it, is in the reciprocal duplicate ratio of the diftance from the center; that is, if a body weighs a pound at the furface of the earth, whofe diftance from the center is four thoufand miles, it will at double that diftance weigh only a quarter of a pound, and at triple the diftance, only the ninth part of a pound, and fo on, whatever be the diftance the force of gravity will be reciprocally as the fquare of the diftance. For is it not highly rational that the power of gravity, whatever it be, fhould exert itfelf more vigoroufly in a fmall fphere, and weaker in a greater, in proportion as it is contracted or expanded? and if fo, feeing that the furfaces of fpheres are as the fquares of their *radii*, this power at feveral diftances muft be as the fquares of thofe diftances reciprocally. Tho' ftrictly fpeaking, this be the law of gravity, yet where the diftances from the furface are inconfiderable with refpect to the earth's radius

the

the force of gravity may be looked upon as equal
at all thofe diftances; thus for inftance, the gravity of a body at the diftance of half a mile from the
earth may be looked upon as equal to the gravity
thereof at the diftance of a quarter of a mile; or
at the very furface; becaufe the difference is fo
fmall, that if it be rejected it will not occafion any
error in calculations. And indeed on this fuppofition are founded moft of the reafonings of GAL-
LILÆO, TORRICELLIUS, HUYGENS, and other
naturalifts concerning the defcent of heavy bodies;
and by the help of the fame fuppofition have the
feveral theorems been formed relating to the acceleration of falling bodies, the fpaces defcribed, the
times of the fall, and the velocities thereby acquired; as I fhall now fhew you.

If the force of gravity whereby a body defcends
remains unvaried, the motion of a body falling by
fuch a force will be accelerated, and that uniformly; that is, the velocity will increafe, and the increments thereof in equal times will be equal. For
let us fuppofe the time of the defcent to be divided
into a number of equal parts indefinitely fmall, in
each of which by fuppofition, the force of gravity
makes equal impreffions on the body to carry it
down; whatever velocity therefore the body receives from the impreffion of gravity in the firft
portion of time, it muft receive as much in every
other portion; fince therefore fetting afide all outward lets and obftacles the effect of every impreffion
remains, the velocity given in the firft portion of
time, will be doubled in the fecond, tripled in the
third, quadrupled in the fourth, and fo on continually thro' the feveral portions of time. So that
the velocity of a body falling by the force of gravity will conftantly increafe in the fame proportion
with the time of the defcent. Or in other words,
the motion of a body carried down by the force of
gravity will be uniformly accelerated: and the ve-

C   locities

locities acquired will be as the times of the defcent from the beginning of the fall.

Fig. 3.
From what has been faid it follows, that if a right line as AB be fuppofed to denote the time of a body's fall, and another right line as BC fet at right angles to the former, to exprefs the velocity acquired by the falling body in the time denoted by AB. The triangle ABC being compleated, and another right line as DE drawn parallel to BC, then will DE denote the velocity acquired by the falling body in a portion of time, which is to the time denoted by AB, as AD to AB. For from the nature of fimilar triangles, AB is to AD as BC to DE; but BC exprefſes the velocity acquired where the time is as AB, confequently, fince the velocities are as the times of the defcent, DE will exprefs the velocity acquired in the time denoted by AD.

And what has been thus proved of the line DE, is in like manner true of any other right line, as FG, or HI, drawn within the triangle parallel to the bafe; for FG and HI will exprefs the velocities acquired in the times denoted by AF and AH.

The fpaces defcribed by bodies falling from a ftate of reft by the force of gravity are to one another as the fquares of the times from the beginning of the fall. In the triangle ABC, let AB exprefs the time of a body's fall, and BC the velocity acquired at the end of the fall, let AB be divided into a number of equal parts indefinitely fmall; and from each of thofe divifions fuppofe lines, as DE drawn parallel to BC; it is evident from what has been faid, that thofe lines will exprefs the velocities of the falling body in the feveral refpective points of time; which velocities, inafmuch as the body is given and the portions of time are indefinitely fmall, will be as the refpective fpaces defcribed in thofe times : but the fum of the fpaces defcribed in all the fmall portions of time is equal

Fig. 4.

to

to the fpace defcribed from the beginning of the fall ; and the fum of all the lines, as D E taken in- definitely near each other conftitute the area of the triangle. And therefore the fpace defcribed by a falling body in the time expreffed by A B, and where the velocity acquired at the end of the fall is denoted by B C, will be as the area of the triangle A B C. And for the fame reafon the fpace defcribed by a falling body in the time expreffed by A D will be as the area of the triangle A D E. But from the nature of fimilar triangles thefe areas are to one an- other as the fquares of their homologous fides ; that is, as A B^q to A D^q, or as B C^q to D E^q. But A B and A D exprefs the times of the fall; and B C and D E the velocities acquired by the fall; where- fore the fpaces defcribed by a falling body are as the fquares of the times from the beginning of the fall, or as the fquares of the velocities at the end of the fall. And what has been thus demon- ftrated from the nature of gravity is likewife con- firmed by experiments. For if a weight of eleven Exp. 8. hundred grains be let fall from the height of three inches, fo as to ftrike one end of a balance, its force will be juft fufficient to raife a pound weight at the other end of the balance to the height of about the eighth or tenth part of an inch ; whereas if the fame body be required to raife a weight of two pounds to the fame height, it muft be let fall from the height of twelve inches ; and if the weight to be raifed, be three pounds, then muft the moving body fall from the height of twenty feven inches, for leffer heights will not fuffice, as will appear from the experiment.

The forces wherewith the defcending body ftrikes the end of the balance are meafured by the weights that are raifed ; which in this cafe are as one, two, and three ; but the forces wherewith one and the fame body ftrikes, are as the velocities of the body ; wherefore in the cafe before us the velocities acquired

by

by the falling body are as one, two, and three; but the heights from which it defcends in order to acquire thofe velocities are as one, four, and nine; that is, as the fquares of the velocities.

Exp. 9.   If this experiment be repeated with a body double in weight to the former, to wit, with one of twenty two hundred grains; the weights raifed by the ftrokes will be two, four, and fix pounds, to wit, double the former.

From this experiment appears the truth of that rule, which collects the quantity of motion in any body by multiplying the velocity of the body into its quantity of matter. For the force of a ftroke is, *cæteris paribus*, always proportional to the quantity of motion in the ftriking body; confequently in like circumftances the motions of bodies may be meafured by the force of their ftrokes; but it has appeared from the experiment that where the ftriking body is as unity, and the velocities wherewith it moves at the times of the ftrokes, as one, two, and three; the forces of the refpective ftrokes are likewife as one, two and three. But where the body is as two, the ftrokes are as two, four and fix: that is, in both cafes the ftrokes are as the products arifing from the multiplication of the quantities of matter in each body into the refpective velocities; wherefore the quantities of motion are as thofe products. Whence as a corollary it follows, that if the weight of one body multiplied into its velocity gives an equal product to what arifes from the multiplication of the weight of another body by its velocity, the motions of thofe two bodies are equal; and this will ever be where the weights of the bodies are reciprocally proportional to their velocities. Thus when the body whofe weight was as unity, was let fall from the height of twelve inches, and thereby acquired a velocity which was as two; it raifed a two pound weight, which was likewife raifed by the body whofe weight was as two, when by fall-

ing

ing from the height of three inches, it had acquired a velocity which was as unity.

From what has been proved concerning the spaces
described by falling bodies it follows, that if the
time of a body's fall be divided into a number of
equal parts, the spaces thro' which it falls in each
of those parts of time taken separately and in
their order, beginning from the first, are as the
odd numbers taken likewise in their order, beginning from unity. For instance, if the time of the
fall be four seconds, the space described in the first
of those seconds will be as one, in the second as
three, in the third as five, and in the fourth as seven; for where the times of the fall are as one,
two, three and four; the spaces described are as
one, four, nine and sixteen; and therefore if from
the space described in two seconds, to wit, four, be
subducted the space described in the first second, to
wit, one, the remainder, to wit, three, will be the
space described in the next second. And if from
nine, which is the space described in three seconds,
be taken four, which is the space described in two
seconds, the remainder, which is five, will be the
space described in the third second. In like manner
subducting nine, the space described in three seconds,
from sixteen, which is the space described in four
seconds, the remainder, to wit, seven, will be the
space described in the fourth second; and so on according to the number of parts into which the time
of the fall is divided.

From what has been said it likewise follows, that
the velocity acquired by a falling body at the end of
the fall is such as with an equable motion would in
the same time in which the body fell, carry it thro'
a space double that of the fall. That the truth of this
may be made appear, it is necessary that some things
be premised concerning the spaces described by bodies carried with an equable motion. And first, if
the velocity of a body moving uniformly be given,

the

the fpace defcribed will be as the time of the motion; for if a body with a given velocity moves
thro' a certain fpace a foot, for inftance, in a fecond
of time, it will in two feconds, with the fame velocity, move thro' two feet, and thro' three feet in
three feconds, and fo on, whatever be the time,
the fpace defcribed will be proportional thereto.
On the other hand, if the time be given, the fpace
defcribed will be as the velocity; for if a body in
a given time moves thro' the fpace of a foot with
a certain velocity, with double the velocity it will
pafs thro' the fpace of two feet, and with triple the
velocity thro' the fpace of three feet, and fo on,
whatever be the velocity, the fpace defcribed will
be in the fame proportion.   But if neither the time
of a body's motion, nor the velocity wherewith it
moves be given, the fpace defcribed will be as the
time and velocity conjointly; for if a body moving
with a certain velocity runs thro' a certain fpace in a
certain time, it follows from what has been faid, that
if the time be increafed or diminifhed in any proportion, in the fame alfo will the fpace be increafed
or diminifhed, fuppofing the velocity to remain the
fame, but if that likewife be changed, it is plain
that the fpace will be changed in the fame proportion; and therefore univerfally the fpace defcribed
by a body moving equably is as the time and velocity conjointly.   For which reafon, if in the rec-

Fig. 5.
tangle, one fide, as A B, be fuppofed to denote the
time wherein a body moves equably, and BC the
velocity wherewith it moves, the rectangle ABCD
will be as the fpace defcribed; but the triangle
ABC of the fame figure, is as the fpace defcribed
by a falling body in the time denoted by AB, and
BC is as the velocity acquired at the end of the
fall; and the rectangle ABCD is double the triangle ABC, confequently the velocity acquired by
a falling body is fuch as will carry the body with an
equable motion in the time of the fall thro' double
the fpace of the fall.   As

As the motion of bodies falling from a state of rest is uniformly accelerated; so likewise the motion of bodies thrown upward is uniformly retarded; for the same force of gravity which conspires with the motion of descending bodies, acts in direct opposition to the motion of such as ascend; and therefore in whatever manner it accelerates·the one, in the very same manner must it retard the other. Whence it follows, that if a body be thrown directly upward, the time of its rise will be equal to that wherein a body falling freely from a state of rest, acquires the same velocity wherewith the body is thrown up. For since the action of gravity is constant and uniform in whatever time it generates any velocity in a falling body, in the same time must it destroy that velocity in a rising body; and therefore the time of the rise must be equal to that of the fall. It likewise follows that the height to which a body thrown upward rises is equal to that from which a body falling freely does at the end of the fall acquire a velocity equal to that wherewith the body is thrown up. For since the times in which the velocity of the falling body is generated, and that of the rising body is destroyed, are equal; and since of the two equal velocities one is generated and the other destroyed by the constant uniform action of one and the same power; it is manifest, that whatever be the space thro' which the falling body moves in order to acquire its velocity, the rising body must ascend thro' an equal space in order to lose its velocity; that is, it must rise to the same height from which the other falls.

The force of gravity at the surface of the earth is such as, setting aside the resistance of the air, makes a body falling from a state of rest to descend thro' a space of sixteen feet and an inch in a second of time. For the time wherein a pendulum performs its smallest vibrations is to the time in which a body falls thro' half the length of the pendulum as the

circum-

circumference of a circle to its diameter (as shall be shewn when I come to treat of the pendulum) wherefore since the spaces described by falling bodies are as the squares of the times, and since the diameter of a circle expresses the time which a body takes to fall thro' half the length of a pendulum vibrating seconds, when the circumference expresses a second; it follows, that as the square of the diameter is to the square of the circumference, so is half the length of the pendulum to the space thro' which a body falls in a second of time. So that putting D to denote the diameter of a circle, which is as unity, P the periphery which is as 3,1416, L the length of the pendulum vibrating seconds, which is $39\frac{1}{8}$ inches, and S to denote the space sought; we shall have this analogy, $D^2 : P^2 :: \frac{L}{2} : S$.

Consequently $S = \frac{P^2 \frac{1}{2} L}{D^2}$, or rejecting the divisor as being equal to unity, $S = P^2 \frac{1}{2} L = 193$ inches, or sixteen feet and an inch.

Before I quit this subject I must observe to you, that bodies do not every where descend at the rate of sixteen feet and an inch in a second of time, but in such places only as are in or near the latitude of forty nine degrees; in places more distant from the line the descent is quicker, and more slow in those less distant. For the force of gravity is less towards the æquator than towards the poles, as has been collected from observations made on pendulums; for they have been found to vibrate more slowly near the line than in places farther removed; insomuch that a pendulum which in the latitude of Paris vibrates seconds, must be shortened one sixth of an inch French measure in order to its vibrating seconds under the line. And the length of a pendulum which in the latitude of Paris performs its vibrations in a second, is to the length of a pendulum whose vibrations are performed in the same
time

time under the line as 220 to 219. Since therefore
the forces of gravity which actuate pendulums that
vibrate in equal times are to one another as the
lengths of the pendulums (as shall be shewn when I
come to treat of pendulums) it is evident that the
force of gravity in the latitude of Paris is to the
same force under the line as 220 to 219. And indeed it has appeared from a great number of observations, that the force of gravity is least at the
æquator, and that it continually increases as we
recede from thence and approach the poles, under
which it is greatest of all. And the chief cause of
this difference is the rotation of the earth about its
axis, whereby all bodies on or near the surface of the
earth are indued with a centrifugal force, which acts
in opposition to that of gravity, and of course must
lessen the same; and the diminution of gravity
arising from this cause must be greatest under the
æquator, and grow less and less in the approach
to the poles: and that for two reasons, first, because the centrifugal force is greatest at the æquator, and from thence towards the poles is continually diminished so as at last to vanish in the polar
points. For all parts of the earth's surface with the
bodies thereto adjacent revolve in the same time
either in the æquator or in circles parallel thereto;
but the æquator is the largest of all those circles,
and the others grow less and less as they are more
and more distant from the æquator. Now the
centrifugal forces of bodies revolving in the same
time in different circles being to one another as the
radii of the circles (as shall be shewn when I come
to treat of those forces) it follows that the centrifugal force must be greatest at the æquator, and
thence be continually diminished towards the poles.
To illustrate this, let A B be the axis of the earth, Fig. 6.
C K the radius of the æquator, D I, E H and F G
the radii of so many circles parallel to the æquator,
the centrifugal forces in the points K, I, H, G, are
as

as thofe radii; fo that the centrifugal force is greateft in the point K, that is at the æquator, and at I
it is lefs than at K, and at H lefs than at I, and lefs
again at G, and fo on till at length it vanifhes at
the polar point where there is no rotation. Whence
it is evident that the force of gravity muft be
fmalleft under the line, and muft increafe towards
the poles, inafmuch as the force which acts in oppofition to it is greateft under the line, and leffens
in the approach to the poles. The force of gravity
muft likewife be lefs under the æquator than in
any other place, becaufe under the line the centrifugal force acts in direct oppofition to the force of
gravity, whereas in other places it acts in an oblique
direction to that of gravity, and of confequence
muft act lefs powerfully againft it. Thus in the
point K the force of gravity pulleth from K towards C, whilft the centrifugal force pulleth directly contrary from C towards K; whereas in the
point L gravity pulleth from L towards C, whilft
the direction of the centrifugal force is from O towards L. Let the centrifugal force in the point L
be expreffed by the line LM, and to CL continued to N let fall the perpendicular MN. The force
LM, according to the known method of refolving
forces, of which I fhall fpeak hereafter, may be
refolved into two forces denoted by the lines NM,
and LN; whereof the latter only acts in oppofition to gravity, as pulling directly againft it; the
other no way affecting the fame: confequently, fuppofing the centrifugal force at L to be the fame as
at K, yet will the force of gravity be lefs diminifhed by it at L than at K, becaufe at L part only of
the centrifugal force refifts that of gravity, whereas at K the whole centrifugal force acts in oppofition thereto.

From what has been faid it follows, that the force
whereby gravity is leffened in the æquator is to the
force whereby it is leffened in any other part of the

earth's

earth's furface as the fquare of radius to the fquare of the fine of the compliment of latitude. For the centrifugal force in the point K, the whole of which acts in oppofition to gravity, is to the centrifugal force in the point L, as CK or CL to OL; but the whole centrifugal force in L is to that part of it which oppofes gravity, as LM to LN, that is, becaufe the triangles LNM and COL are fimilar, as CL to OL; wherefore the centrifugal force, or the force which oppofes gravity in the point K, is to that part of the centrifugal force which oppofes gravity in the point L in the duplicate ratio of CL to OL, that is, as the fquare of radius to the fquare of the fine of the compliment of latitude.

LECTURE III.

Of Repulsion and Central Forces.

AS experience has convinced us that there are Powers in nature, whereby not only the larger fyftems and collections, but likewife the fmaller parcels and particles of matter are in fome cafes made to tend to one another; the fame experience will inform us of other powers in nature, whereby the parts of matter do in fome circumftances recede and fly from each other. For if the difagreeing pole of a loadftone be moved towards a magnetical needle floating on water, the needle will recede; and the nearer the ftone is brought to it, with the greater violence and precipitation will it fly off; the repelling power, like the attractive, exerting itfelf with greater vigor at fmaller diftances.

This repelling power is likewife evident from the experiments which were made relating to electrical attraction: for it was obfervable that upon holding the glafs tube, when heated by friction, nigh fmall pieces of brafs-leaf; fome of thofe pieces which by the

the attraction had been raised towards the tube, were, before they could reach it, driven back again with great precipitation : and of those which adhered to the tube some were thrown off with a velocity much greater than could possibly arise from the force of gravity in such light bodies, and consequently must have been driven down by some repelling power in the glass. And in the experiments of the glass-globe and woollen threads; when the threads were, by the attractive force of the globe, made to extend themselves towards its surface, upon moving one's finger towards them, they were observed to recede and fly off, and that at considerable distances from the finger ; which plainly argues a repelling power interceding the finger and the threads, when under the circumstances of those experiments. . From this power it is, that the leaves of the sensitive plant shrink and retire from the touch of an approaching hand. And to the same power we are to attribute the elasticity of the air; as also the shaking off of the particles of light from the sun and other luminous bodies.

Besides the forementioned principles of attraction and repulsion, whereby nature seems to perform most of her operations, and which for that reason are very properly stiled active principles ; there is another of a passive nature, commonly called the *vis insita* and *vis inertiæ* of matter, a force arising from the inertness or inactivity of matter ; which force in any body is proportional to its quantity of matter. From this force result three passive laws of motion, usually called by modern naturalists the three LAWS OF NATURE[*].

The

---

[*] By virtue of the *vis inertiæ* it is, that the motion of a body produced by a force impressed upon it, is measured by the quantity of matter in the body and its velocity, taken together. For the body by its *vis inertiæ*, resists the force impressed upon it which causes its motion, in proportion to its quantity of matter ; and consequently, to produce a given tendency in the body forward,

The firſt of theſe laws is, That every body, in proportion to its quantity of matter, perſeveres in its preſent ſtate, whether it be of reſt or uniform motion ſtraight forward in a right line. For as every particle of matter is with reſpeĉt to itſelf perfeĉtly unaĉtive, it is utterly impoſſible it ſhould produce any alteration in its own ſtate; for which reaſon (ſetting aſide all impreſſions from external cauſes) if it be at reſt, it muſt continue ſo for ever; or if in motion, it muſt for ever continue its motion without any change either as to direĉtion or velocity: ſo then the continuation of motion in bodies projeĉted, (the cauſe whereof very much perplexed the naturaliſts of old) is to be attributed to the paſſive nature of matter, which makes it as impoſſible for a body of itſelf to ſtop its own motion when once begun, as it is for it to move itſelf originally, or of itſelf to change its figure.

As a conſequence of this law it follows, that all motion is of itſelf equable and reĉtilineal. For firſt whatever be the velocity wherewith a body begins to move, the ſame velocity muſt continue during the motion, unleſs a change be made therein by ſome cauſe from without; wherefore the body

forward, by which it moves at a given rate or with a given velocity, the force impreſſed muſt be proportional to the reſiſtance ariſing from its *vis inertiæ*, that is, to its quantity of matter; and if the quantity of matter in the body, and conſequently the reſiſtance ariſing from its *vis inertiæ*, be given, the force impreſſed will be proportional to the tendency forward which it communicates to the body, that is to its velocity; and if neither the quantity of matter in the body nor its velocity be given, the force impreſſed will be in a ratio compounded of the quantity of matter and velocity; that is, putting F for the force impreſſed, Q for the quantity of matter in the body, and V for its velocity, F will be as $Q \times V$. But the motion of the body is the effeĉt produced by the force F, and is proportional to it, that is, putting M for the motion of the body, M is as F. And therefore, by proportion of equality, M will be as $Q \times V$; or the motion of the body will be meaſured by its quantity of matter and velocity taken together.

muſt

muſt in equal times move thro' equal ſpaces with an
uniform velocity ; that is, the motion muſt be equable. And as motion is by virtue of this law in itſelf equable ; ſo is it likewiſe rectilineal : for motion cannot otherwiſe be conceived than as directed
and determined towards ſome place or other ; and
it muſt by the foregoing law keep the direction
which it had at firſt, until it be hindered or put out
of its way by ſome extrinſic cauſe, that is, it muſt
move on in a right line. If therefore a body moves
in a curve, that curvature muſt of neceſſity proceed from ſome external force continually acting on
the body ; and whenever that force ceaſes to act,
the body will move forward in a right line touching the curve in that point wherein the body is at
the inſtant of time when the force ceaſes to act.
Thus for inſtance, if a ſtone, moved about in a
ſling, be ſet at liberty by ſlipping one end of the
ſling ; it will not continue its circular motion, but
go on in a right line touching the circle made by the
circumvolution of the ſling in that point where the

Fig. 8.
ſtone is let go. If the circle B C D E be the curve
deſcribed by the revolution of the ſling A B about the
center A ; and if the ſtone be let off at the point
B, it will move on in the right line B G, which
touches the circle in B. For by the law, the natural
tendency of the ſtone in the point B is along the
line B G, tho' by the force of the ſling it be made to
revolve in the curve. And what has been ſaid of
the ſtone in the point B, is in like manner true of
the ſame at any other point, as C, D, or E ; for in
thoſe points its tendency is along the lines C F,
D H, and E K.

Another conſequence of the foregoing law is that
all bodies, which revolve about a center, muſt endeavour to recede from the center : for ſince bodies,
that are moved round in a curve, do of themſelves
in every point of the curve tend to move in the
tangents to each point ; and ſince all the parts of
the

the tangents are more diſtant from the center of motion than are the parts of the curve, as is evident from the figure; it is manifeſt that bodies ſo moved muſt perpetually endeavour to fly off from the center of motion, which endeavour of receding is commonly called the centrifugal force; and it is oppoſed to the centripetal force, or that force which by drawing the bodies towards the center makes them to revolve in a curve.

Theſe two forces are by one common name called the central forces: and they are in all caſes equal the one to the other. For let us ſuppoſe a body to revolve in the orbit E A C, and that being in the point A, the centripetal force ceaſes to act; it will then move forward in the direction of the tangent A B, and B C will be the ſpace thro' which the body recedes from the orbit by means of the centrifugal force; and if A B be in its naſcent ſtate, the centrifugal force will be as B C; but if the centripetal force acts at A, it will make the body deſcribe the arc A C in the ſame time that it would deſcribe the tangent A B, in caſe it were not acted upon by the centripetal force; conſequently, the ſpace B C is deſcribed by means of the centripetal force; and the arc A C being in its naſcent ſtate, the centripetal force will be as B C, and of conſequence equal to the centrifugal.

In treating of theſe central forces I ſhall proceed in the following manner. Firſt, I ſhall conſider two equal bodies moving uniformly in two different circles; and thence deduce one general expreſſion for the central forces in the terms of the circle. Secondly, By ſubſtituting other proportional quantities in the place of thoſe which conſtitute the general expreſſion, I ſhall form other general expreſſions for the ſame forces. Thirdly, by a proper application of thoſe expreſſions I ſhall determine the laws of central forces in particular caſes, and

Fig. 9.

and at the fame time confirm each law by an experiment.

Fig. 10.
11. As to the firft, if two equal bodies moving uniformly in the circles marked 1, and 2, do in the fame
portion of time taken indefinitely fmall defcribe the
nafcent arches A C ; and if from the points C be
drawn the lines C B, perpendicular to the tangents
A B, thofe lines will exprefs the proportion of the
central forces. For fince the time in which the
arches A C are defcribed is indefinitely fmall, the
bodies will be carried thro' the fpaces B C, by one
fingle impulfe of each central force ; for which reafon the motions of the bodies thro' thofe fpaces
will be uniform ; confequently, fince the time of
the motion is the fame, and the bodies equal, the
motions will be as the fpaces defcribed, that is, as
the lines B C ; but forces which generate equable
motions are to one another as the motions generated ; that is, in this cafe, as the lines B C ; which
lines being equal to the verfed fines A D of the
arches A C, muft be equal to the fquares of the
arches A C, divided by their refpective diameters
A E. For from the nature of the circle, the verfed
fine of any arch is equal to the fquare of the chord
divided by the diameter ; but as in this cafe the
arches A C are fuppofed to be nafcent, they do not
differ from their chords ; and therefore in each
circle the verfed fine of the arch A C, (which
verfed fine exprefes the central force) is equal to the
fquare of the arch divided by the diameter : confequently, the central forces are as the fquares of the
nafcent arches applied to their refpective diameters ;
and forafmuch as thofe nafcent arches, are to one
another as any other two arches which are defcribed
by the revolving bodies in a given time, the central
forces of two equal bodies revolving uniformly in
different circles, are to one another as the fquares
of the arches defcribed in a given time applied to
their

their refpective diameters; or becaufe the diameters are as the radii, as the fquares of the arches applied to their refpective radii. Wherefore putting A to denote the arch of a circle defcribed in a given time, D for the radius, and F for the central force. F is as $\frac{A^2}{D}$, as it ftands in the firft place of the firft rank of fymbols.

F is as $\frac{A^2}{D}$.	F is as $\frac{QA^2}{D}$.
F is as $\frac{V^2}{D}$,	F is as $\frac{QV^2}{D}$.
F is as $\frac{D}{P^2}$.	F is as $\frac{QD}{P^2}$.
F is as DN^2.	F is as QDN^2.

As the bodies are fuppofed to move uniformly in the circles, it is evident that the arches defcribed in a given time are as the velocities of the revolving bodies; and therefore in the general expreffion for the central force, the velocity of the body may be fubftituted in the place of the circular arch; whence putting V for the velocity of the body, F is as $\frac{V^2}{D}$, as in the fecond place of the firft rank of fymbols, which is a fecond general expreffion for the central force.

Again, the velocity of a body moving uniformly in a circle, is as the radius applied to the periodic time, or the time of one intire revolution. For if the velocity of the body be given, the periodic time muft be proportional to the circumference of the circle, inafmuch as a body, which with a given velocity defcribes a certain fpace in a certain time, will require a double or triple time to defcribe a double or triple fpace; and univerfally whatever be the magnitude of the fpace, the time in which it is

D defcribed

deſcribed will be proportional to it. If the circumference of the circle be given, the periodic time will
be inverſly as the velocity with which the body
moves; for if a body moves thro' a given ſpace
with a certain velocity in a certain time, it will
with double the velocity move through the ſame
ſpace in half the time, and with a triple velocity in
one third of the time; and in general, in whatever
proportion the velocity is increaſed, in the ſame
proportion will the time be leſſened; that is, the
periodic time will be inverſly as the velocity. If
therefore neither the circumference of the circle,
nor the velocity of the body be given, the periodic
time will be directly as the circumference, and inverſly as the velocity; that is, as the circumference
applied to the velocity; or (becauſe the circumference is as the radius) as the radius applied to the
velocity. Wherefore putting P for the periodic
time of a body revolving in a circle, P is as
$\frac{D}{V}$, and conſequently V is as $\frac{D}{P}$. If therefore in the

ſecond general expreſſion $\frac{D}{P}$ be ſubſtituted in the

place of V, we ſhall have a third general expreſſion

for the central force, wherein F is as $\frac{D}{P^2}$, as in the

third place of the firſt rank of ſymbols.

Again the periodic time of a body revolving
uniformly is inverſly as the number of revolutions
performed in a given time. For if the periodic
time of a body be ſuch, as that in a given time it
can perform a certain number of revolutions; if the
periodic time thereof be doubled, it will perform
but half the number of revolutions in the ſame time;
and if the periodic time becomes thrice as great,
it will perform but one third of the number of revolutions in the given time; and ſo on, as the periodic time is inlarged the number of revolutions
will

will be diminifhed in the fame proportion, fo that putting N for the number of revolutions in a given time, P will be as $\frac{I}{N}$. Confequently, if in the third general expreffion $\frac{I}{N}$ be fubftituted in the room of P, we fhall have a fourth general expreffion for the central force, wherein F is as DN^2, as it ftands in the laft place of the firft rank of fymbols.

In collecting thefe general expreffions, I have all along fuppofed the quantity of matter in the revolving body to be given; and for that reafon have not made it a part of thofe expreffions, inafmuch as it may be denoted by unity; and as fuch, whether it be taken in, or left out, it will not vary the expreffions. But the cafe will be different, if the quantity of matter in the revolving body varies; becaufe the central forces, and confequently the expreffions for thofe forces will likewife vary; fo as to be greater *cæteris paribus* in larger quantities of matter than in fmaller. For the whole central force of any body, is made up of the forces of each particle whereof the body confifts; and therefore the more numerous the particles of matter are in any body, the greater will its central force be; fo as to be double in a double quantity of matter, triple in a triple quantity; and fo on in proportion to the quantity of matter. In order therefore to render the expreffions yet more general, let Q be put for the quantity of matter in the revolving body, and let it be multiplied into each of the four expreffions, as in the fecond rank of fymbols.

Before I apply thefe expreffions to the feveral particular cafes, I fhall offer an experiment in confirmation of what I juft now proved, *viz.* that the greater the quantity of matter in any body is, the greater is the central force.

Let

Let three glafs tubes half full, one with mercury and water, and another with water and fmall leaden bullets, the third with water and a piece of cork, be ftopped clofe, and made faft to an inclined plane; and let the plane be fo fixed to a table moveable about its center by means of a wheel and axle, as that the lowermoft part of the plane may reft upon the center of the table. As long as the table continues at reft, the liquors and folids contained in the tubes, will by reafon of their gravity poffefs themfelves of thofe parts of the tubes which lie next the center of the table, leaving the remoter parts empty: and of the two bodies included in each tube, that which is heavieft will be neareft the center; but upon turning the table about, the feveral bodies will by reafon of their centrifugal forces, whereby they are carried from the center of motion, fly to the uppermoft parts of the tubes; and in each tube, the heavier body will poffefs the uppermoft place as being indued with the ftronger centrifugal force.

If bodies moving in equal circles perform their revolutions in equal times, or in other words, if the velocities of bodies revolving in circles be equal, and their diftances from the center likewife equal, their centrifugal forces are as their quantities of matter. For in the fecond general expreffion, fince V and D are given, F is as Q; that is, the central force is as the quantity of matter; which is con-

firmed by the following experiment. Let two fmall troughs be fo fixed to two moveable tables, as that the centers of the troughs may lie upon the centers of the tables, and let the centers of the tables be fixed to two axles, on each of which is a grooved wheel, with equal diameters; let the two wheels be turned by means of one and the fame chord going round them: it is manifeft, that as the wheels are equal, they, and confequently the tables with their affixed troughs, muft perform their revolutions in the fame time; and the parts of the

tables

tables and troughs, whofe diftances from their refpective centers are equal, will revolve equally
fwift; and fo likewife muft all bodies that are placed
in the troughs at equal diftances from the centers;
fo that by this contrivance, if two bodies be placed
one in each trough at equal diftances from the centers, they will revolve equally fwift. Let then two
balls, whereof one is double the other, be laid one in
each trough, and let each ball be faftened to one end
of a chord, whofe other end paffing thro' an hole in
the center of the table is made faft to a weight,
which refts upon the floor; and let the lengths of
the chords be fuch, as that being ftretched, and the
weights not raifed, the balls in the troughs may be
equally diftant from the centers. This being done
if the weights be to one another as the balls, and if
the tables be turned about with fuch a velocity as
that the centrifugal forces of the balls may be fufficient to raife the weights, they will be lifted up precifely at the fame time. Whence it appears, that in
this cafe the centrifugal forces are as the quantities
of matter, inafmuch as they overcome refiftances
which are in that proportion.

If equal bodies moving in unequal circles perform their revolutions in equal times; or in other
words, if the quantity of matter in the revolving
bodies be given, as alfo the number of revolutions
performed in a given time, their centrifugal forces
are as their diftances from the center. For in the
fourth general expreffion fince Q and N are given,
F is as D; that is, the force is as the diftance. For
the confirmation whereof, let two equal balls be
placed in the troughs at diftances from the centers,
which are as one and two, and when the tables are
turned about, that ball, whofe diftance from the
center is double, will raife a double weight.

Exp. 3.

If equal bodies move in unequal circles with equal
velocities; or more generally, if the quantity of
matter in the revolving bodies be given, as alfo the

D 3 velocity

velocity wherewith they revolve; their central forces are inverfly as their diftances from the center. For in the fecond general expreffion fince Q and V are given, F is as $\frac{1}{D}$, that is, the force is inverfly as the diftance. Before I mention the experiment whereby this law is confirmed, I muft obferve to you, that to the axle of one of the tables is fixed a fecond wheel, whofe diameter is but one half of the diameter of the other wheel; and therefore when the chord goes round the fmaller wheel, the table muft turn as faft again as when it goes round the larger wheel; fo that the table which is moved by means of the fmaller wheel, will revolve twice in the fame time, that the other table which is turned by means of the larger wheel, performs one revolution.

Exp. 4.   This being premifed, let two equal balls be fo placed in the troughs, as that the diftance of that ball which is to revolve by means of the fmaller wheel, may be but one half of the other's diftance from the center; in which cafe their velocities will be equal: for tho' the peripheries of the circles which the two balls defcribe, are as one and two; yet will the leffer periphery be defcribed twice in the fame time that the larger is defcribed once; and therefore the fpaces thro' which the bodies move in a given time will be equal, and of confequence their velocities will be fo too. If then two weights be made faft to the chords of the balls in the manner of the former experiments; the tables being turned about, the ball whofe diftance from the center is as one, will raife twice the weight, that is raifed by the ball whofe diftance is as two; fo that the weights raifed, and confequently the forces which raife them, will be inverfly as the diftances of the balls from the center.

If equal bodies revolve in equal circles with unequal velocities, their central forces are as the fquares of the velocities, or becaufe the velocities are as the
number

number of revolutions in a given time, the forces are as the fquares of the numbers of revolutions performed in a given time. For by the fourth general expreffion fince Q and D are given, F is as N^2, that is, the force is as the fquare of the number of revolutions in a given time. To confirm this law, Exp. 5. let two equal balls be placed in the troughs at equal diftances from the centers; and let that table, whofe axle has two wheels, be turned about by means of the fmaller, fo that it may perform two revolutions in the fame time that the other table performs one: in this cafe the numbers of revolutions performed by the two balls in a given time being as one and two, their fquares will be as one and four, in which proportion the weights raifed will likewife be.

If unequal bodies revolve in equal circles with unequal velocities, their central forces are as the products of their quantities of matter into the fquares of their refpective velocities; or, which is the fame thing, as the products of their quantities of matter into the fquares of the numbers of revolutions in a given time. For by the fourth general expreffion, D being given, F is as QN^2. Let therefore two balls, whereof one is double the other, be placed at equal diftances from the centers; and let the larger revolve twice in the fame time that the fmaller re- volves once. In this cafe, the quantity of matter in Exp. 6. the leffer ball, which is as unity, being multiplied into the fquare of its number of revolutions in a given time, which is likewife as unity, gives one for the product. And the quantity of matter in the larger ball, which is as two, being multiplied into the fquare of its number of revolutions in the given time, which fquare is as four, gives eight for the product: fo that the weights raifed by the two balls will be as one and eight.

If unequal bodies revolve in unequal circles with unequal velocities, their forces are as their quantities of matter multiplied into the fquares of their re-

fpective

spective velocities, and that product divided by their respective distances from the centers; or what amounts to the same thing, their forces are as the products arising from the continued multiplication of their quantities of matter into their respective distances from the centers, into the squares of their numbers of revolutions in a given time; or to use the mathematical phrase, their forces are in a ratio compounded of their quantities of matter, of their distances from the center, and of the squares of their numbers of revolutions in a given time. For by

Exp. 7. the fourth general expression, F is as QDN^2. To confirm this law by an experiment, let two balls, whereof one is double the other, be placed in the troughs, so as that the distance of the smaller from the center may be to the distance of the larger as two to one; and let the larger revolve twice in the same time that the smaller revolves once. In this case the quantity of matter in the smaller body, which is as one, being multiplied into the distance from the center, which is as two, and the product being multiplied into the square of the number of revolutions performed by the smaller body in a given time, which is as one, gives two for the product. In like manner the quantity of matter in the larger body, which is as two, being multiplied into the distance from the center, which is as one, and the product of that multiplication being again multiplied into the square of the number of revolutions performed by the larger body in the given time, which square is as four, gives eight for the product; consequently, the weights which are raised, as also the forces which raise them, are as two and eight, or one and four.

If equal bodies revolve in unequal circles in such a manner as that the squares of their periodical times are as the cubes of their distances from the center, their central forces are inversly as the squares of their distances from the center. For since the
quantity

quantity of matter in the revolving bodies is given, and the cubes of the diftances are as the fquares of the times; if in the third general expreffion the cube of D be fubftituted in the room of the fquare of P, F will be as D divided by the cube of D, or as one divided by the fquare of D; that is, the force will be inverfly as the fquare of the body's diftance from the center. To confirm this law, let two equal balls be placed in the troughs, fo as that the diftance of one from the center may be as two, and the diftance of the other as three and one fixth; and let that which is at the fmalleft diftance revolve twice in the fame time that the other revolves once; fo that their periodical times may be as one and two, the fquares of which being one and four, are very nearly proportional to the cubes of the diftances; for the cube of the fmaller diftance is eight, and that of the larger thirty two very nearly; confequently, the balls muft raife weights which are to one another inverfly as the fquares of the diftances from the centers; that is, the weight raifed by the ball, whofe diftance is as two, muft be to the weight raifed by the ball whofe diftance is as three and a fixth, as the fquare of the laft diftance to the fquare of the former, that is, as ten to four, or five to two very nearly.

If the fquares of the periodical times be proportional to the cubes of the diftances, and the revolving bodies unequal, the central forces are directly as the quantities of matter in the bodies, and reciprocally as the fquares of their diftances from the center.' For in the third general expreffion, if the cube of D be fubftituted in the room of the fquare of P; F will be as $\frac{Q}{D^2}$. If therefore all things remain as in the laft experiment, excepting that the body which is at the greater diftance from the center is to the body lefs diftant, as two to one; the weight which is raifed by the former, will be to the weight

raifed

raised by the latter, as two, to two and a half; that
is, the weights raised, will be as the products arising
from the multiplication of the quantity of matter
in one body into the square of the other body's
distance.

Among the several laws of central forces, that
which obtains in nature, and by virtue whereof the
heavenly bodies are made to revolve in their several orbits, is, where the forces are to one another
inversly as the squares of the distances of the revolving bodies from the center. For it has been found
by observation, that all the planets as well primary
as secondary revolve either in circular orbits or such
as are nearly so. And that the six primary planets
move about the sun as their center in such a manner, as that the cubes of their mean distances from
the sun are very nearly proportional to the squares
of their periodical times. And the same thing has
been discovered with regard to the four secondary
planets or satellites that move about JUPITER, as
also with respect to the other five that revolve about
SATURN. And therefore the forces whereby they
are retained in their orbits must be in the inverse
ratio of the squares of their distances from the central bodies about which they revolve.

Exp. 9.   If two bodies are by means of their mutual attraction made to revolve about each other, and also
about a fixed point; and if their distances from that
fixed point, be reciprocally proportional to their
quantities of matter, that is to say, if as much as
one body exceeds the other in quantity of matter,
so much is its distance from the fixed point exceeded
by the other's distance from the same point; or what
amounts to the same thing, if the product arising
from the multiplication of one body into its distance
from the fixed point, be equal to the product arising
from the like multiplication of the other body into its distance from the fixed point, their central
forces are equal. For as the two bodies must of
necessity

neceſſity perform their revolutions in the ſame time; the number of their revolutions in a given time is given : and therefore by the fourth general expreſſion, F is as Q D, that is, the central force is as the product ariſing from the multiplication of the quantity of matter into the diſtance from the center, or fixed point ; but by ſuppoſition the product of one of the bodies into its diſtance from the fixed point, is equal to the product of the other into its diſtance, conſequently, their central forces are equal; for which reaſon neither of them can fly off from the fixed point ſo as to draw the other after it; for however ſtrongly either of them endeavours to recede by virtue of its own centrifugal force, it is with equal ſtrength drawn the contrary way by the centrifugal force of the other. But if the diſtances of the bodies from the fixed point be not reciprocally proportional to their quantities of matter ; that body, whoſe diſtance with regard to the diſtance of the other is greater than in the forementioned proportion, will fly off and draw the other after it; for in this caſe, the product of the former body into its diſtance from the fixed point is greater than the product of the latter into its diſtance ; which products being as the centrifugal forces of the bodies, the former body will have a greater centrifugal force than the latter, and of courſe muſt recede from the fixed point, and drag the other after it ; all which is fully confirmed by the following experiments.

Let two equal balls be tyed together by a ſmall Exp. 10, chord ; and let them be laid in one and the ſame trough, one at each end, ſo as that the chord being ſtretched may have its middle point juſt over the center of the table ; let then the table be turned about, and the balls will revolve about the center without flying off either way ; and continue ſo to do as long as the motion of the table laſteth. And the ſame thing will likewiſe happen tho' one ball be double the other, provided its diſtance from the center

center of the table be but one half of the diftance of the fmaller. But when equal balls are made ufe of, if one of them be placed at a greater diftance from the center than the other, upon turning the table it will fly off and draw the other after it. So likewife when unequal balls are made ufe of, fhould that which is double the other be placed at a diftance from the center greater than one half of the diftance of the fmaller, it will fly off and draw the fmaller after it. And on the other hand, if the diftance of the larger be lefs than half the diftance of the fmaller, the fmaller will in that cafe fly off and draw the larger after it.

LECTURE IV.

OF THE COMPOSITION AND RESOLUTION OF MOTION.

LECT. IV. THE fecond LAW OF NATURE, refulting from the inertnefs of matter, is, that whatever motion, or change of motion is produced in any body, it muft be proportional to, and in the direction of the force impreffed. For fince a body cannot by reafon of its inactivity contribute to the production of its own motion, or of any change therein, it is plain, that whatever motion or change of motion is generated in any body, it muft intirely proceed from the force impreffed on the body; and of confequence, fince effects are ever proportionate to their adequate caufes, muft be proportional thereto. And it muft likewife be directed and determined towards the fame part with the generating force. Wherefore if the body whereon the impreffion is made, was in motion before the impulfe, that motion will be retarded or accelerated according as the force impreffed oppofes it, or confpires therewith, or if it acts obliquely to the fame, the direction thereof will be changed, and the body

will

will move in a direction fituated between the direction of its former motion, and that of the impreffed force.   For inftance, if a body moving from
A towards B, be impelled at the point A by a
force acting in the direction AC, it will move along
a line as AD placed between AB and AC, the
fituation of which may be thus determined.   Let
the lines AB denote the velocity wherewith the
body moves in the direction AB; and let AC denote the velocity wherewith the body would move
by virtue of the impulfe along the line AC, fuppofing it had no other motion: that is, let AB be
to AC as the fpace defcribed by the body in a
given time in the direction AB, to the fpace defcribed by it in the direction AC, each of the motions being confidered fingly and apart; then compleating the parallelogram ABDC, and drawing
the diagonal AD, that diagonal is the line in which
the body moves; for the proof of which, let us
fuppofe a fmall inflexible wire equal in length to the
line AB, to pafs thro' the center of a ball, and
that whilft the ball moves uniformly on the wire
from A towards B, with a velocity which is as AB,
the wire is alfo moved uniformly from AB towards CD, with a velocity which is as AC, and
in fuch a manner as to be always parallel to AB,
and with its extremities to defcribe the lines AC
and BD.   Then, forafmuch as the fpaces defcribed
in a given time where the motions are uniform, are
to one another as the velocities of the motions; it
is evident, that in whatever time the ball moves
the length of the wire, in the fame time will the wire
move the length of AC, to wit, from AB to CD;
confequently, at the end of that time the ball will
be found in D at the extream point of the diagonal AD.   From any point in the diagonal taken
at pleafure as E, let the line EF be drawn parallel
to DB, and from the nature of fimilar triangles,
AF will be to FE, as AB to BD, that is, as the
velocity

velocity of the ball to the velocity of the wire; consequently, in the same time that the ball moves the length of A F along the wire, the wire will move the length of F E, from A B to K L; and the point F, which is the place of the ball on the wire, will be found in E. And what has been thus proved in relation to the two points D and E of the diagonal, may in the very same manner be demonstrated of any other point in the same line; wherefore the ball will by virtue of its own motion, and that of the wire, whereof it partakes, be carried in such a manner as to be always found in the diagonal A D; that is, it will by virtue of its compound motion describe the diagonal line. This being so, it plainly follows, that if the wire be taken away, and the ball at A have two motions impressed upon it at once; one in the direction A B, the other in the direction A C; and if the motions impressed, or, which is the same thing, if the forces impressing those motions be to one another in the proportion of A B to A C, the ball will by virtue of the double impression, move along the diagonal A D. For as to the effect it matters not whether the motion which the ball has in the direction A C arises from a force impressed on it at the point A, or whether it be communicated by a wire supporting the ball, and carrying it along with it in that direction.

Exp. 11.   To confirm this by an experiment, let three ivory balls of equal size, be suspended from three pins by strings of equal lengths, and let the middle ball rest over one angle of a wooden square; then let each of the extream balls be let fall separately from the same height, in such manner as to strike the middle ball in the direction of one side of the square, and the middle ball will by each of the strokes made separately, be moved along over that side of the square, which correspondeth to the direction of the stroke; but if the two balls be at the same instant

of

of time let fall from equal heights, fo as that they
may ftrike the middle ball at once, and in the directions of the two fides of the fquare, the middle
ball will by the double ftroke, be driven over the
diagonal of the fquare.

As a COROLLARY it follows, that a body will in
the fame time defcribe the diagonal A D of a parallelogram with two forces conjoined, that are to
one another as the fides A B and A C, that it would
the refpective fides with each of thofe forces feparately. As alfo, that the velocity wherewith a
body moves along the diagonal, is to the velocity
wherewith it is carried along the fides when acted
upon by each force fingly, as the diagonal to each
fide refpectively : confequently, if the two forces
be given, the velocity along the diagonal, which
arifes from the conjunction of both forces, will be
fo much the greater, by how much the angle B A C
is lefs ; for as that angle is diminifhed, the diagonal
which in this cafe denotes the velocity, is lengthened, till at laft the angle vanifhing by the coincidence
of the fides, the diagonal becomes equal to both the
fides taken together ; and the velocity of the body
equal to the fum of the velocities wherewith the
body would move, were each of thofe forces impreffed upon it in the fame direction. Thus the
lines A B and A C being placed at three different
angles, fo as to conftitute the fides of three different
parallelograms, (the diagonals whereof are reprefented by the pricked lines) it is evident to fight, that
as the angle B A C grows lefs, the diagonal grows
longer ; and that when the angle vanifhes by the
coincidence of A B with A C, the diagonal A D
becomes equal to A C and C D, that is, to A C and
A B ; and the velocity denoted by A D, is in that
cafe a *maximum*, or the greateft that can arife from
the conjunction of thofe two forces. On the other
hand, as the angle inlarges, the velocity along the
diagonal muft decreafe, till at length the angle vanifhing

Pl. 2.
Fig. 2.

Pl. 2.
Fig. 3.

nifhing by the two fides becoming one right line,
the velocity becomes equal to the difference of the
velocities, arifing from the impreffion of each force
Pl. 2.
Fig. 4.
when made fingly and feparately. Thus the lines
AB and AC being as before placed at three different angles BAC, it is evident that the diagonals
AD reprefented by the pricked lines, grow fhorter
as the angle BAC inlarges; till at laft the angle,
and with it the diagonal vanifhing the two fides
Pl. 2.
Fig. 5.
BA and AC conftitute one right line as BAC,
wherein the body is, as it were, carried two contrary ways, to wit, from A towards B by the force
which acts in the direction AB, and from A towards C by the force acting in the direction AC;
and the difference of the velocities, which arife
from the impreffions of the two forces when they
act feparately, is the velocity wherewith the body
actually moves in the direction of the ftronger
force, which velocity is a *minimum*, or the leaft velocity that can arife from the joint action of thofe
two forces.

As a fecond COROLLARY it follows, that a body
may be moved thro' one and the fame line by numberlefs pairs of forces acting upon it. For if in-
Pl. 2.
Fig. 6.
ftead of the force, whofe direction is AB, we fuppofe another, the direction whereof is AE; and if
inftead of the force acting in the direction AC, we
fuppofe one to act in the direction AF, and that
thofe forces are to one another as AE to AF:
then compleating the parallelogram AEDF, the
line AD will be the diagonal of this parallelogram,
as well as of the former; and therefore the body
will from the joint action of thefe two forces defcribe the fame line AD which it did before: and
as AD may be made the diagonal of numberlefs
parallelograms, it is evident that it may be defcribed by a body acted upon by numberlefs pairs
of forces in different directions. And not only fo,
but it may likewife be defcribed by a body, where-

2

on a great number of forces act at the same time; for as the forces acting upon the body in the directions AB and AC make it to move along the diagonal AD, so may the direction along AB arise from the directions of two other forces, and each of those from the directions of two others, and so on without number. Hence we see, that all forces and motions whatever may be resolved into innumerable forces and motions; and any simple direct force or motion may be looked upon as compounded of innumerable oblique forces or motions. For the line and direction of the motion is the same, whether that motion be compounded of two motions arising from forces impressed in the directions AB, AC, or in the directions AE, AF, or arise from the impression of a single force in the direction AD; and therefore the motion along the line AD, tho' it be simple arising from one single force acting in that direction; yet may it be considered as compounded of two or more motions in other directions, such as AB and AC, or AE and AF, since the very same motion would arise from such a composition.

Pl. 2.
Fig. 6.

This composition and resolution of motions and forces is of singular use in mechanicks; for by the help thereof, the effects of powers acting in oblique directions are readily determined, as will appear hereafter.

The third LAW OF NATURE arising from the inertness of matter is, that reaction is always equal to action, and contrary thereto; or in other words, that the actions of two bodies, one upon another, are constantly equal, and in directions contrary to each other; so that whatever change is made in the state of one body, whether at rest or in motion, by the action of another; the same change is produced in the state of the other by the reaction of the former; but the tendencies or directions of those changes are contrary ways. Thus, when one presses

E   a stone

a ſtone with his finger directly downward, the finger is equally preſſed by the ſtone, and that directly upward. And when a horſe draws a load, he
is equally drawn back by the load; for as much as
he promotes the progreſs of the load, ſo much is
he retarded in his own motion; that is, he is in effect drawn back; for the ſame force of muſcles and
ſinews, which he exerts in order to drag on the
load, would, if he was freed from the incumbrance,
carry him forward to a diſtance much greater than
what he reaches in the ſame time whilſt tied to the
load; and conſequently, as far as his progreſs fall-
eth ſhort of that diſtance, ſo much is he in effect
drawn back; and whatever motion he communicates to the load, ſo much does he loſe of his own,
the load reacting upon him with the ſame force
that he acts upon it; for which reaſon, if by addition of weight the load be ſo far increaſed as to require the whole ſtrength of the horſe to move it,
no motion will enſue, the whole power of the
horſe, wherewith he endeavours to go forward, being but juſt equal to the reaction of the load whereby he is drawn back. This equality of action and
reaction obtains in all kinds of attractions whatever.
When a loadſtone attracts a piece of iron, it is
equally attracted by it; as will appear from the
following experiment. Let a piece of iron and a
loadſtone equal in weight, be ſuſpended by two cords
of an equal length, and let the diſtance between
them be ſo ſmall, as that they may not be out of
the reach of each other's attraction; then will they,
from a ſtate of reſt, begin to move towards each
other, and that with equal velocities, ſo as to meet
at the middle point of their firſt diſtance: if they
be again ſeparated, and the loadſtone fixed, the
iron being ſuſpended at the ſame diſtance from it
as before, will move towards it, ſo as at length to
touch it, and adhere thereto. And on the other
hand, if the iron be fixed, and the ſtone moveable,
the

Exp. 12.

the ſtone will approach the iron in the ſame manner, as the iron did the ſtone; all which plainly ſhews that the attraction between the loadſtone and iron is mutual, the one drawing the other as much as it is drawn by it : ſo that the reaction of the iron upon the ſtone is exactly equal to the action of the ſtone upon the iron.

The equality of action and reaction with reſpect to attractions is likewiſe manifeſt from hence, that if a man placed in a boat, draws another boat by means of a rope faſtened thereto; the boat wherein the man is placed will be equally drawn with the other, and the two boats will approach one another with equal quantities of motion; ſo that if they be equal in weight, and of the ſame ſize and ſhape, they will approach with equal velocities, and meet at the middle point : but if one be heavier than the other, then by how much it exceeds the other in weight, by ſo much will it be exceeded by the other in the velocity of its motion; for inſtance, if the weight of one be to the weight of the other as one to two, then will the velocity of the former be to the velocity of the latter as two to one; that is, their velocities will be reciprocally proportional to their weights. To confirm this by an experiment, let a cord be made faſt to one end of a ſmall boat, and let it paſs over a pulley fixed to the end of another ſmall boat of the ſame ſhape and ſize, and let a weight be tied to the end of the cord, and hang in the water; this being done, let the boats be placed at ſuch a diſtance as that the cord may be ſtretched, then letting go the boats the weight will deſcend, and in deſcending draw the boat to whoſe end the cord is faſtened towards the other, and at the ſame time the other will move towards it; and when they come together, the ſpace deſcribed by the boat whoſe weight is as one, will be to the ſpace deſcribed by the boat, the weight whereof is as two, as two to one; that is, if the

Exp. 13.

diſtance

diftance between the two boats be divided into three
equal parts, that boat which is double in weight to
the other, will move thro' one of thofe parts in the
fame time that the lighter moves thro' the other
two.

As action and reaction are equal with regard to
attractions, fo are they likewife in refpect of ftrokes
or impulfes made by bodies one upon another; the
force of two bodies, ftriking each other equally,
affecting the motions of both, and producing equal
changes therein towards contrary parts. On this
equality of action and reaction do the feveral laws
which have been collected concerning the collifion
of folid bodies in a great meafure depend; which
laws, as they relate to bodies void of elafticity, I
fhall now explain; in doing of which, I fhall lay
down one general PROPOSITION concerning the
collifion of fuch bodies, whence I fhall deduce the
laws of particular cafes, and at the fame time confirm each law by an experiment.

The PROPOSITION is as follows: *If two bodies
void of elafticity move in one right line, either the fame
or contrary ways, fo as that one body may ftrike directly againft the other; let the fum of their motions
before the ftroke when they move the fame way, and
the difference of their motions when they move contrary
ways, be divided into two fuch parts as are proportional to the quantities of matter in the bodies; and
each of thofe parts will refpectively exhibit the motion
of each body after the ftroke.* For inftance, if the
quantities of matter in the bodies be as two and one,
and their motions before the ftroke as five and four,
then the fum of their motions is nine, and the difference is one; and therefore when they move the
fame way, the motion of that body, which is as
two, will after the ftroke be fix, and the motion of
the other three: but if they move contrary ways, the
motion of the greater body after the ftroke will be
two thirds of one, and of the leffer one third of one.

For

For since the bodies are fuppofed to be void of elafticity, they will not feparate after the ftroke, but move together with one and the fame velocity; and of confequence, their motions will be proportional to their quantities of matter; and from the equality of action and reaction it follows, that no motion is either loft or acquired by the ftroke when the bodies move the fame way, becaufe whatever motion one body imparts to the other, fo much muft it lofe of its own; confequently the fum of their motions before the ftroke is neither increafed nor diminifhed by the ftroke, but is fo divided betweeen the bodies, as that they may move together with one common velocity, that is, it is divided between the bodies in proportion to their quantities of matter; but it is otherwife, where the bodies move contrary ways; for then the fmaller motion will be deftroyed by the ftroke, as alfo an equal quantity of the greater motion, becaufe action and reaction are equal; and the bodies after the ftroke will move together equally fwift, with the difference only of their motions before the ftroke; confequently, that difference is by means of the ftroke divided between them in proportion to their quantities of matter.

The feveral particular cafes concerning the collifion of bodies may be reduced to four general ones. For, 1ft, it may be that one body only is in motion at the time of the ftroke. Or, 2dly, they may both move one and the fame way. Or, 3dly, they may move in direct oppofition to each other, and that with equal quantities of motion. Or, laftly, they may be carried with unequal motions in directions contrary to each other. As the bodies may be either equal or unequal, each of thefe four general cafes may be looked upon as confifting of two branches; and as fuch I fhall confider them, and treat of them in the order, wherein I have laid them down.

As

As to the firſt, if a body in motion ſtrikes another equal body at reſt, they will by the propoſition move together, each of them with one half of
the motion that the body had which was in motion
before the ſtrokë ; and ſince the quantity of motion
in any body, is as the product ariſing from the
multiplication of its quantity of matter into its velocity ; the common velocity of the two bodies,
will be but one half of the velocity of the moving

Exp. 14. body before the ſtroke. For the confirmation
whereof, let two equal balls of clay be ſuſpended
from two pins of an equal height, by threads of an
equal length, and in ſuch a manner, as that when
they hang freely they may juſt touch one another,
and that their centers and point of contact may lie
in a right line parallel to the horizon. This being
done, and one of the balls being at reſt, let the
other be removed to any diſtance from it, and then
let fall ; it will in its deſcent deſcribe the arch of a
circle, and by the time it arrives at the loweſt point
of the arch, that is, when it comes to touch the
quieſcent ball, it will have acquired ſuch a velocity
as would carry it to the ſame height from which it
fell, as ſhall be ſhewn when I come to treat of pendulums ; and conſequently, if the other ball was
removed, would actually aſcend to that height ; but
upon ſtriking the other ball, which is of equal ſize,
it will communicate one half of its motion to it,
and they will move together with half the velocity
that the moving body had at the time of the
ſtroke, ſo as to aſcend to one half only of the
height from which the ſtriking body fell.

That the nature of this and the other experiments relating to the colliſion of bodies may be
more readily comprehended ; I ſhall lay down ſome
things concerning the motion of bodies thro' the
arches of circles, the truth whereof ſhall be demonſtrated in my lecture upon pendulums. And
firſt, all the arches of a circle, provided they be

4   not

not large, are defcribed in equal times by bodies
defcending along them ; and therefore if two bodies be let fall at the fame time, one from C and
the other from E, or from D and F, they will
·both arrive at the loweft point B at one and the
fame time ; and the ftroke of the fubfequent body
upon the preceding will be made at B : and for the
fame reafon if one be let fall from C, and the
other from D or F, or one from E, and the other
from D or F, they will meet and ftrike one another at B.

2dly, The velocity which a body acquires in
falling thro' the arch of a circle, is as the chord of
the arch ; that is, the velocity of a body which has
fallen from C to B, is to the velocity of a body
that has fallen from E to B, as the chord CB to the
chord EB. And here I muft obferve to you, that
when in the following experiments I fpeak of a
body falling from, or rifing to any height, as four,
fix, or ten inches, I would be underftood to mean
it of a body's falling thro' or moving up an arch,
whofe chord is of fuch a length.

3dly, The velocity wherewith a body begins to
rife up thro' the arch of a circle, is as the chord of
the arch which the body defcribes in its afcent.
Thus the velocity wherewith a body begins to move
from the point B towards D, if it afcends as high
as D, is as the chord BD ; but if it rifes only to F,
the velocity is as the chord BF. So that in the
experiments the chords of the arches thro' which
the bodies defcend, exprefs the velocities of the bodies in the point B at the time of the ftroke ; and
the chords of the arches thro' which the bodies afcend after the ftroke exprefs the velocities of the
bodies immediately after the ftroke.

Thefe things being laid down, I fhall now proceed to determine the laws of the four general cafes.
As to the firft, it has been already fhewn, that

E 4

where

LECT.
IV.

where the moving body is equal to the quiefcent, the common velocity of the two bodies after the ftroke, is but one half of the velocity of the moving body before the ftroke ; and of confequence, the motion of each body after the ftroke, is equal to one half of what the moving body had before the ftroke. But if the quiefcent body differs in fize from the moving body, then the common velocity after the ftroke will be fo much lefs than the velocity of the moving body before the ftroke, by how much the fum of the two bodies exceeds the body which was firft in motion. Thus, if the moving body be to the quiefcent as two to one, the common velocity after the ftroke will be to the

Exp. 15.

velocity of the moving body before the ftroke, as two to three ; wherefore if a ball of clay, falling from the height of nine inches, ftrikes another at reft, and of one half the magnitude, they will afcend together to the height of fix inches only : and on the other hand, if the larger be quiefcent, and the fmaller falls from the height of nine inches, they will afcend to the height of three inches only ; and the quantity of motion in each body immediately after the ftroke will be had, by multiplying each of them into the common velocity.

As to this and all other experiments of this nature, it muft be obferved, that they do in fome meafure vary from the theory, and that for two reafons. Firft, becaufe clay or any other body, wherewith thefe experiments can be made, is not perfectly void of elafticity. Secondly, becaufe the air refifts the motions of the balls, and by fo doing diminifhes their velocities.

As to the fecond general cafe, where both the bodies are in motion before the ftroke, and move one and the fame way : In order to find their common velocity after the ftroke, let the fum of their motions before the ftroke be divided by the fum of the bodies,

bodies, and the quotient will exprefs the common
velocity. Wherefore, if two equal balls of clay
be let fall at the fame time, one from the height
of three inches, and the other from the height of
fix, after the ftroke they will afcend to the height of
four inches and an half ; for as in this cafe the bodies are equal, their motions are as their velocities,
that is, as fix and three, the fum of which being
divided by two, the fum of the bodies, gives four
and an half for the common velocity after the
ftroke.

Where the bodies are unequal, let us fuppofe the
preceding body to be as one, and to fall from the
height of three inches as before, fo that its quantity
of motion will be as three ; and let the fubfequent
body be as two, and fall from the height of fix inches, fo that its quantity of motion will be twelve ;
and the fum of the two motions will be fifteen,
which being divided by three, the fum of the two
bodies, gives five in the quotient ; fo that in this cafe,
after the ftroke, the balls will afcend to the height
of five inches, and the motion of the greater will
be as ten, and that of the fmaller as five.

As to the third general cafe, where the bodies
move in direct oppofition to each other, if they
have equal quantities of motion, they will upon
the ftroke lofe all their motion, and continue at
reft ; for by the propofition, the bodies after the
ftroke will be carried with the difference of their
motions before the ftroke ; which difference is fuppofed to be nothing. Wherefore, if two equal
balls of clay be let fall at once from equal heights,
upon the ftroke they will ceafe to move ; and the
fame thing will happen where the balls are unequal,
provided the heights from which they fall are reciprocally proportional to their quantities of matter ;
for inftance, if the balls be as one and two, let the
former fall from the height of fix inches, and the
latter from the height of three, and upon their
meeting

meeting they will ftand ftill, for in this cafe, the quantities of motion, wherewith they oppofe each other, will be equal.

When two bodies meet with unequal quantities of motion, if the difference of their motions be divided by the fum of the bodies, the quotient will exprefs their common velocity after the ftroke; for by the propofition, the difference of their motions before the ftroke, is equal to the fum of their motions after the ftroke; confequently, that difference divided by the fum of the bodies muft give the velocity. Wherefore, if two equal balls of clay be let fall at the fame time, one from the height of three inches, and the other from the height of fix, after the ftroke they will afcend together to the height of an inch and an half; for fince the balls are equal, their motions will be as their velocities, that is, as fix and three, the difference whereof is three, which being divided by two, the fum of the bodies gives one and an half in the quotient. If the balls be unequal in the proportion, for inftance, of two to one; and if that which is as two falls from the height of fix inches, and the other from the height of three; after the ftroke they will afcend together to the height of three inches; for the greater ball being as two, and its velocity as fix, its motion is as twelve; whereas the fmaller being as one, and its velocity as three, its motion is likewife as three, which being fubducted from the greater motion leaves a remainder of nine; and this being divided by three, the fum of the bodies, gives three for the common velocity, or the height to which the bodies will rife.

In order to difcover the quantity of motion communicated by one body to the other, I fhall lay down four rules adapted to the four general cafes. And Firft, if one of the bodies be quiefcent at the time of the ftroke, let that body be multiplied into the common velocity after the ftroke, and the product

Exp. 20.

Exp. 21.

duct

duct will exprefs the communicated motion.   For
fince that body had no motion before the ftroke, it
is manifeft, that whatever motion it has after the
ftroke muft be communicated to it by the ftriking
body; but that motion is as the product arifing
from the multiplication of the quantity of matter in
the body into the common velocity, confequently,
that product expreffes the communicated motion.

Since the body which is at reft before the ftroke
has no motion, but what is imparted to it by the
ftriking body; and fince the motion of the ftriking
body is by the propofition to be divided between
the two bodies in proportion to their quantities of
matter; it follows, that where the ftriking body is
greater than the quiefcent, it will communicate lefs
than half its motion, and where it is equal to it, it
will impart one half; and where it is lefs, more
than one half: and if the quiefcent body be infinitely great with refpect to the ftriking body, which
is in effect the cafe where the quiefcent body is fixed,
fo as not to give way to the ftroke, the ftriking
body will impart all its motion to the other; for
as the quiefcent body is fuppofed to be infinitely
greater than the ftriking body, the motion, which
it receives from the ftriking body, muft bear an infinite proportion to the motion remaining in the
ftriking body; but as the motion communicated is
a finite quantity, it cannot bear an infinite proportion to the remaining motion, unlefs that remaining
motion be in its evanefcent ftate, and reduced to
nothing.

When both the bodies are in motion before the
ftroke, and their motions are directed the fame
way, which was the fecond general cafe; the rule
for determining the quantity of motion communicated is as follows.   Let the preceding body be
multiplied into the common velocity after the ftroke,
and from the product let the motion which it had
before the ftroke be fubducted, and the remainder
will

will be the motion communicated. For the product arising from the multiplication of the preceding body into the common velocity, gives the whole motion of that body after the stroke, and therefore, if from thence be taken the motion which it had before and independent of the stroke, the remainder must be the motion acquired by the stroke.

When the bodies move towards one another with equal quantities of motion, as in the third general case; the motion communicated is equal to the motion of either before the stroke. For as in this case, both their motions are destroyed by the stroke; it is plain, that whichever of the bodies is considered as giving the stroke (and either of them may) it must communicate just as much motion to the other, as the other has at the time of the stroke; for by this means the motion communicated, as it is directly opposed to the former motion of the body, will be just sufficient to destroy the same, and by so doing cause the body to rest.

When the quantities of motion in two bodies moving directly towards each other are unequal, which is the fourth general case; the motion communicated is determined by the following rule. Let the body which had the lesser motion before the stroke be multiplied in the common velocity after the stroke, and to the product let the motion which it had before the stroke, be added, and the sum will be the motion communicated. For as the body, to which the motion is communicated, does after the stroke move in a direction contrary to what it did before, it is evident, that besides the motion, wherewith it is carried in that contrary direction, it must have received as much more in the same direction, as was sufficient to withstand the motion it had before the stroke in an opposite direction; for till that motion was destroyed by an equal motion opposed thereto, the body could not change its direction, and move backward.

LECTURE

LECTURE V.

OF THE COLLISION OF ELASTICK BODIES.

HAVING given you an account of the collifion of bodies void of elasticity, I come now
to confider the effects thereof in fuch as are elaftick;
by which I mean bodies that confift of fuch parts
as yield and give way when preffed, and which reftore themfelves upon the removal of the preffure:
if the force wherewith they reftore themfelves be
exactly equal to the preffure whereby they are bent
inward, then are the bodies faid to be perfectly
elaftick ; and fuch are all thofe bodies fuppofed to
be, wherewith experiments are ufually made for
confirming the theory relating to the collifion of
elaftick bodies ; but as there is not perhaps in nature
any body perfectly elaftick, if among the experiments that are now to be made, any fhall be found
to vary a little from the theory, fuch variation muft
be looked upon as rifing rather from the want of
perfect elafticity in the bodies, than from any error
in the theory itfelf, or in the calculations grounded
thereon.

The method which I fhall obferve in treating of
the percuffion of elaftick bodies is this ; Firft, to
lay down one general propofition concerning fuch
percuffion, and then, Secondly, to deduce the laws
relating to the four general cafes mentioned in my
laft lecture, and to confirm each of thofe laws by
experiments.

Before I lay down the propofition, I muft obferve
to you, that wherever I mention the ftriking body,
I thereby mean that body which is in motion where
one of the two is quiefcent, as alfo that body which
moves fwifteft when they both move the fame way ;
and laftly, that body which has the greateft quan-

4   tity

tity of motion, when they move in oppofition to one another, or in this cafe, if their motions be equal, then either of them may be taken indifferently for the ftriking body.

This being premifed, the PROPOSITION is as follows.

If of two bodies perfectly elaftick, one be at reft, and the other in motion ; or if they both move either the fame or contrary ways, fo as that one fhall ftrike the other ; let them be confidered as void of elafticity, and by the propofition laid down in my laft lecture, let the motion of each body after the ftroke be found, and by one of the four rules laid down in the fame lecture, let the motion communicated by the ftriking body to the other be likewife found ; and let this motion be fubducted from the motion of the ftriking body after the ftroke, and added to that of the body which received the ftroke, and the refidue will be the motion of the ftriking body, and the fum the motion of the other body after reflexion. For, fince the bodies are fuppofed to be perfectly elaftick, their parts which are bent in by the ftroke will reftore themfelves with a force equal to that which bends them in ; but the force which bends them in, is meafured by the quantity of motion communicated by the ftriking body to the other, and therefore the parts of each body which are bent inward will reftore themfelves with fuch a force, as is fufficient to generate a motion equal to that which is communicated ; confequently, the bodies will by virtue of their elafticity throw one another contrary ways, each with a quantity of motion equal to that which the ftriking body communicates to the other ; for which reafon, if that motion be fubducted from the motion remaining in the ftriking body after the ftroke, as being contrary thereto, and added to the motion of the other body after the ftroke, as confpiring therewith, the refidue and fum will give the true motions of the bodies after reflexion.

To

To apply what has been faid to the four general L E c т. cafes, the firft whereof is where one of the bodies V. is at reft at the time of the ftroke. If a body per- fectly elaftick ftrikes another of the fame kind and of equal magnitude at reft, the ftriking body will communicate all its motion to the other and remain at reft; for by the firft of the four rules laid down in my laft lecture, the ftriking body will upon the ftroke, communicate half its motion, and by the propofition now laid down, a quantity of motion equal to that which is communicated, muft be fub- ducted from the motion remaining in the ftriking body, and be added to the motion of the body which receives the ftroke, by which means the ftriking body will have no motion left; but the other body will have a quantity of motion equal to what the ftriking body had before the fhock. For the confirmation of which, let two equal ivory balls Exp. 1. be fufpended as were thofe of clay; and let one of the balls fall from any height, and fo as to ftrike the other at reft, the ball which receives the ftroke will afcend to the fame height from which the other fell, and will leave the other at reft.

If inftead of one there be two, three, or more Exp. 2. quiefcent balls contiguous to one another, that which is fartheft removed from the ftriking ball will fly off with the velocity of the ftriking ball, and all the intermediate balls together with the ftriking ball will quiefce; for as the ftriking ball imparts all its motion to the firft of the quiefcent balls, fo does that in like manner to the ball which lies next be- yond it, and that again to a third, and fo on; till at length the laft ball meeting with none other to refift it, flies off with all the motion of the ftriking ball, leaving that and the intermediate ones at reft.

If two balls be let fall together contiguous to one Exp. 3 another, upon the ftroke the two fartheft will fly off, leaving the others at reft; for as the foremoft

of

LECT.
V.

of the two moving balls is carried equally fwift with the fubfequent, it cannot during its motion receive any impreffion from the fubfequent ball; confequently, when it makes the ftroke, it will produce the fame effect in the quiefcent balls as if the fubfequent ball was away; that is, it will by means of the intermediate balls communicate all its motion to the laft, and make that fly off; but no fooner has it made the ftroke, and thereby parted with its own motion, but the fubfequent ball impels it, and imparts to it all its motion, and this motion being propagated thro' the feveral intermediate balls as before, makes the laft but one to fly off, and that in fuch a manner as to keep pace with, and clofely purfue the other; becaufe in the fame inftant of time that the foremoft of the two moving balls makes its ftroke, it likewife receives the ftroke from the hindmoft ball, and of confequence, the flying off of the two laft balls, which is the effect of the double ftroke, muft happen at one and the fame time.

Exp. 4. For the fame reafon that two balls fly off where the number of ftriking balls is two, three will fly off when there are three ftriking balls, and four, where there are four, and fo on, whatever be the number of ftriking balls, an equal number will conftantly go off.

Exp. 5. If two elaftick balls be unequal, for inftance, if one be double the other, and if the greater be let fall from the height of nine inches, and ftrike the fmaller at reft; they will both move forward after the ftroke, the ftriking body with one third of the motion which it had before the ftroke, and the other with two thirds; and the ftriking body will afcend to the height of three inches, and the other to the height of twelve. For fince the ftriking body is to the quiefcent as two to one, it will by the firft of the four rules laid down in my laft lecture, communicate one third of its motion to it, and on account

of

of the elasticity a quantity of motion equal to what L e c t.
is communicated, muſt be taken from the motion V.
remaining in the ſtriking body, and added to the
motion of the other; conſequently, the ſtriking
body will retain one third only of its motion, the
other two thirds being communicated to the body
which receives the ſtroke; wherefore ſince the ſtriking body is as two, and the height from which it
falls as nine, its motion muſt be as eighteen, one
third of which, to wit, ſix, it will retain after the
reflexion; and the other two thirds, to wit, twelve,
will be the motion of the other body, and theſe
motions being divided by the bodies, will give three
and twelve for the quotients; which quotients are
as the velocities of the bodies after reflexion, or as
the heights to which they aſcend.

On the other hand, if the larger ball be quieſcent, Exp. 6.
and the ſmaller be let fall from the height of nine
inches, its motion will be as nine, whereof two
thirds will by the firſt of the four rules be communicated by the ſtroke to the greater, and one third
only will remain in the ſtriking ball, from which
on account of the elaſticity muſt be taken as much
as was communicated to the larger ball, that is, two
thirds, but upon ſubducting two thirds from one
third, there will remain one third negative; which
ſhews, that the ſtriking ball will be reflected with
one third of the motion it had at the time of the
ſtroke, ſo as to aſcend backward to the height of
three inches; and the quieſcent ball, to which two
thirds of the ſtriking ball's motion was communicated by the ſtroke, will likewiſe on account of
the elaſticity receive two thirds more, ſo as to be
carried forward with a motion equal to what the
ſtriking ball had at the time of the ſtroke, and one
third more; that is to ſay, with a motion which is
as twelve, which being divided by two, the quantity
of matter in the ball gives ſix for the velocity, or
the height to which that ball muſt aſcend.

F

From

From what has been said it follows, that when the quiefcent ball is fmaller than the ftriking ball, there can be no reflexion, becaufe in that cafe the ftriking ball will by virtue of the ftroke communicate lefs than half its motion, and the motion which is to be taken from the ftriking ball on account of the elafticity being equal to the motion communicated, will upon the fubduction always leave fome motion in the ftriking ball to carry it forward, confequently, it cannot be reflected. Where the two balls are equal there will likewife be no reflexion, but the ball which was quiefcent will go forward with all the motion of the ftriking ball, and the ftriking ball will become quiefcent; as is evident from what was faid concerning that cafe. But where the ftriking ball is lefs than the quiefcent, it will be reflected, and there will likewife be an augmentation of motion in the greater ball; for the fmaller ball muft upon the ftroke communicate more than half its motion to the greater ball, and there muft likewife, on account of the elafticity, as much motion be fubducted from the fmaller ball, and added to the larger, as is communicated; wherefore, fince two equal quantities of motion, each of which exceeds half of the fmaller ball's motion, are to be fubducted from the fmaller ball, and given to the larger; it is plain, that the fmaller muft lofe all its motion and fomething more, that is, it muft be carried backward or reflected; and the greater ball muft go forward with more motion than was in the fmaller at the time of the ftroke, that is, there will be an augmentation of motion; and the excefs of motion in the greater ball, above the motion which the fmaller ball has at the time of the ftroke, is ever equal to the motion wherewith the fmaller ball is reflected after the ftroke, as is evident from what has been faid. If therefore motion be communicated from a fmaller elaftick body to a larger, by means of feveral intermediate bodies each

larger

larger than the other, the motion will be augmented in each of them, and the motion of the laſt will greatly exceed that of the firſt ; and this augmentation of motion is greateſt when the bodies are in a geometrical progreſſion ; for inſtance, if there be two bodies which are as one and four, and if the ſmaller communicates motion to the larger by means of one intermediate body, the motion will be greater in the larger body, if the middle body be as two, that is, a geometrical mean between the two, than if it be as one and an half, or two and an half, or three, or in ſhort in any other proportion whatever but that of the geometrical mean. For the proof of which, let the leſſer body be expreſſed by unity, and the larger by the ſquare of a, and the geometrical mean will be expreſſed by a ; ſo that the three bodies taken in their order from the leaſt, will be expreſſed by the ſymbols in the firſt ſtep, and the motion produced in the ſecond body by the ſtroke of the firſt, will be expreſſed by the ſecond ſtep ; and the motion produced in the third by the ſtroke of the ſecond, will be expreſſed by the third ſtep. Again, let another body greater or leſs than a be ſubſtituted in the room thereof, and let the difference between that body and a be called x, in this caſe, the bodies will be expreſſed by the ſymbols in the fourth ſtep, and the motion produced in the ſecond by the ſtroke of the firſt, will be expreſſed by the fifth ſtep ; and the motion produced in the third by the ſtroke of the ſecond, will be expreſſed by the ſixth ſtep ; but this fraction of the ſixth ſtep is leſs than that of the third ſtep, for if from the product ariſing from the multiplication of the denominator of this fraction into the numerator of that be ſubſtracted, the product which ariſes from the multiplication of the numerator of this into the denominator of that ; that is, if from the ſeventh ſtep the eighth be ſubducted, there will remain the

Lᴇᴄᴛ.   the quantity which is expreſſed in the ninth ſtep.
  V.   Whence it appears, that the former product is
greater than the latter; and therefore, by the 2d
Cᴏʀᴏʟ. of the 19th Pʀᴏᴘ. of the 7th book of the
elements, the numerator of the former fraction
bears a greater proportion to its denominator, than
that of the latter fraction does to its denominator,
that is, the fraction in the third ſtep which expreſſes
the motion of the greateſt body when the intermediate one is a geometrical mean, is greater than the
fraction in the ſixth ſtep, which expreſſes the motion of the greateſt body when the middle body is
not a geometrical mean; conſequently, the motion
is more augmented when the intermediate body is
a geometrical mean, than when it is greater or leſs
in any proportion.

1ſt. $1, a, a^2.$

2d. $\dfrac{2a}{1 + a}.$

3d. $\dfrac{a^3}{a^3 + 2a^2 + a} \times 4.$

4th. $1, a \pm x, a^2.$

5th. $\dfrac{2 \times \overline{a \pm x}}{1 + a \pm x}.$

6th. $\dfrac{a^3 \pm a^2 x}{a^3 + 2a^2 + a \pm a^2 x \pm 2ax + x^2 + x} \times 4.$

7th. $a^6 + 2a^5 + a^4 + a^5 x + 2a^4 x + a^3 x + a^3 x^2.$

8th. $a^6 + 2a^5 + a^4 + a^5 x + 2a^4 + a^3 x.$

9th. $- - - - - - - - - - - a^3 x^2.$

   To give you an inſtance, how prodigiouſly motion may be augmented, by being ſucceſſively communicated to ſeveral bodies in a geometrical pro-
                                              greſſion;

greſſion; if twenty elaſtick bodies be placed one after another, each ſucceeding body exceeding the foregoing in the proportion of twenty to one; and if motion be communicated thro' the ſeveral intermediate bodies from the firſt to the laſt, it will be ſo far augmented, as to be two hundred thouſand times greater in the laſt body than in the firſt; ſo that if we ſuppoſe the firſt to be a cannon-ball, moving with the ſame velocity wherewith it flies from the mouth of a cannon, which, from the obſervations of Mr. DERHAM, I ſhall ſuppoſe to be at the rate of 612 feet in a ſecond, tho' there are pieces of cannon which diſcharge their balls with double that velocity, the motion of the laſt body will be ſo great as if applied to the ball would carry it at the rate of above twenty three thouſand miles in one ſecond of time; which velocity is five thouſand times as great as the velocity of a body revolving about the earth, by the force of gravity at a ſmall diſtance from its ſurface; for a body ſo revolving will not come round in leſs than an hour and twenty four minutes.

From the increaſe of motion in elaſtick bodies, a reaſon may be drawn for the augmentation of ſound in ſpeaking trumpets; for as the ſpeaking trumpet is narroweſt at the mouth-piece, and thence widens and inlarges continually to the extremity, the air within it, which is an elaſtick fluid, as ſhall be ſhewn hereafter, may be conſidered as divided into a great number of cylindrical bodies, of very ſmall but equal altitudes, the baſis of the firſt being equal to the mouth of the trumpet, and the baſis of the reſt increaſing one above another as they are more and more removed from the mouth; upon which account the motion that is impreſſed by the force of the voice on the firſt cylindrical body of air, grows larger in the ſecond, and larger ſtill in the third, and ſo on, till at length at the exit of the tube it becomes ſo large as to magnify the ſound to a great degree;

degree; and of the feveral kinds of trumpets, thofe
magnify the found moft, that are of fuch a figure as
arifes from the revolution of the logarithmick curve

Pl. 2.
Fig. 8.
about its axis; that is, let AG be the logarithmick
curve, and HO its axis, the figure arifing from the
revolution of AG about HO, is fuch as a fpeaking
trumpet ought to have in order to give it the greateft advantage poffible. For from the natu.e of the
curve, if HI, IK, KL, LM, and fo on, be taken
equal; the ordinates HA, IB, KC, LD, nd
fo on, are in geometrical proportion; wherefore'
if HI, IK, and fo on, be taken very fmall, they
will reprefent the equal altitudes of the cylindrical bodies of air in the trumpet; and the ordinates HA, IB, and fo on, will be the radii of their
bafes, and the bodies of air being of equal heights
will be to one another as their bafes, that is, as the
fquares of their radii; but the radii being to one
another in a geometrical proportion, their fquares
will be fo too; confequently, the little cylindrical
bodies of air will be in a geometrical progreffion,
the fmalleft whereof lies next the mouth, and the
largeft at the exit of the tube; for which reafon
the augmentation of found will be greater, *cæteris
paribus*, in a trumpet of fuch a form than of any
other form whatever.

But to proceed to the fecond general cafe, wherein both the bodies move one and the fame way, but
the fubfequent more fwiftly than the preceding.

If two equal elaftick bodies move in the fame direction, and in fuch a manner as that one may over-
take and ftrike the other, upon the ftroke they will
change their quantities of motion with each other;
for inftance, if the motion of the fubfequent body
before the ftroke be double the motion of the preceding body, then will the preceding body after the
ftroke have double the motion of the fubfequent
body after the ftroke; and the preceding body after the ftroke, will move with the fame velocity
where-

wherewith the fubfequent body moved before the L e c t
ftroke; and the fubfequent body will after the V.
ftroke be carried with the velocity of the preceding
body before the ftroke; fo that upon the ftroke the
bodies will change their motions and velocities. For
fince by fuppofition the fum of the motions is three,
and fince the bodies are equal, the motion of each
after the ftroke, fetting afide the elafticity, muft be
one and an half; and by the fecond rule for determining the quantity of motion communicated by
the ftriking body to the other; the motion communicated in this cafe will be as one half, and fo
likewife will the motion arifing from the elafticity,
which being deducted from the motion which remains in the ftriking body after the ftroke, and
added to that of the preceding body, leaves the
motion of the former as one, and of the latter as
two; fo that upon the ftroke the motions will be
changed. Wherefore if two ivory balls of an equal
fize be let fall at the fame time, one from the height
of fix inches, and the other from the height of
three, after the ftroke, the preceding ball will rife
to the height of fix inches, and the fubfequent to
the height of three only.

If the bodies be unequal and move the fame way,
their motions and velocities after the ftroke may in
like manner be difcovered by the help of the propofition. For inftance, if the fubfequent body be as
two, and have twelve parts of motion, and the preceding body as one, and its motion as three; the
motion of the fubfequent body after the ftroke will
be as eight, and that of the preceding body as feven, and the velocity of the former will be as four,
and that of the latter as feven; for the fum of the
two motions before the ftroke being fifteen, and
the bodies being as one and two, the motion of the
leffer body after the ftroke, fetting afide the elafticity, will be as five, and that of the greater as ten,
but the motion of the leffer body before the ftroke

F 4   was

was at three, confequently, the communicated motion is as two ; wherefore adding fo much on account
of the elafticity to the motion of the leffer body,
and fubducting as much from that of the greater
body, which in this cafe is the ftriking body, we
fhall have eight for the motion of the greater, which
being divided by two, the quantity of matter in the
greater, gives four for its velocity; and we fhall
have feven for the motion of the leffer body, which
becaufe the quantity of matter in the leffer is as
one, will likewife exprefs the velocity. Wherefore

Exp. 8. if two ivory balls, one double of the other, be let fall
at the fame time, the larger from the height of fix
inches, and the fmaller from the height of three,
after the ftroke the leffer will afcend to the height
of feven inches, and the greater to the height of

Exp. 9. four.   On the other hand, if the fmaller ball be let
fall from the height of fix inches, and the greater
from the height of three ; after the ftroke the leffer
will afcend to the height of two inches, and the
greater to the height of five, and the motion of
the former will be two, and that of the latter ten ;
for fince the fmaller ball is as unity, and falls from
the height of fix inches, its motion at the time of
the ftroke is fix ; and fince the larger ball is as two,
and falls from the height of three inches, the motion thereof at the time of the ftroke is likewife
fix ; and the fum of thofe two motions, which is
twelve, being divided between the bodies in proportion to their quantities of matter gives eight for
the motion of the greater, and four for the motion
of the leffer, which motions they would have after
the ftroke, fuppofing they were not elaftick ; and
fince the motion of the greater body before the
ftroke was fix, the motion communicated to it by
the ftroke is two, which by reafon of the elafticity
being fubducted from four, the motion of the
ftriking body, and added to eight, the motion of
the other body, gives two and ten for the motions
of

of the two bodies, which motions being divided by the refpective bodies, give two and five for the velocities.

If two equal bodies meet one another with equal quantities of motion, which is one branch of the third general cafe, they will rebound with the fame motions and fame velocities wherewith they approached; for were they void of elafticity they would upon the ftroke ftand ftill, becaufe they communicate to one another a quantity of motion equal to that which each of them has at the time of the ftroke, and that in a contrary direction; but by the propofition, each of them muft on account of the elafticity receive as much motion as was communicated by the ftroke; and the motions which are thus received by the bodies being equal, and contrary to the motions wherewith the bodies met, and which were deftroyed by the ftroke, muft carry the bodies backward with the fame velocities wherewith they approached. Wherefore, if two Exp. 10. equal ivory balls be let fall at the fame time from equal heights, fo as to meet one another, upon the ftroke they will be reflected back to the heights from which they fell.

If the balls be unequal, for inftance, if one be Exp. 11. double the other; let the larger fall from one half only of the height from which the fmaller defcends, by which means when they meet their motions will be equal, and upon the ftroke they will be reflected each to the height from which it fell.

Where the bodies meet one another with unequal motions, which is the fourth general cafe, if the bodies be equal, they will both be reflected, and each of them will recede with the motion and velocity wherewith the other approached; that is, they will change their motions and velocities; for let us fuppofe the motions of the two bodies to be as fix and three; if they were void of elafticity the body which has the fmalleft quantity of motion would

LECT. V.

would upon the ftroke be turned back, and the two bodies would be carried with the difference of their motions divided equally between them, that is, the motion of each would be as one and an half, and the motion communicated would by the fourth rule be as four and an half; but a quantity of motion equal to what is communicated, muft be fubducted from the motion remaining in the ftriking body, and added to the motion of the other, that is, four and an half muft be fubducted from one and an half, and likewife added thereto; whereby there will be three negative for the motion of the ftriking body, which fhews that it will be carried back with a motion which is as three; and there will be fix pofitive for the motion of the other body, which fhews that it will be carried with a motion which is as fix, in the direction of the ftriking body before the ftroke; that is, it will be reflected; fo that each of them will be carried back with the

Exp. 12. motion wherewith the other approached. Wherefore, if two equal balls of ivory be let fall at the fame time, one from the height of fix inches, and the other from the height of three, upon the ftroke they will return back; but that which fell from the height of fix inches will rife only to the height of three, whereas that which fell from three inches will rife to fix.

If the balls be unequal, and meet one another with unequal quantities of motion, their motions after the ftroke may in like manner be determined by the help of the rule laid down in the propofition;

Exp. 13. for inftance, if two ivory balls which are as one and two be let fall at the fame time, the greater from the height of fix inches, and the fmaller from the height of three; in this particular cafe, the greater ball will upon the ftroke lofe all its motion, and the fmaller will be reflected with the difference of their motions, fo as to rife to the height of nine inches; for fince the larger ball which defcends from the

height

height of fix inches is as two, its motion is as twelve, whilft the motion of the fmaller ball, which is as unity, and defcends only from the height of three inches, is as three, the difference of which motions is nine; and this being divided between the bodies in proportion to their quantities of matter, gives fix for the motion of the larger, and three for that of the fmaller; and with thefe motions the bodies would be carried after the ftroke, fuppofing they were void of elafticity; but becaufe of the elafticity, a quantity of motion equal to what is communicated by the ftriking body to the other, which in this cafe is fix, muft be taken from the motion of the greater body, and added to that of the fmaller, which two motions being fix and three, the remainder after fubduction, which expreffes the motion of the greater body, will be nothing; and the fum arifing from the addition, which expreffes the motion of the fmaller ball, will be nine.

LECTURE VI.

OF THE CENTER OF GRAVITY, BALANCE, AND LEVER.

MY defign in this lecture is to give you an account of the firft and fecond of the mechanick powers, commonly called the balance and the lever; but I fhall firft take notice of fome things relating to heavy bodies, the knowledge of which is in a great meafure neceffary to the right underftanding of what fhall be faid concerning the mechanick powers in general. And Firft, in every body there is a certain point, commonly called by the writers of mechanicks, the center of gravity; the nature of which will beft appear from its chief properties, which are thefe.

1ft, If a body be fufpended by its center of gravity, it will continue in any pofition whatever
wherein

LECT. wherein it is placed ; whereas if it be fufpended by
VI.   any other point, it will not reft in any other pofition
       but where the center of gravity is either directly
       above, or directly beneath the point of fufpenfion ;
Exp. 1. thus, if two beams be fupported, the one by an
       axle paffing thro' its center of gravity; the other
       by an axle which doth not pafs thro' the center of
       gravity, but thro' fuch a point, as when the beam
       is parallel to the plane of the horizon, lies directly
       above the center of gravity ; the former will reft
       in any pofition, whether it be perpendicular, paral-
       lel, or inclined to the horizontal plane ; but the
       latter will reft in the parallel pofition only ; and
       fhould it by any force be removed from that pofi-
       tion, it will, upon the removal of the force, begin
       to move in order to recover the parallel pofition, and
       after feveral vibrations will at length fettle therein.

       A fecond property of the center of gravity is, that
       where that is fupported the whole body is likewife
       fuftained ; for which reafon the whole weight of a
       body may be looked upon as applied to that fingle
       point, and as centered therein.

       A third property of this center is, that it continu-
       ally endeavours to move downward towards the cen-
       ter of the earth, and where all lets and impediments
       are removed does actually defcend ; and therefore if
       in any cafe a body feems to move upward by the force
       of gravity, it will be found that the center of gravity
       defcends notwithftanding any appearance to the con-
Exp. 2. trary. Thus, if two rulers be fo placed as to meet
       in an angle at one of their ends, and there to reft
       upon an horizontal plane, whilft at their other end
       they are raifed a little above the plane ; and if a
       body confifting of two equal fimilar cones united
       at their bafes, be laid upon the rulers in fuch a man-
       ner, that the edge of their bafes may lie between
       the rulers, it will when left to itfelf begin to roll
       towards the elevated extremities of the rulers, and
       upon that account appear to afcend, whereas in re-
                                                    ality

ality it moves downward ; for if a ftring be ftretched horizontally beneath the rulers fo as that it may
touch the edge of the bafes of the cones at the
concourfe of the rulers, it will be found that the
edge of the bafes defcends below the ftring, and
that more and more as the body moves nearer to
the higher end of the rulers.

Whilft the body rolls upon the rulers, the parts
of the cones which reft thereon, do by reafon of the
widening of the rulers grow continually fmaller ;
upon which account, at the fame time that the body afcends along the plane of the rulers, it is as it
were carried down another plane equal in length to
the fide of the cone, and whofe perpendicular altitude is equal to the femidiameter of the bafes of the
cones ; and therefore, if the perpendicular altitude
of the rulers in that part where their diftance is
equal to the length of the double cone, be lefs than
the femidiameter of the bafes, the body will move
up along the rulers, becaufe, by fo doing, it will in
reality defcend, and the defcent thereof will be
equal to the difference between the femidiameter of
their bafes, and the perpendicular altitude of the
rulers in that part where their diftance is equal to the
length of the cones ; but if that perpendicular altitude be equal to the femidiameter, the body will
reft on any part of the rulers, being carried as
much upward on one account, as it is downward
on the other ; and if the altitude of the rulers be a
little increafed, fo as to exceed the femidiameter of
the bafes of the cones, the body will roll down
the rulers, and thereby defcend thro' a fpace equal
to that excefs.

If a cylinder be fo contrived as to have its center Exp. 3.
of gravity near one of its fides, which may be done
by making a wooden cylinder hollow towards one
fide, and then filling it with lead ;. when it is placed
on an inclined plane in fuch a manner as that the
side

fide which is neareft to the center of gravity may lean
towards the upper part of the plane, it will afcend,
provided the inclination of the plane be not too
fmall, but the center of gravity will at the fame
time defcend; for it will fuitably to its nature endeavour to move downward, and thereby caufe the
cylinder to revolve about its axis; and this revolution will make the cylinder, and confequently its
center of gravity, to move up the plane; fo that
the center of gravity will have as it were two motions, one upward arifing from the progreffion of
the cylinder along the plane, the other downward
occafioned by the rotation of the cylinder about its
axis; but the defcent occafioned by the latter motion, will be greater than the afcent arifing from
the former; as will appear by ftretching a line horizontally at the fame height with the center of gravity before the cylinder begins to roll, for after the
rotation ceafes the center of gravity will be beneath
the line; fo that upon the whole, that center will
be found to defcend notwithftanding the afcent of
the cylinder on the plane.

When the elevation of the plane becomes fo great
that the afcent arifing from the progreffion becomes
equal to, or greater than the defcent arifing from
the rotation, the cylinder will in the former cafe
continue at reft, and in the latter roll down the
plane.

A line drawn from the center of gravity of any
body, perpendicular to the plane of the horizon, is
called the line of direction of the center of gravity,
becaufe when the body is carried downward by the
force of gravity, if it meets with no let or obftacle, its center of gravity will defcribe that line.
The chief property of this line is, that as long as
it falls within the bafe of the body, fo long the body ftands, whereas no fooner does it fall beyond the
bafe, but the body tumbles; as will appear from
                                                  the

the following experiment; let a piece of wood be
set on a moveable plane with a plummet hanging
from its center of gravity, and let the plane be gradually elevated, till at length the plum-line (which,
as it is always perpendicular to the horizon, will
represent the line of direction) falls beyond the base;
the wood will not tumble as long as the plummet
line falls within the base, whatever be the elevation
of the plane whereon it stands, but the moment
that line gets beyond the base the body falls.

The reason why a body stands during the continuance of the line of direction within its base is,
that no motion can arise in any body from the force
of gravity, unless the center of gravity can by such
motion be carried downward; but as long as the
line of direction of any body falls within the base,
its center of gravity is supported, and therefore
cannot descend; and consequently, the body will
remain unmoved; whereas upon the removal of the
line of direction beyond the base, the center of
gravity ceases to be supported, and is therefore at
liberty to descend.

From what has been said it appears, why among
bodies descending on inclined planes, some, for instance cubes, only slide, whilst others as globes or
cylinders roll; the lines of direction falling beneath the bases of the former, but not the latter.

The center of motion in any body is a fixed
point or axis about which the several parts of a
body do move, and in moving describe circular
arches.

The direction of any power or weight is, that
strait line wherein it moves or endeavours to
move. And the moment of any power or weight
is, that force wherewith it either moves or endeavours to move, and it is always proportional to the
product arising from the multiplication of the power or weight into the velocity wherewith it moves
or would move if it were not hindred by some opposite

polite power or weight; and therefore if the product ariling from the multiplication of one weight
or power into its velocity, be equal to the product
ariling from the like multiplication of any other
weight or power into its velocity, the moments of
thole two weights or powers mult be equal; and
this will always be where the weights or powers are
to one another reciprocally as their velocities; con-
lequently, two weights or powers may balance, if
as much as one exceeds the other in magnitude, lo
much mult it be exceeded by the other in velocity;
and herein conlilts the whole force and efficacy of
all mechanical engines; for they are lo contrived as
to diminilh the velocity of one weight or power and
to increale that of the other, by which means a very lmall weight or power may become a balance to
one exceedingly great, as will appear from what
lhall be laid concerning the mechanick powers,
which are commonly reduced to lix, namely, the
balance, the *lever,* the *pulley,* the *axle in the wheel,*
the *wedge,* and the *lcrew,* of each of which in their
order.

The BALANCE, ltrictly lpeaking, is a beam lup-
ported by an axle whereon it turns; which axle
therefore is the center of motion; the parts of the
beam which lie on each lide of the axle are called
its arms, and thole parts of the arms to which the
weights are applied are called the points of lulpenlion; concerning which it mult be oblerved, that
the appending weight, whatever be the length of the
cord by which it hangs, acts with the lame force
and in the lame manner as if its center of gravity was
applied to the point of lulpenlion; lo that it matters
not what the diltance is between the weight and point
of lulpenlion, as will appear from the following experiment. Let a weight appended at one arm of a
balance be counterpoiled by a weight at the other,
and let it by means of a cord be hung at different
diltances below the point of lulpenlion; the polition

Exp. 5.

of

of the balance will remain unvaried, and the weights L ᴇ ᴄ ᴛ. will continue to counterpoise each other at all those VI. distances.

The moment of any weight appended at the arm of a balance, is proportional to the product arising from the multiplication of the weight into the distance of the point of suspension from the axis of the balance; for as was before said, the moment of a weight is proportional to the product of the weight into its velocity, and in this case the velocity of the weight is as the distance of the point of suspension from the axis; for since the weight acts in the same manner as if its center of gravity was applied to the point of suspension, whatever be the velocity wherewith that point moves round the axis, the same will the velocity of the weight be; but the velocities wherewith the several points in the arm of a balance move round the axis, are as the spaces, that is, as the circular arches, which they describe in the same time, which arches from the nature of the circle are to one another as their respective radii, that is, as the distances of the points from the axis. Thus, if A B represents the arm of the balance Pl. 2. moving round the axis at A, the velocities of the Fig. 9. points B and D, which describe the arches B C and Exp. 6. D E, will be as those arches, because they are described in the same time; but from the nature of the circle, those arches are to one another as their radii A B, and A D, that is, as the distances of those points from the axis; consequently, the moment of a weight appended at the arm of a balance, is as the product of the weight into the distance of the point of suspension from the axis. Whence it follows, that if two weights be appended at the arms of a balance in such a manner, as that the distances of the points of suspension from the axis shall be reciprocally proportional to the weights, those weights will counterpoise each other, and the balance will be in *æquilibrio*; for instance, if two equal weights

G   be

be applied at equal diſtances from the axis, the balance will not incline to either ſide,-but remain parallel to the horizon, the weights in this caſe coun-

Exp. 7. terpoiſing one another. Again, if one weight be
larger than the other in any proportion, for inſtance,
in the proportion of three to one, if the point at
which the ſmaller is applied be thrice as far diſtant
from the axis as the point at which the larger is applied, the balance will be in *æquilibrio*.

On this *æquilibrium* ariſing from the ſuſpenſion of
Exp. 8. weights at diſtances reciprocally proportional to the
weights, is founded the *Statera Romana*, otherwiſe
called the ſteel-yard, which conſiſts of two arms
very unequal in length, but equally poiſed by means
of a weight annexed to the ſhorter, from which
likewiſe hangs a ſcale in order to receive ſuch things
as are to be weighed; the longer arm is divided
into a number of equal parts beginning from the
axis, and ſuſtains a weight with ſlides from one
end to the other; which weight being applied to
the ſecond diviſion, will counterpoiſe double the
weight in the ſcale of the ſhorter arm, that it will
when applied to the firſt diviſion; and triple when
applied to the third diviſion; and ſo on, whatever
be the diviſion to which it is applied, the weight in
the ſcale of the ſhorter arm muſt be proportional
thereto; otherwiſe the products ariſing from the
multiplication of the weights into their reſpective
diſtances from the axis would not be equal, and
conſequently would not balance each other.

Exp. 9. On the ſame *æquilibrium* is likewiſe founded the
deceitful balance, which is ſo contrived, as tho' one
arm be longer than the other, yet is the ſhorter
made ſo much thicker than the longer, as thereby
exactly to poiſe the ſame; upon which account the
balance appears to be juſt, and conſequently ſuch
weights as counterpoiſe are judged equal, whereas
in truth that which is appended at the longeſt arm
is leſs than the other, and that in the proportion of
the

the length of the shorter arm to that of the longer for instance, if the longer arm be to the shorter as ten to nine; a weight of nine ounces applied at the longer arm, will counterbalance ten ounces append-ed at the shorter.

Several weights appended at several distances from the axis in one side of a balance, will counter-poise several others appended likewise at several dis-tances on the other side; provided the sum of the products which arise from the multiplication of the weights on one side into their respective distances from the axis, be equal to the sum of the products arising from the like multiplication of the weights on the other side into their respective distances. Thus, if on one side a weight of one ounce be ap-pended at the distance of two inches from the axis, and another of two ounces at the distance of three inches, and a third of three ounces at the distance of four inches; and if on the other side be append-ed one weight of five ounces at the distance of an inch from the axis, and another of three ounces at the distance of five inches; the two latter will ba-lance the three former; for the product of five in-to one, being added to the product of three into five, gives the sum of twenty; as does likewise the addition of the three products of one into two, two into three, and three into four.

Exp. 10.

The chief use of the balance, commonly called a pair of scales, is to compare the weights of different bodies together; and that this machine may be as exact and perfect as possible, it is requisite, 1st, that the center of gravity of the beam be placed a little below the axis, because in this case, when there is an *æquilibrium*, the beam will not rest in any position but the parallel; consequently, the weights which are compared together will appear to be equal, as they really are; whereas if the axis be placed beneath the center of gravity, should the center of gravity be moved out of the perpendicu-

lar

lar line, which can fcarcely be avoided, it will not
return, but from its tendency downward will be
carried lower, fo as to give the beam an inclined pofition; for which reafon the weights will appear to
be unequal, tho' in reality they are not fo; and the
fame inconvenience will arife if the axis paffes thro'
the center of gravity, for in that cafe it has been
already fhewn, that the beam, notwithftanding the
equilibrium, will reft in any pofition.

Secondly, the arms of the beam ought to be exactly equal both as to weight and length, the reafon of which is evident, from what was faid concerning the deceitful balance.

Thirdly, the points from which the fcales are
fufpended, ought to be in one right line paffing
thro' the beam's center of gravity; for by this contrivance the weights will act directly againft each
other, fo that no part of either will be loft on account of any oblique direction.

Fourthly, the friction of the beam againft the
axis ought to be as little as poffible; becaufe, fhould
the friction be great, it will require a confiderable
force to overcome it; upon which account, tho'
one weight fhould a little exceed the other, it will
not preponderate, the excefs not being fufficient to
overcome the friction, and bear down the beam.

That the friction may be as little as poffible, the
parts of the beam which play upon the axis, as alfo
the axis itfelf, fhould be well polifhed, and the axis
fhould be made as fmall as the ufes of the balance
will admit; but as friction cannot be entirely pre-
vented, to remedy the inconveniences arifing from
it as much as poffible, the arms of the beam ought
to be made as long as they conveniently can; becaufe the longer the arms are, the lefs will the
weight be that is requifite to overcome the friction;
the moments of weights increafing in proportion to
their diftances from the center of motion, as has
been already fhewn.

I fhall

I shall close what I had to say concerning the balance, by laying before you one property of it, which is somewhat singular and surprising; tho' it has not that I can find been taken notice of by any of the mechanick writers *, namely, that if a man standing in one scale and counterpoised by a weight in the other, lays his hand to any part of the beam, and presses it upward, he will thereby destroy the balance, and make the scale wherein he stands to preponderate.

In order to account for this property, let A B represent the beam of a pair of scales playing on the axis at C, and let a man standing in the scale D, and counterpoised by a weight in the scale E, lay his hand to some part of the beam, either on the same side of the axis with himself as at H, or on the other side as at K, and press the same upward; inasmuch as action and reaction are always equal, it is manifest that with whatever force the hand presses upward against the point H or K, with the same the hand, and consequently the man's whole body, is pressed downward; and therefore the scale D wherein he stands bears the same pressure from his feet that the point H or K does from his hand; but the pressure upon the scale D may be looked upon as applied to the beam at the point A from which the scale hangs; consequently, the same force which presses up the point H or K, presses down the point A; wherefore putting F to denote that force, $F \times HC$ will express the moment wherewith the arm A C is pressed upward when the hand is applied at H, and $F \times KC$ the moment wherewith the arm

---

* The property here mentioned, had not been taken notice of by any of the Mechanick Writers, when the Author composed this Lecture; but has been published since, both in the Philosophical Transactions for the year 1729, and in a course of experimental Philosophy, by Dr. DESAGULIERS, to whom our Author communicated it, as he told me and many others, about thirteen or fourteen years ago when he was in London.

BC is preſſed upward, the hand being applied at K ;
and in both caſes $F \times AC$ will expreſs the moment
wherewith the arm AC is preſſed downward by
means of the reaction ; if therefore the hand be
applied at H, it is manifeſt that as the arm AC is at
one and the ſame time preſſed upward by a force
which is as $F \times HC$, and downward by a force
which is as the ſame $F \times AC$, and as HC is ever
leſs than AC, the arm AC muſt deſcend with the
difference of thoſe forces, that is, with a force equal
to $F \times AH$, which is the diſtance of the hand from
the point A ; if the hand be applied at K, the arm
CB is preſſed upward, and conſequently AC downward, with a force equal to $F \times KC$, and upon account of the reaction AC is likewiſe preſſed downward with a force equal to $F \times AC$; and therefore
it muſt deſcend with a force equal to the ſum of
thoſe two forces, that is, with a force equal to
$F \times AK$ the diſtance of the hand from the point A ;
ſo that the ſcale D muſt preponderate whether the
hand be applied to that part of the beam which
lies on the ſame ſide of the axis with the man, or to
that which lies on the other ſide ; and if D be put
to denote the diſtance of that point to which the
hand is applied from the point A, the force wherewith the preponderating ſcale deſcends will be univerſally as $F \times D$, that is, as the force which the
hand exerciſes againſt the beam, multiplied into
the diſtance of the hand from the point A. And
if the force wherewith the hand preſſes the beam
be required, it may be diſcovered by throwing in
as much weight into the ſcale E as is ſufficient to
balance the force of the hand, and to prevent the
deſcent of the ſcale D ; for putting W to denote that
weight, its moment is as $W \times BC$ or AC, which being equal to $F \times D$ the moment of F, F will be
found equal to $W \times \dfrac{AC}{D}$, that is, to the weight
multiplied into half the length of the beam, and
divided

divided by the diſtance of the hand from A. For
inſtance, if the balancing weight be twenty pounds,
and the diſtance of the hand from A be to half the
length of the beam as one to two, the force wherewith the hand preſſes the beam is equal to twenty
pounds multiplied by two and divided by unity,
that is, it is equal to forty pounds ; from what has
been ſaid it follows, that when the hand is applied
to that part of the beam which lies on the ſame ſide
of the axis with the man, the force of the hand
upon the beam is greater than the weight which bal-
lances it in the ſcale E, and leſs than the ſame when
the hand is applied to that part of the beam which
lies on the other ſide of the axis with reſpeſt to the
man; for in the firſt caſe, $W \times \dfrac{AC}{D}$ is greater than
W, and in the latter leſs, inaſmuch as A C is in the
former caſe always greater, and in the latter leſs
than D.

The ſecond, and indeed the moſt ſimple of all the
mechanick powers is the LEVER; an engine chiefly
made uſe of to raiſe large weights to ſmall heights.
By the writers of mechanicks, it is ſuppoſed to be
an inflexible line void of all gravity; tho' ſuch as
are in common uſe are both flexible and weighty.
In every lever there is one immoveable point, about
which as a center all the parts of the lever turn;
and whatever ſupports that point is called the prop;
and with regard to the different ſituations of the
moving power, and the weight to be moved in reſpeſt to the prop, the lever is divided into three
kinds; the firſt of which is where the prop is placed
between the moving power and the weight to be
raiſed; which kind of lever is repreſented, wherein C denotes the prop, B the weight, and A the
power. In this lever there will be a balance between the power and the weight, provided they be
to one another reciprocally as their diſtances from
the prop; that is to ſay, if the power at A be to

Pl. 2.
Fig. 11.

the

the weight at B, as CB to C A; for upon the motion of the lever round its fixed point C, the power
at A will defcribe the arch AD in the fame time
that the weight at B defcribes the arch BE; confequently, the velocity of the power will be to the
velocity of the weight, as the arch AD to the arch
BE; that is, becaufe the arches are fimilar, as is evident from the manner wherein they are generated,
as AC to CB. That therefore the product arifing
from the multiplication of the power into its velocity, may be equal to the product of the weight into its velocity; or in other words, that their moments may be equal, the power muft bear the fame
proportion to the weight, that BC the diftance of
the weight from the prop bears to AC the diftance
of the power from the prop. For inftance, if BC be
to AC as one to two, and if a man's ftrength be

fuch as that without the help of a machine he can
fupport an hundred weight, he will by the help of
this lever be enabled to fupport two hundred; becaufe as BC is to AC, which by fuppofition is as
one to two, fo muft the power at A be to the
weight at B; but the power at A is fuppofed to be
equal to one hundred, confequently the weight muft
be equal to two.

As in this lever the prop may be placed either
at the middle diftance between the moving power
and the weight, or nearer to one than the other, it
is evident that there may be a balance between the
power and the weight, either when they are equal,
or when the one exceeds or is exceeded by the
other according to the different fituations of the
prop.

To this kind of lever may be reduced feveral
forts of inftruments, fuch as fciffars, pincers, fnuf-
fers, each of which may be confidered as made up
of two levers, whofe prop is the fame with the pin
which rivets them together. Quarry crows are
likewife levers of this kind, concerning which it

2

muft

muſt be obſerved, that the larger and more ponderous they are, provided they are not ſo big as to become unmanageable, the more uſeful they muſt be, becauſe the weight of that part of a crow which lies on the ſame ſide of the prop with the power, and which uſually far exceeds the other part in length, acts in conjunction with the power, and thereby facilitates the raiſing of the ſtones.

If the arms of this lever, inſtead of lying in a right line, meet each other at the prop in a right angle, where A C and B C repreſent the arms of a lever united at the prop C, in ſuch a manner as to conſtitute a right angle A C B ; if to one arm as C B placed horizontally, a weight be appended at B, and to the other as A C ſtanding perpendicularly a power be applied at A acting in the direction A D. In order to a balance the power muſt be to the weight as B C to A C, that is, the power and weight muſt be in the inverſe *ratio* of the lengths of the arms to which they are applied. For as the arms turn together upon the prop C, in the ſame time that the point B deſcribes any arch as B K, the point A muſt deſcribe a ſimilar arch as A H ; conſequently, the velocity of A will be to the velocity of B as A C to B C ; but as the moment of the power at A is ſuppoſed equal to the moment of the weight at B, the power muſt be to the weight, as the velocity of the latter to the velocity of the former, that is, as B C to A C.

To confirm this by experiment, let B C be one fourth of A C, and a weight of twelve ounces be appended at B ; to the cord A D F made faſt to the point A and paſſing over a pulley at D, let a weight of three ounces be hung at F ſo as to pull the arm A C in the direction A D, and there will be a balance. And if B C be one third or one half of A C, then a weight at F, which in the former caſe is one third, and in the latter one half of P, will balance·

the

Pl. 3.
Fig. L

Exp. 13.

the fame; and if AC and BC be equal, the balancing weights muft be fo too.

From the experiments, and what has been faid
concerning them, it is evident, that the greater the
proportion is which AC bears to BC, the greater is
the force of the lever, or the lefs the power at A
requifite to balance a given weight at B. And forafmuch as the hammer when made ufe of in drawing nails is a lever of this kind, it is manifeft,
that the longer the handle is in proportion to
that part of the hammer which lies between the
handle and that portion of it which gripes the nail,
the lefs will the force be that is requifite to draw the
nail.

The fecond kind of lever has its prop at one end,
the power at the other, and the weight between, as
where C is the prop, A the power, and B the
weight; in this lever, in the fame time that the
power at A moves thro' the arch of a circle whofe
radius is AC, the weight at B moves thro' a fimilar arch of a leffer circle whofe radius is BC; confequently, the velocity of the power is to the velocity of the weight as AC to BC; in order therefore
to a balance, the power muft be to the weight as
BC to AC; that is, as much as AC, the diftance
of the power from the prop, exceeds BC, the diftance of the weight from the prop, fo much muft
the weight exceed the power.

As in this lever the diftance of the weight from
the prop is always lefs than the diftance of the
power from the prop, it is evident that there cannot be a balance in any cafe but where the weight
exceeds the power.

To this kind of lever may be reduced the oars
and rudders of fhips, cutting-knives fixed at one
end, and doors moving upon hinges.

If in this lever we fuppofe the power and the
weight to change their places, fo as that the power
may

Pl. 3.
Fig. 2.

Exp. 14.

may be applied at B between the weight at A and the prop at C, it will become a lever of the third kind; wherein in order to a balance, the power at B muſt ſo far exceed the weight at A, as BC the diſtance of the power from the prop, is leſs than AC the diſtance of the weight from the prop.

It is evident, that the moving power receives no advantage from this kind of lever, and therefore it is never made uſe of but in caſes of neceſſity, and where the weights to be raiſed cannot be managed in a more convenient manner; as is the caſe of ladders, which being fixed at one end are by the force of a man's arms reared againſt a wall.

As levers are of ſervice in raiſing weights, ſo are they likewiſe in carrying and ſupporting the ſame; concerning which it is to be obſerved, that when two powers ſupport a weight by help of a lever, the ſum of the powers muſt equal the weight; and the weight being placed between them, their reſpective diſtances therefrom muſt be reciprocally as the powers; thus, if a weight reſting on the lever at B be ſupported by two powers, one at A and the other at C, the diſtance of A from B muſt be to the diſtance of C from B, as the power at C is to the power at A. For in this caſe the lever is of the ſecond kind, where each of the powers is in its turn to be looked upon as the prop, and then the other power muſt be to the weight as the diſtance of the weight from the prop to the diſtance of the power from the prop; that is, when A is conſidered as the prop, the power at C muſt be to the weight at B, as AB to AC; and when C is conſidered as the prop, the power at A muſt be to the ſame weight at B, as CB to CA. Conſequently, ſince the power at A is to the weight, as BC to AC; and ſince the ſame weight is to the power at C, as AC to AB, the power at A muſt be to the power at C, as BC to BA, that is, the powers muſt be to one another inverſly as their diſtance from the weight; and thus

Pl. 3.
Fig. 3.

it

it will appear to be from experiments. For if from the point B of the lever A C a weight as D be suf-

Exp. 16. pended, and if two other weights as E and F be suspended from the extream points A and C by

Pl. 3. cords passing over pullies, so as that they may draw
Fig. 4. the lever directly upward; they will support the weight D provided the sum of those two weights be equal to the weight D, and the weight E be to the weight F as BC to BA.

Exp. 17. The same thing will happen, if the three weights be made to pull the lever horizontally, which may be done by passing the cords over small wheels or pins placed on a level with the lever.

In shewing what the proportion ought to be be-tween two powers which support a weight placed upon a lever, I have supposed the position of the lever to be parallel to the plane of the horizon; what the proportion ought to be, and in what man-ner such proportion is determined in inclined posi-tions of the lever, shall be shewn, when I come to treat of powers acting in oblique directions.

If instead of a single lever, several be combined together in such a manner, as that a weight being appended to the first lever, may be supported by a

Pl. 3. power applied to the last, as in the machine, which
Fig. 5. consists of three levers of the first kind, and is so
Exp. 18. contrived as that a power applied at the point L of the lever C, may sustain a weight at the point S of the lever A. The power must be to the weight, in a *ratio* compounded of the several *ratios*, which those powers that can sustain the weight by the help of each lever when used singly and apart from the rest, have to the weight; for instance, if the power which can sustain the weight P by help of the lever A alone, be to the weight as one to five; and if the power whereby the same weight can be sustained by the help of the lever B alone, be to the weight, as one to four; again, if the power which can sup-port the same weight by the help of the lever C

alone,

alone, be to the weight as one to five; the power
which supports the weight by means of those three
levers joined together will be to the weight in a
ratio compounded of one to five, one to four, and
one to five, that is, it will be as one to an hundred.
For since in the lever A, a power equal to one fifth
of the weight P pressing down the lever at L, is
sufficient to balance the weight; and since it is the
same thing whether that power be applied to the
lever A at L, or the lever B at S, the point S bear-
ing on the point L, a power equal to one fifth of
the weight P being applied to the point S of the
lever B, and pressing the same downward, will support the weight; but one fourth of the same power being applied to the point L of the lever B, and
pushing the same upward, will as effectually depress
the point S of the same lever, as if the whole power
was applied at S; consequently, a power equal to
one fourth of one fifth, that is, to one twentieth
part of the weight P, being applied to the point L
of the lever B, and pushing up the same, will support the weight; but it matters not whether that force
be applied to the point L of the lever B, or to the
point S of the lever C, since if S be raised, L
which rests thereon must be so too; but one fifth
of the power applied at the point L of the lever C,
and pressing it downward, will as effectually raise
the point S of the same lever, as if the whole power
was applied at S and pushed up the same; consequently, a power equal to one fifth of one twentieth,
that is, to one hundredth part of the weight P, being applied to the point L of the lever C, will balance the weight at the point S of the lever A; that
is, a power which is to the weight, in a *ratio* compounded of the three *ratios*, which the powers have
to the weight in each lever taken separately, will be
a balance to the weight, when the three levers are
used jointly.   And by the same way of reasoning it
will

will be found, that in all machines of this kind, the power requifite to fuftain the weight, is to the weight, in a *ratio* made up of the feveral *ratios* of the power to the weight in each lever taken fepa- rately, whatever be the number of levers.

In all that has been hitherto faid concerning the lever, the power and the weight are fuppofed to act in direct oppofition to each other; and on this fup- pofition, the power muft be to the weight in each of the three kinds of levers, in the reciprocal *ratio* of their diftances from the prop, as has been fully proved with regard to each kind; but where the directions of the power and weight are inclined to each other, the proportion will vary from what has been here determined, as fhall be fhewn, when I come to treat of powers acting in oblique di- rections.

LECTURE VII.

OF THE PULLEY.

L E C T.
VII.

IN this lecture I fhall give you an account of the *Pulley*, the *Axle in the Wheel*, the *Wedge*, and the *Screw*. The PULLEY is a fmall wheel that turns about its axis, and which has a drawing rope paffing over it. It is made ufe of in raifing large weights to confiderable heights; and is of two kinds, fixed and moveable; the fole ufe of the fix- ed pulley, is to change the direction of the moving power; which in all cafes where weights are to be raifed to great heights, is exceedingly convenient,

Exp. 1.
Pl. 3.
Fig. 6.

and very often of abfolute neceffity; for inftance, if the weight P is to be raifed by the force of a man's hand to any height as A above the reach of the hand, the man muft quit his place and afcend in order to carry up the weight, which for the moft

part

part is found to be inconvenient, and sometimes im-
practicable; whereas if to a rope as PAF paffing
over the fixed pulley at A, the weight be made faft
at one end as P, and the hand applied to the other
end at F, the man by drawing the rope AF downward, will without moving from his place raife
the weight as effectually, as if his hand was applied
to it and moved upward from P to A ; fo that in
raifing weights to great heights the fixed pulley is
of fingular fervice, in as much as by changing the
direction of the power, it takes off the neceffity
that a man would otherwife lie under of afcending
along with the weight, and by fo doing leffens his
labour; befides, it has this farther convenience at-
tending it, that by means thereof the joint ftrength
of feveral perfons may be made ufe of to raife one
and the fame weight, which in many cafes cannot
be done, at leaft not fo conveniently, where the
weight is raifed by the immediate application of
the hands; but this pulley does not in the leaft af-
fift the power, by increafing its moment; becaufe
it neither leffens the velocity wherewith the weight
rifes, nor augments that of the power; for whatever be the fpace thro' which the power moves by
drawing the rope AF, the weight muft in the fame
time be drawn up thro' an equal fpace ; the rope
AP conftantly fhortening in the fame proportion,
that the rope AF is lengthened; and therefore,
wherever any power fupports a weight by means of
a fixed pulley, that power muft be equal to the
weight.

When a pulley rifes and falls along with the
weight, as does this pulley, it is faid to be move- Pl. 3.
able, and with regard to its ufe, it is juft the re- Fig. 7.
verfe of the fixed pulley ; for it adds to the moment of the power, but caufes no change in its direction: for if the hand be applied at F to the
rope D, in order to raife the weight P appended to
the moveable pulley E, it muft move directly upward

ward in the very fame manner, as if it was applied immediately to the weight; confequently, the direction of the hand which raifes the weight is no way altered by this pulley, but the moment thereof is doubled, becaufe it is made to rife twice as faft as the weight; for in the fame time that the hand moves upward from F to G, thro' the fpace F G equal in length to the two equal ropes D and C, the pulley, and confequently the weight annexed, will be drawn up thro' the fpace E H, whofe length is equal to one of the ropes only.

In machines confifting of feveral pullies, whereof fome are fixed and fome moveable, and which have one common rope that goes round them all; if one end of the rope be fixed, as is the cafe in the Pl. 3.
Fig. 8, 9,
10. machines reprefented by thefe figures, in order to a balance, the moving power muft be to the weight, as one to twice the number of moveable pullies; becaufe the velocity wherewith the power moves in raifing weights by the help of fuch engines, is to the velocity of the rifing weight, as twice the number of moveable pullies to unity; as I fhall now Pl. 3.
Fig. 8. fhew you in the machine, which confifts of one fixed pulley as A, and another moveable as E. Since it is one and the fame rope that is continued from G to F, the part A F which lies beyond the fixed pulley, cannot be drawn down and thereby lengthened, unlefs the two parts D and C, which lie on each fide of the moveable pulley, be at the fame time drawn up and fhortened, and that equally; whence it is evident, that the part A F will be lengthened as faft again as either D or C is fhortened, inafmuch as what each of thofe parts lofe of their length is added to the length of A F; but the point F to which the power is applied, defcends as faft as A F is lengthened, and the point E to which the weight is faftened, afcends as faft as D or C is fhortened; confequently, the velocity of the power is to the velocity of the weight, as two

to

to one, that is, as twice the number of moveable pullies to unity ; if therefore a weight appended at F, be to a weight appended at E, as one to two, they will balance each other, as being to one another in the reciprocal *ratio* of their velocities.

Pl. 3.
Fig. 9, 10.
Exp. 3.

In the machines, each of which confifts of two fixed and as many moveable pullies, and which differ only in this, that in one the pullies of the fame kind move upon one and the fame axis, and in the other upon different axes ; I fay, in thefe machines, the velocity wherewith the power moves is to the velocity wherewith the weight rifes, as four to one, that is, as twice the number of moveable pullies to one ; for as the part of the rope A F is drawn down and lengthened, the four parts B, C, D, H, which lie on each fide of the two moveable pullies are drawn up and fhortened, and that equally ; and what each of them lofes of its length is added to the length of A F ; confequently, A F is lengthened four times as faft as each of the other parts fhortens ; but the power moves as faft as A F lengthens, and the weight rifes as faft as the other four fhorten ; and therefore, the velocity of the power at F is to the velocity of the weight at E, as four to one, or as twice the number of moveable pullies to unity : for which reafon, if a weight be appended at F, which is to the weight at E as one to four, that is, in the reciprocal *ratio* of their velocities, there will be a balance.

What has been thus proved with regard to the three laft machines, namely, that the velocity wherewith a power moves in raifing a weight is to the velocity wherewith the weight rifes, as twice the number of moveable pullies to unity, is in the fame manner demonftrable with regard to any other machine of the fame kind, whatever be the number of pullies whereof it confifts ; and therefore, in all machines confifting of feveral pullies, whereof fome are fixed and others moveable, and round

H   which

which goes one common rope, fixed at one end, it may be laid down as a general rule, that in order to a balance between the moving power and the weight, the former muſt be to the latter, as one to twice the number of moveable pullies.

Exp. 4.   If the rope which goes round the pullies, inſtead of being fixed at one end, be faſtened to the weight or to the block which ſupports the moveable pul-

Pl. 3.   lies, ſo as to riſe therewith, as in this machine,
Fig. 11. which conſiſts of five pullies, whereof three are fixed and two moveable, and in which the end of the rope is joined at G to the block which ſupports the two moveable pullies ; the velocity of the power is to the velocity of the weight, as the ſum of twice the number of moveable pullies increaſed by unity to one ; for in this caſe, the parts of the rope which are equally ſhortened in order to lengthen the part A F, are more in number by one than the ſum of the moveable pullies when doubled ; conſequently, ſince the power at F moves as faſt as A F is lengthened, whilſt the weight at E riſes in proportion only to the ſhortening of the ropes B, C, D, H, K, the velocity of the power bears the ſame proportion to the velocity of the weight, as the ſum of twice the number of moveable pullies increaſed by unity does to one ; and therefore, if the power be to the weight in the inverſe *ratio*, that is, as one to twice the number of moveable pullies added to unity, there will be a balance. Thus, if

Pl. 3.   in the machine a weight appended at F be to an-
Fig. 11. other at E, as one to five, they will balance, and remain unmoved.

Exp. 5.   If to any of the forementioned machines be added a runner, that is, a ſingle moveable pulley, which has its own rope diſtinct from that which is common to the other pullies, one end whereof is

Pl. 3.   fixed as at L, the other being faſtened to the
Fig. 12. block at E, and the weight appended at M, the force of the former machines will be doubled by this
addi-

additional pulley; for fince the point E moves with twice the velocity of the point M, as I fhewed when fpeaking of the fingle moveable pulley, whatever be the proportion which the velocity of the power at F bears to the velocity of the weight when appended at E, it will be doubled if the weight be appended at M; confequently, the power will by the help of the runner be able to fuftain twice the weight that it did before.

Pl. 3.
Fig. 13,
Exp. 6.

If a machine be combined of one fixed and feveral moveable pullies, put together in fuch a manner as that each of the moveable pullies has a feparate rope, one end whereof being fixed, the other either paffes over the fixed pulley, as does that of the firft moveable pulley E, or is joined to the moveable pulley which lies next above it, as is the cafe of the ropes B, C, D, which belong to G, H, and I, the fecond, third and fourth moveable pullies; B being joined at N to the firft moveable pulley, C at K to the fecond, and D at L to the third; the weight being appended to the laft moveable pulley at H. The velocity wherewith the weight rifes in fuch a machine is to the velocity of the power, as one to the laft term of a duple progreffion, whereof the firft term is unity, and the number of terms more by one, than the number of moveable pullies.

For as I proved when fpeaking of the fingle moveable pulley, the velocity of the power at F is to the velocity wherewith the pulley E rifes, as two to one; and fo likewife is the velocity of E, to that of G, and that of G, to that of I, and fo on, whatever be the number of moveable pullies, the velocity of each fucceeding pulley is but one half of the velocity of the preceding; wherefore, if the velocity of the laft pulley, which is the fame with the velocity of the weight, be put equal to unity, the velocity of that which immediately precedes it, to wit H, will be as two; and the velocity

city of G, as four, and of E, as eight, and so on;
if there be more moveable pullies, the velocity
will be continually doubled, and since the velocity
of the laft pulley is expreffed by unity, that of the
firft will be expreffed by the laft term of a duple
progreffion whofe firft term is unity, and the number of terms equal to the number of moveable pullies; and confequently, since the velocity of the
power is double that of the firft moveable pulley,
if the duple progreffion be continued to one term
more, that term will exprefs the velocity of the
power, the velocity of the weight being as unity;
thus, in this machine, the number of moveable
pullies being four, the velocity of the weight at
M is to that of the power at F, as one to fixteen;
if therefore a weight appended at F be to the weight
at M, as one to fixteen, there will be a balance.

Tho' this engine be of greater force than any
other wherein there is the fame number of moveable pullies, yet inafmuch as it does for that very
reafon raife weights more flowly; men for the
fake of difpatch choofe rather to make ufe of fuch
combinations of pullies as are reprefented in the
9th and 10th Figures, and where they have occafion
to raife very large weights, they double the force
of thofe machines by the addition of a runner.

The fourth mechanick Power is called the
AXLE IN THE WHEEL; which is a fimple engine
confifting of one wheel fixed to the end of an axle
that turns along with the wheel; the manner of
raifing weights by the help of this machine is thus;
the power being applied to fome part of the wheel's
circumference, turns the wheel and together with
it the axle, by which means a rope that is tied to
the weight at one end, and made faft to the axle
at the other, is wound about the axle, and thereby
the weight drawn up; and for as much as the wheel
and its axle revolve together, in whatever time the
power moves thro' a fpace equal to the circumference

of

of the wheel, the weight muſt in the ſame time be raiſed up thro' a ſpace equal to the circumference of the axle, conſequently, the velocity of the power is to the velocity of the weight, as the circumfe- rence of the wheel to the circumference of the axle ; that is, from the nature of the circle, as the diame- ter of one to the diameter of the other ; if there- fore the power be to the weight in the inverſe *ratio* of thoſe diameters ; that is to ſay, if the power be to the weight, as the diameter of the axle to the diameter of the wheel, there will be a balance ; the power in that caſe being juſt ſufficient to ſupport the weight. For inſtance, if the diameter of the wheel be five inches, and that of the axle one, a weight of one ounce hanging from any point in the cir- cumference of the wheel, will ſupport a weight of five ounces hanging at the axle ; and if the diame- ter of the axle be but half an inch, then will ten ounces at the axle be ſupported by one at the wheel.

Where the parts of the axle differ in thickneſs, if weights be hung at the ſeveral parts, they may be ſuſtained by one and the ſame power applied to the circumference of the wheel, provided the pro- duct ariſing from the multiplication of the power into the diameter of the wheel be equal to the ſum of the products ariſing from the multiplica- tion of the ſeveral weights into the diameters of thoſe parts of the axle from which they are ſuſ- pended. Thus a weight of five ounces hanging from the part of an axle whoſe diameter is one inch, and another of ten ounces from a part whoſe diameter is half an inch, will be balanced by a weight of two ounces hanging from the circumfe- rence of the wheel whoſe diameter is five inches ; for the ſum of the products of five into one, and of ten into one half, which expreſs the moments of the weights, is equal to ten, as is alſo the pro- duct of two into five, which expreſſes the moment of the power.

If

If to the axle in the wheel be added one or more
wheels with teeth, fo that motion may be communicated from the firft wheel to the laft; the weight
being hung from the axle of the laft wheel, whilft
the moving power is applied to the circumference
of the firft wheel; in order to a balance, the power muft be to the weight in a *ratio* compounded of
the inverfe *ratio* of the diameter of the firft wheel
to the diameter of the laft axle, and of the inverfe
ratio of the number of revolutions made by the firft
wheel, to the number of revolutions made by the
laft axle in a given time; for if the firft wheel and
the laft axle revolved in the fame time, the *ratio* of
the diameter of the wheel to that of the axle, would
exprefs the *ratio* of the velocity of the power, to
the velocity of the weight; but if the wheel revolves oftener than the laft axle in a given time, it
is evident, that the *ratio* of the velocity of the
power to that of the weight, will be greater in that
proportion; confequently, the velocity of the power
muft be to the velocity of the weight in a *ratio*
compounded of the *ratio* of the diameter of the
firft wheel to the diameter of the laft axle, and of
the revolutions of the firft to thofe of the laft axle in
a given time; and therefore, that there may be a
balance between the power and the weight, the former muft be to the latter inverfly in the fame compounded *ratio*. For inftance, in a machine confifting of two wheels with their axles, wherein the
diameter of the firft wheel is four inches, and that
of the fecond axle a quarter of an inch, and wherein the cogs or teeth of the firft axle, by applying
themfelves fucceffively to the teeth of the fecond
wheel, turn it about, and therewith its axle; but
the teeth of the firft axle being in number but
one fourth of the teeth of the fecond wheel, that
axle, and confequently the firft wheel, muft revolve four times in order to turn the fecond wheel
and its axle once; fo that the revolutions of the
firft

Exp. 9.

firſt wheel in a given time are to the revolutions of the ſecond axle, as four to one : in this machine, in order to a balance, the power muſt be to the weight inverſly in a *ratio* compounded of ſixteen to one, and of four to one; that is, it muſt be to the weight inverſly as ſixty-four to one; ſo that a weight of one ounce at the circumference of the firſt wheel, will ſupport a weight of ſixty-four ounces faſtened to the ſecond axle.

Again, in a machine compoſed of three axles, the two laſt having wheels with teeth, and the firſt a perpetual ſcrew, which in each revolution of the firſt axle moves one tooth only of the wheel of the ſecond axle; which wheel having twenty-eight teeth, moves round once in the ſame time that the firſt axle turns twenty-eight times; and there being a ſmall wheel with fourteen teeth at the other end of the ſecond axle, and theſe teeth applying them- ſelves continually to the teeth of a wheel fixed on the third axle, which are twenty-eight in number; the wheel of the third axle muſt revolve but once in the ſame time that the wheel of the ſecond axle revolves twice, and of conſequence the third wheel and its axle move round but once whilſt the firſt axle performs fifty-ſix revolutions; and the diame- ter of the firſt axle is to that of the laſt as two to one; in order therefore to a balance between the power which is applied to the firſt axle, and the weight which is applied to the laſt; the power muſt be to the weight inverſly in a *ratio* compounded of two to one, and of fifty-ſix to one; that is, the power muſt be to the weight as one to an hundred and twelve; ſo that one ounce hanging from the firſt axle will ſupport an hundred and twelve ounces hanging from the laſt axle.

In order to exhibit the force of the w e d g e, which is the fifth mechanick power, let A D repre- Pl. 3. Fig. 14 ſent the baſe of a wedge, from whoſe middle point B let the line B E be drawn perpendicular to the

ſide

LECT.
VII.
side DC, and the line BC at right angles to AD, and consequently, bisecting the angle ACD made by the concourse of the wedge's sides.

In cleaving timber with a wedge, the force of the mallet which strikes the wedge is to be looked upon as the moving power, and the cohesion of the parts of the timber, as the resistance or weight to be moved; now, whilst the wedge is driven by the repeated strokes of the mallet from B to C, (for I suppose the edge of the wedge to be placed on the top of a piece of timber at B in order to rend it) the space described by the wood as it yields on each side of the wedge in lines perpendicular to those sides, is equal to BE. Consequently, that the moment of the mallet may be equal to the resistance of the wood, the absolute force of the mallet must be to the force wherewith the parts of the wood cohere, as BE to BC, that is, as the sine of the angle BCD to radius; whence it follows, that all similar wedges are of equal force, for in such the angle BCD is given; it likewise follows, that the powers of dissimilar wedges are inversly as the sines of the angles BCD, or in other words, that the forces requisite to rend timber with such wedges, are directly as the sines BE, which is confirmed by the following experiment.

Exp. 11.
Let a machine be so contrived, as to consist of two equal cylinders, rolling upon their axles in an horizontal position along the edges of two rulers, and let them be drawn and kept together by a weight of 2000 grains, hanging freely by a rope, fastened at each end to the cylinders, and let the edge of a wedge be placed between the cylinders, so as that when a sufficient weight is hung to it, it may be drawn down between the cylinders; in this machine the force wherewith the cylinders are drawn together, added to the attrition of their axles in rolling upon the rulers, may be looked upon as the resistance of the timber, and the weight of the
wedge

wedge together with the appending weight where-by it is pulled down between the cylinders, as the force of the mallet upon the wedge; now, if three wedges be made ufe of, each three inches long, in which the fines BE are as one, two, and three, their weights likewife being in the fame proportion, the firft will be drawn down by a weight of 300 grains, the fecond by one of 600 grains, and the third by one of 900 grains.

To the wedge may be reduced the axe or hatchet, the teeth of faws, the chizel, the augur, the fpade and fhovel, knives and fwords of all kinds, as alfo the bodkin and needle, and in a word, all forts of inftruments which beginning from edges or points grow gradually thicker as they lengthen; and the manner wherein the power is applied to fuch inftruments, is different according to their different fhapes and figures, and the various ufes for which they were contrived.

The next and laft mechanick power is the SCREW, which confifts of two parts, whereof the firft is called the male or outfide fcrew, being a cylinder cut in, in fuch fort as to have a prominent part going round it in a fpiral manner, which prominent part is commonly called the thread of the fcrew; the other part which is called the female or infide fcrew, and by common workmen the nut, is a folid body that contains an hollow cylinder, whofe concave furface is cut in the fame manner as the convex furface of the male fcrew, fo that the prominent parts of the one may fit the cavities of the other. The chief defign of this machine is to prefs the parts of bodies clofely together, and in fome cafes to break and divide them; when it is made ufe of one part is commonly fixed, whilft the other is turned round, and in each revolution the moveable part is carried in the direction of the axis of the cylinder thro' a fpace equal in length to the interval between two contiguous threads, where-

by

by the parts of the body whereon the preffure is made are forced to move towards-one another thro' a fpace equal to that interval; which interval therefore does exprefs the velocity wherewith the feveral parts of the body give way to the preffure, whilft the circular periphery, which is defcribed by the power whereby the moveable part of the fcrew is turned round, expreffes the velocity of the power; for the moveable part of the fcrew is ufually turned by means of an handle or handfpike, to fome part of which the power is applied, and by moving round with that part defcribes the circumference of a circle; if therefore the moving power be to the refiftance of the body which is preffed, as the diftance between two contiguous threads of the fcrew to the circular periphery defcribed by the power, there will be a balance; and if the power be ever fo little increafed beyond that proportion, it muft overcome the refiftance, and move the fcrew; and thus it would conftantly be, provided there was no refiftance from the attrition of the parts of the fcrew one againft another; but as that is very confiderable, there is an addition of power requifite to overcome it, over and above what is neceffary to overcome the refiftance of the body whereon the preffure is made: for which reafon fuch experiments as are made to fhew the force of the fcrew, muft vary more from the theory, than thofe which have been made concerning the other mechanick powers, wherein the attrition is far lefs confiderable; however, it will appear from the following experiment, that fmall powers are fufficient by the help of the fcrew to overcome great refiftances in the bodies which are preffed.

Exp. 12.   Let a wheel whofe diameter is four inches, be fixed at its center to the head of a male fcrew in an horizontal pofition, and let the end of a rope, which is wound about the groove of the wheel, pafs over a pulley in fuch a manner as that having

a weight

a weight faftened to it, it may be drawn in a line,
that is a tangent to the wheel, by which means the
intire gravity of the weight will be employed in
turning the wheel; to one end of a lever, fupported
by a prop at the middle, let a weight of feven pounds
be hung, and let the bottom of the male fcrew
reft on the other end of the lever; and let the diftance between the threads of the fcrew be equal to
one fifth of an inch, and a weight of three ounces
and 250 grains being hung to the end of the rope
which paffes over the pulley, will juft turn the wheel,
and thereby thruft down the fcrew, and with it the
end of the lever whereon it refts, and by fo doing
raife up the weight at the other end.

　In this cafe the power which moves the fcrew, is,
to the weight raifed whereby the refiftance that is,
made to the preffure is meafured, as one to 24
nearly; whereas it ought not to exceed the proportion of one to 63; for the diameter of the wheel being four inches, the circumference is twelve and an
half nearly, but 12.5 is to $\frac{1}{5}$, which is the interval
between the threads of the fcrew, as $62\frac{1}{2}$ to one;
confequently, if the power which turns the fcrew be,
to the weight that is to be raifed in the inverfe *ratio*
of thofe numbers, that is, as one to $62\frac{1}{2}$, it ought
to balance the weight, and if it be increafed ever
fo little, it fhould overpower and raife the weight:
fince therefore the force that is requifite to turn the
wheel is nearly three times as great as what is neceffary to overcome the refiftance of the weight to
be raifed, it is evident, that almoft two thirds of
that force is employed in overcoming the refiftance
arifing from the attrition of the parts of the fcrew
one againft another; what the nature of this refiftance is, and in what proportion it varies, fhall be
fhewn hereafter.

L E C T U R E

LECTURE VIII.

OF COMPOUND ENGINES.

THE mechanick powers, which for the moſt part are made uſe of ſeparately, may in many caſes be combined together, and engines thereby formed of ſuch efficacy, as that by the help thereof exceeding great weights may be raiſed by very ſmall powers. In all ſuch compounded machines the proportion which the moving power bears to the weight when they balance each other, is compounded of the ſeveral *ratios* which thoſe powers have to the weight which balance it in each ſimple machine, whereof the compounded engine conſiſts. Thus when a machine is compoſed of an axle in the wheel and a pulley, by faſtening the drawing rope of the one to the axle of the other ; the power which balances the weight in ſuch a machine, muſt be to the weight, in a *ratio* compounded of the *ratio* which that power has to the weight which balances it by means of the axle in the wheel alone, and of the *ratio* which that power has to the weight, which balances the weight by means of the

Exp. 1. pulley alone. For inſtance, if the nature of the pulley be ſuch, as that a power equal to one tenth part of the weight balances it ; and if the axle in the wheel be ſuch, as that a power equal to one fifth part of the weight can ſupport it ; the power which balances the weight in the compounded machine, will be to the weight in a *ratio* compounded of one to ten, and of one to five, that is, it will be to the weight as one to fifty ; for, ſince the weight is in effect faſtened to the axle of the wheel by means of the rope which goes round the pullies, it is evident that the axle will be drawn by a force equal to that,

which

which when applied to the drawing rope of the pulley is requifite to fuftain the weight by means of the pulley, which force is by fuppofition equal to one tenth part of the weight; but that force at the axle is balanced by a fifth part thereof applied to the wheel; confequently, the power requifite to balance the weight in this machine, is equal to one fifth of one tenth part of the weight, that is, the power is to the weight, as one to fifty. So that one ounce at the wheel will fupport fifty ounces at the pulley.

If a machine be compofed of the lever, the axle, and the perpetual fcrew; the lever being thirteen inches long, and fixed at its center to an axle, whereon is a perpetual fcrew, the tooth whereof a-dapts itfelf to the teeth of the wheel of an axle, the teeth of that wheel being twenty-four in number, and the diameter of the axle belonging to that wheel equal to fix tenths of an inch; in fuch a machine the power being applied to one end of the lever, and the weight to the axle of the toothed wheel, the former will balance the latter, if it be in proportion thereto, as one to 520; for if the lever to which the power is applied, moved round in the fame time with the axle of the toothed wheel whereunto the weight is faftened, the power would be to the weight, as the diameter of the axle to the length of the lever, that is, as fix tenths of an inch to thirteen inches, or in whole numbers, as fix to an hundred and thirty; but as there are 24 teeth in the wheel of that axle which fuftains the weight, and as the endlefs fcrew moves but one of thofe teeth in each revolution of the lever, the lever muft go round 24 times in order to turn the axle, which fuftains the weight, once; upon which account the power muft be to the weight, as one to 24, which *ratio* of one to 24 being combined with the former of fix to 130, gives a *ratio* of fix to 3120, or of one to 520; fo that an ounce weight being made to act with all its gravity at one end of the

the lever in order to turn it round, which may be done by fixing a wheel to the lever, will balance a weight of 520 ounces at the axle of the toothed wheel.

Exp. 3.   If to the laſt machine one moveable pulley be added, it will conſtitute a machine of double the force ; for the *ratio* of the power to the weight in the foregoing machine, being as one to 520, and in a ſingle moveable pulley, as one to two ; the *ratio* compounded of both, will be as one to 1040 ; ſo that in this machine an ounce will balance 86 pounds 8 ounces ; and if the ſtrength of a man's hand be ſuch, as that it can without the aſſiſtance of an engine ſupport an hundred pounds, it will by the help of this machine ſuſtain 104000 pounds.

In all that has been hitherto ſaid concerning the mechanick powers, the moving force and the weight or reſiſtance have been ſuppoſed to act in direct oppoſition to one another. I ſhall now conſider the effects of powers acting obliquely, and ſhew in what caſes they balance each other.

And firſt, if three powers acting in oblique directions, be to one another, as the reſpective ſides of a triangle formed by the concourſe of three lines drawn parallel to the directions of the powers ; thoſe powers will balance one another. For inſtance, if three powers drawing the point A in the directions A B, A C, and A E, be to one another, as the ſides of the triangle A D E, or A D C, made by the concourſe of the lines A D, A E, and E D ; or A D, C D, and A C, which lines are parallel to the directions of the powers ; they will balance one another, and the point A will remain unmoved.

Pl. 4.
Fig. 1.
Exp. 4.

For if the line A D be ſuppoſed to denote a power equal to that which acts in the direction A B, but contrary thereto ; the power denoted by A D will draw the point A as forcibly towards D, as it is drawn by the oppoſite power towards B ; conſequently, there will be a balance between the two

powers ;

powers ; but the power denoted by A D may be resolved into two powers denoted by A E or C D, and A C or E D ; which two powers acting together upon the point A in their proper directions A E and A C, will draw it as strongly towards D, as it is drawn by the single power denoted by A D ; as is evident from what has been said concerning the resolution and composition of motions and forces ; consequently, two powers which are as A E or C D, and E D or A C, acting in the directions A E and A C, will balance the third power which is as A D acting in the direction A B ; that is, two powers, which are as the two sides of a triangle, acting in directions parallel to those sides will balance a third power, which is as the third side, and which acts in a direction parallel thereto ; and what has been thus proved in particular of two of the powers with regard to the third, is in like manner demonstrable of any two of the powers with respect to the other ; consequently, any three powers which are to one another respectively as the sides of a triangle, and which act in directions parallel to those sides, will destroy each the other's effect, and remain in *æquilibrio*. To confirm this by an experiment ; let the sides of a triangle A B C drawn on an horizontal plane be as two, three, and four ; and let C E be parallel to the side A B, and the side A C continued towards D. Let three small cords be joined together at C, and stretched over three pullies in such a manner, as that one of them may cover the line C D, another the line C E, and the third the line C B ; this being done, if a weight of four ounces be hung to the cord which passes over C D, and one of three ounces to that which covers C B, and one of two ounces to that which covers C E, there will be a balance ; the weights, which in this case are the moving forces, being to one another as the sides of the triangle to which the direction of the weights are parallel.

Pl. 4.
Fig. 2.
Exp. 5.

If

If the weight A hangs freely from one end of a balance, so as to have its line of direction D A perpendicular to the arm of the balance; and if another weight as B, be hung at the other end E, in such a manner, as that its line of direction E C by passing over a pulley at C may be oblique to the arm of the balance, the weight B must be to the weight A when it counterbalances it, as E C to C F, that is, as radius to the sine of the angle C E F made by the oblique direction of B with the arm of the balance; for if the whole force of gravity in the weight B acting in the direction E C, be denoted by the line E C, it may be resolved into two forces denoted by E F and F C, acting in the directions of those lines, of which two forces, the latter only which acts in the direction F C perpendicular to the arm of the balance withstands the force of gravity in the weight A, the other force which acts in the direction E F being entirely employed in pressing the balance against the axis of its motion; since therefore, that part of the weight B which acts in opposition to the weight A, is to the whole weight B, as F C to E C; it is manifest, that in order to make the weight B balance the weight A, it must exceed the weight A in the same proportion that the line E C exceeds the line F C; and thus it is found to be from experiments; for if the pulley be so ordered as that E C may be to F G as three to two, then a weight of three ounces appended at E, will balance one of two ounces appended at D.

As a corollary it follows, that the perpendicular distances of the lines of direction from the center of motion, are to one another inversly as the weights; for, if from G the center of motion be let fall G H perpendicular to E C, that line will be the perpendicular distance of the direction E C from G; and E G, equal to D G, is the perpendicular distance of the direction D A; but the triangles
E F C

EFC and EHG are fimilar, becaufe their angles
at E are equal, and they have each a right angle;
confequently, as EC is to CF, fo is EG to HG;
but the weight B is to the weight A, as EC to FC,
that is, as EG or DG to HG; fo that wherever two
powers, which act in oblique directions, are to one
another in the inverfe *ratio* of the perpendicular or
fhorteft diftances of their lines of direction from the
center of motion, they muft balance one another;
whence it follows, that if two weights as A and B, Pl. 4.
be fufpended from two points as D and E in the Fig. 5.
plane of a wheel placed in a vertical pofition; and
if the line DE which is drawn thro' the two points
of fufpenfion, paffes thro' C the center of motion,
the weights will balance, provided they be to one
another inverfly as the diftances of their points of
fufpenfion from the center of motion, that is, if A
be to B, as CE to CD; for fince the weights hang
freely, their lines of direction DA and FB, will
be perpendicular to the horizon, and of confequence,
parallel to each other; wherefore, if the line HCF
be drawn thro' the center of motion perpendicular
to the two lines DA and FB, the triangles DHC
and ECF will be fimilar, confequently, DC will
be to EC, as HC to FC; but by fuppofition, the
weight A is to the weight B, as CE to CD; that
is, as CF to CH; fo that the weights are to one
another inverfly as the perpendicular diftances of
their lines of direction from the center of motion;
confequently, they muft balance; and tho' the
wheel fhould be turned upon its axis, and the diftances of the lines DA and EB from C be thereby
altered, yet will the fimilarity of the forementioned
triangles continue, and of confequence the balance
between the weights will be preferved; as will appear from the following experiment. Let a weight Exp. 7.
of one ounce be fufpended from the point D, and
another of two ounces from the point E; DC be-

I   ing

ing to EC, as two to one, that is, inverſly as the
weights, there will be a balance, and the wheel
will continue at reſt.   And if by the force of the
hand it be turned about its axis either to the right
from I towards K, or to the left towards M, the
balance will ſtill continue, and the wheel will remain unmoved when the hand quits it, whatever
be its poſition.

Pl. 4.
Fig. 6.
If the points of ſuſpenſion D and E be ſo po-
ſited, as that the right line DE which joins them,
does not paſs thro' C the center of motion ; let that
line be divided any where as in G by another line
as IL paſſing thro' the center C, and there will be
a balance, if the appending weights be to one another inverſly as the parts of the line DE, that is, if
A be to B as EG to DG, provided the poſition of
the wheel be ſuch, as that the line IL may be perpendicular to the horizon ; for ſince the lines EF,
GC, and DH are parallel, FC is to HC, as EG to
DG ; but by ſuppoſition, as EG is to DG, ſo is
A to B ; wherefore A is to B, as FC to HC, that
is, the weights are inverſly as the perpendicular diſtances of their lines of direction from the center of
motion, conſequently, their moments are equal ; but
if by turning the wheel about its axis the line IL
be put out of its perpendicular poſition, the balance will be deſtroyed ; becauſe, in that caſe, one
of the lines of direction will approach nearer to
the center of motion, whilſt the other recedes ; and
of courſe their perpendicular diſtances will not
continue in the inverſe *ratio* of the weights ; for if
the wheel be moved upon its axis from I towards K,
ſo as to have the line SCR perpendicular to the plane
of the horizon ; the line of direction DA will approach towards the center ſo as to become DP, and
its perpendicular diſtance from the center of motion will be NC, whilſt the other line of direction
recedes as far as EQ, and its perpendicular diſtance
from

from C becomes equal to OC; for which reaſon
the weight B muſt preponderate, and move the
wheel about its axis in the direction IKL. And
as the wheel continues to move in that direction,
the direction of the weight A will approach nearer
and nearer to the center of motion, and at length
paſs beyond it, ſo as to be on the ſame ſide with
the direction of the weight B; ſo that the wheel
will then be moved by the joint force of both
weights, and continue ſo to be, till ſuch time as
the direction of the weight B getting on the other
ſide of C, B begins to act in oppoſition to A, and
at length the point I being brought into the place
of L, the weights do again balance each other,
the line BE being divided by the perpendicular
line IL in the reciprocal *ratio* of the weights. To
confirm what has been ſaid by an experiment, let
the line DE in the plane of a wheel be divided in
G by the line IL in ſuch a manner, as that DG
may be double of EG; then ſetting the line IL
perpendicular, let a weight of one ounce be hung
from D, and another of two ounces from E, and
the wheel will remain unmoved; let then the wheel
be turned a little upon its axis, either to the right
hand or to the left; in the former caſe, the two
ounce weight will prevail, and carry the wheel
downward to the right hand, but in the latter the
ſmaller weight will preponderate, and make the
wheel to revolve towards the left.

Exp. 8.

If the line DE be divided in another point as T,
by the line SR, ſo as that DT may be one third of
ET; and if a weight of three ounces be ſuſpended
at D, and another of one ounce at E, the ſame
things will happen as in the former experiment;
for the line SR being placed vertical there will be
a balance; and upon moving it out of that poſition the balance will be deſtroyed.

Exp. 9.

If the crooked lever FCD be ſo placed on its
prop at C, as that the arm CF may be parallel to

Pl. 4.
Fig. 7.

the plane of the horizon, and the arm CD inclined
thereto; if two weights as B and A, appended at
D and F, be in the reciprocal proportion of the
perpendicular diſtances of their lines of direction
from the prop; that is, if B be to A as FC to EC
there will be a balance; for as long as the arm CF
continues parallel to the horizon, the weight B hanging from the point D acts in the ſame manner in
oppoſition to the weight A, as if it hung from E
the extremity of the ſtrait lever FC continued on
to E, in which caſe the weight B that balances the
weight A muſt bear the ſame proportion to it that
FC does to EC; if therefore the arm DC be bent
in ſuch a manner, as that EC may be one half or
one third of FC, in the former caſe a weight of two
ounces, and in the latter one of three ounces hanging from D, will be counterpoiſed by one ounce
hanging from F.

If by moving the lever, the arm FC be put out
of its parallel poſition, the balance will be deſtroyed; for that cannot be preſerved, unleſs the diſtance of B's direction from the prop continues to
bear the ſame proportion to the diſtance of A's direction, that EC does to FC; which in this caſe is
impoſſible; for firſt, if the point F be moved upward towards H, and of courſe the point D downward towards G, it is manifeſt, that the diſtances
of both directions will be leſſened; but the decreaſe
of EC in a given time will bear a greater proportion to the decreaſe of FC than EC does to FC;
for by that time the point D has moved from D to
G thro' the arch DG, which meaſures the angle of
CD's inclination, EC will vaniſh; whereas FC
cannot vaniſh till ſuch time as the point F has moved
from F to M thro' the quadrantal arch FM; but in
the ſame time that the point D moves from D to
G thro' the arch DG, the point F can move only
from F to H thro' the arch FH ſimilar to DG;
which arch being always leſs than the quadrant, the
                                    perpen-

perpendicular diftance of A's direction from the prop,
to wit F C, will not vanifh upon the arrival of the
point F at H, that is, it will not vanifh fo foon as
E C; confequently, the decreafe of E C in a given
time muft bear a greater proportion to the decreafe
of F C, than E C does to F C: wherefore E C as
diminifhed in any given time, will be to F C as
diminifhed in the fame time, in a lefs proportion
than that of E C to F C; or in other words, the
perpendicular diftance of B's direction from the
prop will bear a lefs proportion to the perpendicular diftance of A's direction, than E C does to F C;
and therefore, the weight A will preponderate. If
the point F be moved downward, and confequently
D upward, it is manifeft from the infpection of the
figure, that the diftance of A's direction from the
prop continually diminifhes, at the fame time that
the diftance of B's direction increafes; and therefore the weight B muft in that cafe overbalance
the weight A.

If F C D be a crooked lever placed as the laft,
and if a weight, inftead of being hung from the
arm D C, be laid thereon at D, and by a vertical
plane, as H K, fet clofe to it, be hindered from
falling off; from the point D whereon the weight
refts, let the line D E be drawn perpendicular to the
arm F C continued on towards G; the weight at D
will be balanced by the weight A hanging freely
from F, provided the weight D be to the weight
A, in a *ratio* compounded of E C to C D, and of
F C to C D; that is, as a rectangle under E C and
F C the perpendicular diftances of the directions of
the two weights from the prop, to the fquare of
C D the inclined arm of the lever. For whatever
be the moment wherewith the weight A preffes
down the arm F C, the arm D C muft with an
equal moment be preffed upward, and with it the
weight D in the direction D G perpendicular to

Pl. 4.
Fig. 8.

C D;

CD; and forafmuch as the fame weight preffes perpendicularly againft H K the vertical plane, it muft
be preffed backward by the fame in an horizontal
direction; and at the fame time it muft have a tendency downward from the force of gravity in the
direction ED; fo that it is acted upon by three forces
in the directions DG, GE and ED; in order
therefore to a balance, the forces muft be as the
fides of the triangle DGE; and the force of gravity
which preffes it in the direction ED, muft be to the
force preffing it in the direction DG, as ED to
DG, or, becaufe the triangles DGE and CDE
are fimilar, as EC to CD; but as the force which
preffes it in the direction DG is of equal moment
with the weight A, that force muft be to the weight
A, as FC to CD; confequently, the force of gravity in the weight D muft be to the force of gravity in the weight A, that is, the weight D muft be
to the weight A, in a *ratio* compounded of EC to
CD, and of CF to CD, or as the rectangle under
CE and CF to the fquare of CD. To confirm
Exp. 10. this by experiment, let a crooked lever as FCD
confift of equal arms, and let it be bent in fuch a
manner, as that EC may be to CD, as one to two;
and let a weight of one ounce be laid on at D, and
another of two ounces be hung from F, and they
will balance each other; for in this cafe the product of EC which is as one, into CF which is as
two, will be two; and CD being as two, the fquare
thereof will be four; fo that the rectangle under
EC and CF, is to the fquare of CD, as two to
four, or as one to two; in which proportion therefore the balancing weights muft be.

Exp. 11.   All things remaining as in the laft experiment,
excepting that the arm CF is as long again as CD,
fo that EC, CD, and CF are as one, two, and
four; a weight of one ounce at D will be balanced
by one ounce hanging freely from F; for CD being

ing as two, its fquare is four; and the produrt of EC, which is as one into FC, which is as four, is likewife four.

In wheels turned by the force of water falling upon them from an height, and which on that account are commonly called overfhot wheels, the moving power is partly the percuffive force of the water which falls into the uppermoft bucket, and partly the gravity of the water contained in the other buckets, which are lodged on the rim of the uppermoft quarter of the defcending part of the wheel; and the effects which thefe forces have upon the wheel are greater or lefs in proportion to their abfolute quantities, and the diftances of their lines of direction from the center of the wheel. Thus, where AIOP reprefents an overfhot wheel, C its center, K, L, M, N four buckets fixed on the uppermoft quarter of the defcending part of the wheel; AB the direction of the water flowing into the uppermoft bucket K, CB the perpendicular diftance of that line from the center C; DE, FG, and HI, the lines of direction of the centers of gravity of the feveral portions of water contained in the buckets L, M, N; CE, CG, and CI, the perpendicular diftances of thofe lines from the center C. The force of the water flowing into K is proportional to the quantity flowing in in a given time, as alfo to the velocity wherewith it flows, and the diftance of its line of direction from the center; and therefore, where the quantity and velocity are given, the force will be as BC the perpendicular diftance of AB the line of direction, from C the center of motion; confequently, the nearer AB approaches to the tangent in the point A, or the more obliquely the water flows in upon the wheel, the greater will its force be. The portions of water contained in the buckets L, M, N, have different forces according to their different quantities, and the different diftances of their lines of direction from the center

Pl. 4. Fig. 9.

ter

ter C, their quantities being greateft, when the diftances of their directions are leaft, for the buckets
empty as they defcend; fo that their force leffens as
they defcend, by reafon of the diminution of their
quantities, but at the fame time it likewife increafes
on account of the increafe of the diftance of their
lines of direction from the center of motion; fo
that upon the whole, the force in each bucket may
be looked upon as invariable; but whether this be
fo or not, certain it is, that if the wheel be truly
centered, and the buckets be equal and alike, and
if the water flows in uniformly, the whole moving
force muft continue the fame as long as the wheel
continues to move; and fince it acts inceffantly,
the motion of the wheel muft be continually accelerated, and that uniformly; and thus it would
be, were it not that when the wheel arrives at a certain degree of velocity, the refiftance which is given becomes fo great as to deftroy the increments
of motion as faft as they are generated by the
moving force; by which means the wheel is made
to revolve with one uniform velocity, which is the
greateft that can be given it by that moving
power.

Pl. 4.
Fig. 10.
A plane as A B placed obliquely to B C, which
reprefents an horizontal plane, is called an inclined
plane; the angle A B C is called the angle of elevation, and its complement B A C the angle of inclination; the line A C perpendicular to B C is called
the height of the plane, and A B its length. If a
weight as P be laid on an inclined plane as A B, and
be thereon fuftained by a power acting in a direction,
as P F, parallel to the inclined plane; in order to a
balance, the fuftaining power muft be to the weight
as the height of the plane to the length thereof,
that is, as A C to A B, or, putting B A for the radius,
as the fine of the angle of elevation to radius; for
the weight P is acted upon by three powers in different directions, the firft of which is the force of
gravity,

gravity, which preffes it downward in the direction PD perpendicular to BC; the fecond is the power which draws it in the direction PF parallel to BA; and the third is the plane BA, which does as it were prefs it upward in the direction PH perpendicular to BA; for as the weight P preffes the plane in a direction perpendicular thereto, it is reacted upon by the plane in a contrary direction. If therefore the line EG be drawn parallel to PD, the fides of the triangle PEG will be proportional to the three powers, and the force which fupports the weight on the inclined plane, and which acts in the direction PF, will be to the abfolute weight of the body acting in the direction PD parallel to GE, as PG to GE; but inafmuch as the triangles PEG and CBA are fimilar, as PG is to GE, fo is AC to AB; confequently, the power neceffary to fupport a weight on an inclined plane muft bear the fame proportion to the weight fuftained, that the height of the plane does to its length; which is confirmed by experiments; for if a weight of four ounces be laid on a plane whofe length is to its perpendicular height, as two to one, it will be counterbalanced by a weight of two ounces, provided the whole gravity thereof be made to act in drawing the other weight in a direction parallel to the inclined plane, which may be done by faftening one end of a cord to the greater weight, and then ftretching the cord along the plane, fo as to keep it parallel thereto, and paffing it over a pulley at the top of the plane; for the fmaller weight being tied to the end of the cord which lies beyond the pulley will hang freely, and for that reafon act with all its gravity in a direction parallel to the plane.

Exp. 13.

The fame weight of four ounces being laid on an inclined plane, whofe length is to its height as four to one, will be fuftained by a weight of one ounce hanging freely as before.

Exp. 14.

The

The force wherewith a body refting on an inclined plane preffes the fame, is to the weight of the
body, as the fine of the angle of inclination to radius ; for in the triangle P E G, P E denotes the
force wherewith the body preffes the plane, and G E
the weight of the body ; but from the fimilarity of
the triangles, as P E is to G E, fo is B C to B A ; and
putting B A for the radius, B C is the fine of B A C the
angle of inclination ; wherefore as B C the fine of
the inclination is to the radius A B, fo is the force
wherewith the body preffes the plane to the abfolute

Pl. 4.
Fig. 11.

weight of the body. Hence, if upon an inclined lever
as A B, refting on the two props A and B, a weight be
laid any where as at P, it will be eafy to determine
what proportion of the weight each prop bears ;
for drawing the horizontal line A E equal in length
to A B, and from the point P whereon the weight
refts letting fall P D perpendicular to A E, if A E
be fuppofed to denote the whole weight of the body,
A D will denote that part of it which is fuftained
by the uppermoft prop, and D E that part which
is fupported by the lower ; for if the lever was horizontal, fo as that the body might prefs it with all its
gravity, the whole weight of the body would be to
that part of it which preffes the prop B, as B A to
P A, as is evident from what has been faid concerning the fecond kind of lever ; but as in the inclined
pofition of the lever the whole weight of the body
does not prefs upon it, that part of the weight
which the prop B fuftains in the horizontal pofition,
muft be to the part fuftained in the inclined pofition,
in the fame proportion with the abfolute weight of
the body to the force wherewith it preffes the inclined plane, that is, as P A to A D ; for putting
P A for the radius, A D is the fine of the inclination
of the lever ; confequently, the whole weight of
the body muft be to that part which preffes on the
prop B in the inclined pofition of the lever, in a *ratio*
compounded

compounded of BA to PA, and of PA to AD, that is, it muſt be as BA to AD, or becauſe AB and AE are equal, as AE to AD; and of conſequence the part ſupported by the other prop A muſt be as DE.

Hence it follows, that if two perſons carry a load fixed upon a lever, the load being placed between them, which is the caſe of chairmen, upon deſcents the foremoſt man will bear the greateſt burthen, and upon aſcents the hindermoſt. It likewiſe follows, that in coaches and all other fourwheel carriages, which have the foremoſt wheels ſmaller than thoſe behind, the load muſt be thrown more upon the former than the latter; what effect this has upon the draft, ſhall be ſhewn in my next lecture.

LECTURE IX.

Of FRICTION.

IN my laſt lecture I ſhewed you what force is requiſite to ſuſtain a body on an inclined plane. If a body be laid on a plane parallel to the horizon, it does not ſtand in need of any force to ſupport it; for as the direction of gravity is perpendicular to the plane of the horizon, the whole weight of the body muſt be ſuſtained by the horizontal plane whereon it reſts: whence it follows, that if any power endeavours to move a body reſting on an horizontal plane in a direction parallel to the plane, it will meet with no reſiſtance from the weight of the body, that being intirely taken off by the reaction of the plane whereon the body preſſes; but a reſiſtance will ariſe from the attrition of the body againſt the plane; for the ſurfaces of all bodies whatever, even ſuch as are of the fineſt poliſh, being in ſome meaſure rough and unequal, (as is evident from the

obſer-

obfervations that have been made by the help of
microfcopes (when a body is moved upon a plane,
the prominent parts both of the body and plane
muft of neceffity fall into each others cavities, and
thereby create a refiftance to the motion of the body, inafmuch as the body cannot be moved unlefs
the prominent parts thereof be continually raifed
above the prominent parts of the furface whereon
it flides ; and this cannot be done unlefs the whole
body be at the fame time lifted up, and as it were
raifed on an inclined plane equal in height to the
forementioned protuberant parts ; upon which account the moving power muft fuftain fome part of
the weight of the body, even in moving it along an
horizontal plane. But as this is occafioned by the
inequalities in the furface, if thofe were intirely taken off, fo as to leave the furface perfectly fmooth
and even, the refiftance arifing from friction would
likewife be removed ; and fetting afide the refiftance of the medium, the fmalleft force would be
fufficient to move the moft ponderous body along
an horizontal plane. But fince there are not in nature any bodies, whofe furfaces are perfectly equal,
there will ever be fome refiftance arifing from friction ; which refiftance will remain unvaried, whatever be the magnitude of the furfaces that rub one
againft the other, provided the weight which preffes
thofe furfaces together, as alfo the roughnefs of
the furfaces, continue the fame ; for the fame weight
will ever require the fame force to raife it over prominencies of a given height, whatever be the magnitude of the furface whereon the weight refts ;
confequently, the quantity of refiftance will not be
varied by varying the magnitude of the furface ;
which may be confirmed by the following experiment. Let four pieces of polifhed box be laid on
a polifhed horizontal plane, and let each piece be
fo loaded as that its own weight, together with that

Exp. 1.

of

of its load, may be 6685 grains, and let the bafis of one be two inches long, and half an inch broad, and thofe of the other three be each four inches in length, but let their breadths be half an inch, an inch, and an inch and an half, fo that the magnitudes of the bafes may be as one, two, four, and fix; let then a fmall cord be faftened to the end of each piece, and by paffing over a pulley, be kept in a pofition parallel to the plane, and a weight of 2030 grains hanging from the end of the cord which lies beyond the pulley, will juft fuffice to move each piece along the plane; fo that the refiftance arifing from friction is the fame in each piece, notwithftanding the different magnitudes of the furfaces whereon they reft.

If the roughnefs of the furfaces whereon the bodies move be given, the refiftance arifing from friction will vary with the weights of the bodies, and be proportional thereto; for if a certain force be fufficient to raife a certain weight over prominences of a given height, it is manifeft that a double or triple weight will require a double or triple force to raife it to the fame height. If therefore the pieces of box be fo loaded, as that each of them with its load may weigh 13370 grains, that is, as much again as in the laft experiment, a weight of 4060 grains, that is, twice as much as before, will be neceffary to move them along the fame plane. Exp. 2.

If the roughnefs of the furface whereon a body moves be increafed, the refiftance will likewife increafe tho' the weight of the body remains the fame; but as the degree of roughnefs in any furface cannot otherwife be determined than by experiment, fo neither can the refiftance arifing therefrom: if the plane made ufe of in the laft experiments be thinly covered with fine fand, the refiftance will thereby become greater in the proportion of about five to four; for the fame pieces of box which were fet a going by 2030 grains when the Exp. 3.

plane

plane was free from fand, will in this cafe require 2500 grains, that is, about one fourth more.

To avoid as much as poffible the refiftance arifing from friction, which in rough and uneven roads muft needs be very great, WHEEL CARRIAGES have been contrived ; the advantages whereof I fhall endeavour to explain to you, but I fhall firft fhew you from what caufe it is that wheels turn round during their progreffive motion along a plane. If a wheel as ACB playing freely on the axis at A, be lifted off the plane BD by a power applied to the axle, and be carried in any direction whatever, it will not revolve about the axle; for fince in all wheels that are truly made the axle paffes thro' the center of gravity, it is evident, that in this cafe the wheel is fufpended by its center of gravity, and of confequence will not of itfelf change its pofition, but each point thereof will defcribe a line parallel to the direction of the moving power without any rotation about the axle, in the very fame manner as if the wheel was fixed to the axle; but if one point of the wheel as B refts upon the plane BD, and if a power applied to the axle draws the wheel in any direction as AP, fo as to move it along the plane BD; the motion of the point B will be retarded by the refiftance arifing from friction, whilft the point C which meets with no refiftance is carried forward without any retardation of its motion, and confequently muft move forward fafter than the point B; but as all the parts of the wheel cohere, the point C cannot move forward fafter than the point B, unlefs the wheel revolves about its axis from C towards E; and as the feveral points of the wheel's circumference, which are fucceffively applied to the plane, fuffer a retardation in their motion whilft the oppofite points move freely, the wheel during its progreffive motion along the plane, muft continue to revolve about its axle.

Pl. 4.
Fig. 12.

By

By this rotation of wheels about their axles, the refiftance arifing from friction is very much diminifhed, and drafts thereby rendered more eafy; for in plain roads, where the height of the prominent part is inconfiderable with refpect to the diameter of the wheel; the parts of the revolving wheel which apply themfelves fucceffively to the road, may be looked upon in fome meafure as defcending upon the minute prominencies, and of courfe muft pafs over them without any confiderable friction. And fo much is the refiftance arifing from friction diminifhed in wheel carriages, that if upon the fame plane whereon the pieces of box were drawn, a carriage be laid with four equal wheels, each three quarters of an inch in diameter, and loaded in fuch a manner, as that the weight of the carriage and load may amount to 6685 grains, which was the weight of each piece of box with its load; it will be fet a going by a weight of 420 grains drawing it horizontally, whereas 2030 grains were requifite to move the pieces of box along the fame plane.

From this experiment it appears, that the friction is very much leffened by means of wheels; which diminution is not to be attributed to the wheel's touching the plane in a few points, as may poffibly be imagined, but to the rotation of the wheels; for if the wheels of a carriage loaded as before be made faft to the axle, fo as not to revolve in their motion, 2030 grains will be neceffary to fet the carriage a going, that is, juft as much as was requifite to move the pieces of box.

As wheel carriages in general meet with lefs refiftance in their motion than any other, fo thofe of larger wheels, *cæteris paribus*, are lefs refifted than thofe of fmaller; for the proof whereof, it will be neceffary to premife two LEMMAS; the firft of which is, *that the fecants of angles are to one another inverfly as the fines of their complements, that is,* AD, Pl. 4. Fig. 13. *which is the fecant of* BAD, *is to* AC, *which is the fecant*

fecant of BAC, as AF, *which is the fine of the com-plement* of BAC, to AH, *which is the fine of the com-plement* of BAD. For from the nature of fimilar triangles AD is to AC, as AE to AK, that is, as AE to AG; but AE is to AG, as AF to AH; confequently, AD is to AC, as AF to AH.

The fecond ʟᴇᴍᴍᴀ is, *that if two arches of un-equal circles have their verfed fines equal, the arch of the leffer circle is greater in proportion to the whole periphery, than the arch of the greater circle; or in other words, the angle meafured by the arch of the leffer circle, is greater than the angle meafured by the arch of the greater circle.* Let HF and DB be two

Pl. 4.
Fig. 14.
arches of unequal circles, whofe verfed fines FG and BC are equal; I fay, the angle HEF is greater than the angle BAD; for fince EF is lefs than AB, and GF and CB are equal, EF is to GF in a lefs proportion than AB to CB; confequently, EG is to EF in a lefs proportion than AC to AB, that is, the fine of the angle GHE is lefs than the fine of the angle CDA, and of courfe the angle GHE is lefs than the angle CDA; confequently, the angle HEG, which is the complement of the leffer angle GHE, is greater than the angle DAC, which is the complement of the greater angle CDA.

Pl. 4.
Fig. 15.
Thefe two ʟᴇᴍᴍᴀs being premifed, let HM reprefent a plane whereon move the two wheels ABH and KLR, which are of different magni-tudes, but equal in weight, and let BC and LM be two obftacles of equal heights, and of fuch a na-ture as that the wheels cannot otherwife pafs than by furmounting thofe obftacles; the force requifite to draw the larger wheel over the obftacle BC, is lefs than what is requifite to draw the leffer wheel over the obftacle LM equal in height to the former; for fince the wheels revolve in paffing over the points B and L, their centers of gravity A and K may be looked upon as revolving about the fixed points B and L, and defcribing the arches AF and KP;

consequently,

consequently, the forces which move the wheels may be looked upon as drawing them upon inclined planes, whose directions coincide with the directions of the curves in the points A and K, that is, they coincide with the tangents A E and K O; which tangents being parallel to the tangents of the wheels in the points B and L, that is to say, to D B and N L; the centers of gravity of the two wheels, and consequently, the wheels themselves may be looked upon as drawn up the inclined planes D B and N L; but since the wheels are supposed to be equal in weight, the forces which support them on the inclined planes D B and N L, the height whereof is given, must be to one another inversly as the lengths of the planes; that is, the force which supports the larger wheel on the plane D B, must be to the force supporting the smaller wheel on the plane N L, as N L to D B; that is, putting B C or L M for the radius, as the secant of the angle N L M to the secant of the angle D B C; or, because K S and L M, as also A I and B C are parallel, as the secant of the angle K S L to the secant of the angle A I B; but, from the nature of similar triangles, the angle K S L is equal to the angle K L Q, as is also the angle A I B to the angle A B G; and therefore the force which sustains the greater wheel on the inclined plane D B, is to the force sustaining the lesser wheel on the inclined plane N L, as the secant of the angle K L Q to the secant of the angle A B G; but, by the first *Lemma*, the secant of K L Q is to the secant of A B G, as the sine of B A G to the sine of L K Q; and, by the second *Lemma*, the sine of B A G is less than the sine of L K Q; consequently, the force which raises the greater wheel over the obstacle B C, is less than the force which raises the lesser wheel over the obstacle L M equal in height to the former; but the forces requisite to make the wheels surmount the obstacles are the measures of the resistances, and therefore,

K   *cæteris*

cæteris paribus, the greater wheel muſt meet with leſs reſiſtance from the ſame obſtacle than the ſmaller. To confirm this by an experiment, let an obſtacle one tenth of an inch in height be fixed on an horizontal plane, and cloſe behind it let there be placed the carriage with four equal wheels, each three quarters of an inch in diameter, and if it be loaded in ſuch a manner as that the weight of the carriage and load may amount to 6685 grains, it will not be raiſed above the obſtacle by leſs than 2850 grains drawing it in a direction parallel to the plane ; whereas if four wheels, each an inch and an half in diameter, be fitted to the ſame carriage, the weight of the whole being the ſame as before, it will be raiſed above the obſtacle by 2050 grains, that is, by 800 grains leſs than were requiſite to raiſe it with the ſmaller wheels.

From this experiment it appears, that the reſiſtance which larger wheels meet with in ſurmounting obſtacles, is leſs than the reſiſtance given to ſmaller wheels by the ſame obſtacles ; and from what has been demonſtrated it is evident, that the reſiſtance given to the greater wheel is to the reſiſtance given to the ſmaller, as the ſine of an angle meaſured by an arch of the greater wheel, to the ſine of an angle meaſured by an arch of the ſmaller wheel, the verſed ſine of each angle being equal to the height of the obſtacle ; ſo that putting R and r for the radii of the two wheels, and x for the verſed ſine or the height of the obſtacle, it follows from the nature of the circle, that as $\dfrac{\sqrt{2Rx - xx}}{R}$ is to

$\dfrac{\sqrt{2rx - xx}}{r}$, ſo is the reſiſtance given to the larger wheel to the reſiſtance given to the ſmaller, or dividing by $x^{\frac{1}{2}}$, as $\dfrac{\sqrt{2R - x}}{R}$ to $\dfrac{\sqrt{2r - x}}{r}$; but as the

proportion

proportion of thefe fines is not fixed, but varies with the height of the obftacle, fo likewife muft the proportion, which the refiftance given to the greater wheel bears to the refiftance given to the fmaller; and all that can be determined in this cafe is, that larger wheels ever meet with lefs refiftance in furmounting obftacles than fmaller; and that the difproportion between the refiftances fuffered by each wheel, increafes with the height of the obftacle. Indeed where the obftacle vanifhes, which is the cafe when wheels move upon planes, the expreffions for the refiftances, and confequently the refiftances themfelves, are as $\frac{1}{\sqrt{R}}$ and $\frac{1}{\sqrt{r}}$, that is, the refiftances are inverfly as the fquare roots of the femidiameters of the wheels; fo that where the heights of the wheels are as one and two, the forces requifite to draw them along the fame horizontal plane, are as fourteen and ten, that is, inverfly as the fquare roots of one and two, which is confirmed by experiments; for whereas the carriage whofe wheels are three quarters of an inch in diameter, required 420 grains to move it along the horizontal plane, the weight of the carriage and load being 6685 grains; the carriage whofe wheels are $1\frac{1}{4}$ inch in diameter, when loaded in the fame manner, will be fet a going by 300 grains; but 420 is to 300, as 14 to 10, that is, the forces requifite to move the two carriages along the fame plane, are inverfly as the fquare roots of the heights of the wheels.

If the nature of the obftacle be fuch, as to be bore down by the preffure of the wheel, the larger wheel will in this refpect likewife have the advantage over the fmaller, and deprefs the obftacle with greater force. For let L K be continued to T, fo that T L may be equal to A B; and fince the wheels are fuppofed to be equally weighty, let A B and T L ex-

Pl. 4.
Fig. 15.

K 2

TL exprefs the abfolute forces of the two wheels acting againft the obftacles in the directions A B and K L; it is evident from what has been faid concerning the refolution of forces, that the force denoted by A B may be refolved into two forces; one whereof may be denoted by A G, and the other by G B, whereof A G alone acts in depreffing the obftacle B C, inafmuch as it bears directly down upon it; whereas the other force denoted by G B, inafmuch as its direction is perpendicular to the obftacle, may thruft it forward, but can contribute nothing towards preffing it downward from B towards C. In like manner the force denoted by T L, is refolvable into two forces, which may be denoted by T V and V L, whereof T V alone acts in depreffing the obftacle L M; confequently, the force wherewith the greater wheel depreffes the obftacle, is to the force wherewith it is depreffed by the leffer, as A G to T V, or as the fine of the angle A B G to the fine of the angle T L V or K L Q; but by the fecond *Lemma*, the angle B A G, which is the complement of G B A, is lefs than the angle L K Q the complement of K L Q; confequently, the angle A B G is greater than T L V, and A G the fine of the former greater than T V the fine of the latter; but as A G is to T V, fo is the depreffing force of the greater wheel to the depreffing force of the leffer; confequently, the fame obftacle is more eafily depreffed by the larger wheel than the fmaller, and of courfe muft give lefs refiftance to the former than to the latter.

If the obftacle be fuch, as that it can neither be furmounted nor depreffed, but muft be driven forward, then indeed the fmaller wheel has the advantage of the larger; for the forces of the wheels being refolved as before, the lines G B and V L will exprefs the forces which act in driving the obftacle forward; but it has been demonftrated, that G B the fine of the angle G A B, is lefs than V L the fine of the

angle

angle VTL equal to QKL; and therefore the force wherewith the greater wheel propels the obstacle, is lefs than the force wherewith the fmaller wheel propels the fame; befides, as the greater wheel preffes the obstacle directly downward with a greater force than the fmaller, the refiftance made by the fame obstacle to the propelling force of the larger wheel, will be greater than what is made to the propelling force of the fmaller; fo that where the obstacle is to be propelled, the fmaller wheel is preferable to the larger; but as in drafts this is rarely if at all the cafe, the obstacles which are commonly met with in roads being fuch as muft either be furmounted or depreffed by the wheels, fuch wheels are to be preferred as beft ferve both thofe purpofes, and thofe I have fhewn to be the larger wheels; which likewife are attended with other advantages befides what have been already mentioned; for firft, it frequently happens in rough and uneven roads, that two obstacles are placed fo near each other, that before the wheel has quitted one it meets with the other, and refting upon each, hangs between them; in which cafe the fmaller the wheel is, the lower it defcends between the obstacles, and thereby renders the draft more difficult; inafmuch as it muft be raifed to a greater height in order to pafs over the foremoft obstacle, than when the wheel is larger: For the illuftration Pl. 5. of which, let FE and HG reprefent two obstacles Fig. 1. placed at fo fmall a diftance, that the wheel having furmounted the firft but not quitted it, may meet with the fecond, fo as to hang between them; it is manifeft, that as the arch FDH of the leffer wheel, which lies between the obstacles, has a greater curvature than FBH the arch of the greater wheel, which lies between the fame obstacles, the point D muft defcend lower than the point B; confequently, the fmaller wheel muft be raifed to a greater height than the larger, in order to pafs over the fame obstacle;

ſtacle; and therefore a greater force will be neceſ-ſary to pull up the ſmaller wheel, than what is re-quiſite to raiſe the larger; which is confirmed by

the following experiment. Let a carriage with four wheels, each an inch and an half in diameter, and ſo loaded as that the weight of the carriage and load may amount to 6685 grains, be ſo placed on the plane before made uſe of, as that the two foremoſt wheels may hang between two obſtacles whoſe diſ-tance is half an inch, and their height likewiſe half an inch, and a weight of 1150 grains drawing the carriage horizontally, will move the wheels from between the obſtacles; whereas if four ſmaller wheels be made uſe of each three quarters of an inch in diameter, a weight of 2700 grains will be requiſite to draw them from between the obſtacles.

As wheels cannot always run upon the nail, but muſt frequently meet with heavy roads, they will ſink down, and thereby render the draft more difficult; but the larger the wheels are, the leſs _cæ-teris paribus_ will the depth be to which they ſink.

Pl. 5. Fig. 2.

For if ABC denotes the plane of the road, and if it be of ſuch a nature as to ſuffer the ſmaller wheel to ſink down as far as E; it is manifeſt that the gra-vity of the wheel muſt overcome the reſiſtance of as much of the earth whereon it preſſes, as is equal to the ſegment HED; for it cannot otherwiſe ſink, than by forcing ſuch a quantity of the earth out of its place; and ſhould the larger wheel ſink to the ſame depth, the gravity thereof muſt overcome the reſiſtance of as much earth as is equal to the ſegment AEC, that is, it muſt overcome a greater reſiſtance in order to ſink to the ſame depth with the ſmaller; but it cannot poſſibly overcome a greater reſiſtance, becauſe it is ſuppoſed to have the ſame gravity with the ſmaller; conſequently, it will not ſink as deep as the ſmaller, and for that reaſon will make the draft leſs troubleſome.

As

As large wheels have the advantage of small ones with regard to the resistance arising from the obstacles and impediments in the roads, so have they likewise in relation to the resistance occasioned by the friction of the box against the arm of the axle ; not that this resistance is less in greater wheels than in smaller ; for since it is not varied by varying the magnitude of the surface, as has been shewn, if the boxes and arms are truly fitted and of an equal smoothness, and the weights whereby the arms and boxes are pressed together be equal, the quantity of resistance will be given, whatever be the magnitude of the wheels, as also of the arms of the axle whereon they play ; but where the arms of the axles are of equal diameters, (which is commonly the case in one and the same carriage, tho' the wheels be unequal) a less force is requisite to overcome the given resistance in a larger wheel than in a smaller ; for in this case the semidiameter of the wheel may be looked upon as a lever, whose prop or fixed point is at the center of the arm, and the impediment arising from the friction of the box against the arm may be looked upon as a weight placed upon the lever at the distance of the arm's semidiameter from the prop, whilst the moving power is applied to the extremity of the wheel's semidiameter ; and therefore in order to a balance, the power must be to the resistance, as the semidiameter of the arm to the semidiameter of the wheel ; since then the impediment is given, as also the distance thereof from the prop, it is evident, that the larger the lever is, and consequently, the larger the wheel, the less is the force requisite to overcome the resistance. Thus, if BEF represents the circumference of the arm of an axle, whereon the wheels AGH and DIK revolve, C the center of the arm, BC its semidiameter, DC the semidiameter of the smaller wheel, and AC that of the larger ; in the

Pl. 5.
Fig. 3.

K 4   bigger

bigger wheel the length of the lever is A C, and ir
the finaller D C; fince therefore the fame impediment is in both levers placed at the fame diftance
from the prop C, to wit at B, it will be balanced
by a lefs force at A than at D; and the force at A
is to the force at D, as D C to A C; that is, inverfly
as the femidiameters of the wheels; for the force at
A is as B C applied to A C, and the force at D is as
the fame B C applied to D C; that is, the force at
A which balances the refiftance at B, is to the force
at D which balances the fame refiftance, as B C divided by A C, to B C divided by D C, that is, multiplying croffwife, and throwing out B C, as D C to
A C. Whence it follows, that when the femidiameter of the arm is given, the more the wheel is
enlarged, the lefs will the force be that is requifite
to overcome the refiftance arifing from the friction
of the wheel againft the arm; fo that upon this account as well as the former, large wheels are to be
preferred to fmall ones.

In order to leffen the refiftance arifing from the
friction of the box againft the arm of the axle,
there has been a late contrivance, whereby the axle,
contrary to what is ufual in moft carriages, is made
to revolve, and its arms, inftead of preffing againft
the boxes, are made to bear on the circumferences
of moveable wheels, which wheels from their ufe
in diminifhing the friction, are by the author of this
contrivance called *friction wheels*. Now that fuch
wheels, where they can be made ufe of, do take off
much of the refiftance occafioned by friction, will

Exp. 8. appear from the following experiments; from the
axle of the machine called the axle in the wheel, in
which the diameter of the wheel is to the diameter
of the axle, as nine to one, let a weight of 23163
grains be hung, and a weight of 2770 grains hanging at the circumference of the wheel, will turn the
machine, provided the axle turns on the circumferences

rences of two moveable wheels; whereas, if it turns
in the pivets it will be neceffary to add 600 grains
more, fo as to make the whole 3370 grains; confequently the refiftance occafioned by friction in the
latter cafe, is more than four-fold what it is in the
former; for fince the diameter of the wheel is nine
times as great as that of the axle, a weight of 2574
grains at the wheel is requifite to balance the weight
of 23163 at the axle, which balancing weight
being deducted from 2770, and likewife from
3370 grains, leaves 196 grains for overcoming the
refiftance in one cafe, and 796 in the other; but
796 is to 196, as four and a little more to one.

Again, let a fmall cart with friction wheels be fo
loaded, as that its own weight added to that of the
load, may amount to 20000 grains: a weight
of 54 grains drawing horizontally, will move it
along a fmooth level table: whereas, if the friction
wheels be taken off, 322 grains will be neceffary to
fet it a going. If the cart be fo loaded, as that the
weight of the whole may amount to 40000 grains,
then in each cafe, a double force will be requifite to
move it, that is to fay, 108 grains with the friction wheels, and 644 without them; fo that in this
cart the friction wheels take off five parts in fix of
the refiftance; for 54 is but a little more than a fixth
part of 322, as is likewife 108 of 644. And from
thefe experiments it does again appear, that under
like circumftances the refiftance arifing from friction,
is proportional to the weight, whereby the furfaces
which rub one againft the other are preffed together.

Seeing then that great wheels have in fo many re-
fpects the advantage over fmall ones, it will not be
improper in this place to fhew you, on what account
it is, that the wheels of common carts, as alfo the
foremoft wheels of coaches, chariots, and moft
other four-wheel carriages, are commonly made fo
fmall as feldom to exceed two feet and an half in
diameter; and the firft reafon of this contrivance

2

is

is for the convenience of turning; for as in most
roads, but more especially such as are narrow, there
are windings of such a nature as to allow but a small
space for carriages to turn in, it is necessary to make
use of such wheels as can turn in the narrowest compass, and such are small ones; for it is a thing well
known to carters and all others who are used to
drive wheel carriages, that the larger the wheels are,
the greater compass do they require in order to turn
with ease and safety; and should they at any time
attempt to turn carriages with large wheels as short
as those which have smaller, the wheels will drag,
and thereby render the draft very difficult, and
sometimes endanger the oversetting of the carriage.

But the second, and indeed the principal reason
for the use of small wheels is, that upon ascents, and
in passing over obstacles in rough and hilly roads,
as little of the horse's force may be lost as possible;
if roads were level and smooth without risings or impediments, the most convenient size for wheels, setting aside the necessity of turning, would be where
the axle is upon a level with the breast of the horse;
for since the whole force of the horse in drawing is
applied to that part of the tackle which lies upon
the breast, and to which the traces are joined; and
since the traces are fastened to the carriage in such
a manner, as that being continued they must pass
thro' the axle of the foremost wheels, it is manifest,
that if that axle be of an equal height with the chest
of the horse, the traces, in whose plane the line of
direction lies, will be parallel to the road whereon
the carriage is drawn; consequently, the whole force
of the horse will be employed in drawing the carriage directly forward, without any loss or diminution; whereas if the wheels be of such a size as that
the height of the axle is either greater or less than
that of the horse's chest, the whole force of the horse
will not be employed in the direct draft; but in the
former

former cafe, fome part of the force will be fpent in prefling the carriage directly downward, and in the latter, in lifting the fame directly upward. For the proof and illuftration whereof, let the firft of Pl. 5. the three wheels be of fuch a fize, as that its axle A Fig. 4. may be of an equal height with the horfe's breaft at B ; and let the fecond wheel be fo large as that its axle A may ftand higher than the horfe's cheft at B, and in the third, let the axle be lower than the breaft of the horfe ; and in each wheel let the lines of direction of the horfe's draft, to wit, A B be taken equal, and let each of thofe lines exprefs the force of the horfe ; it is manifeft, that in the firft wheel, the whole force denoted by A B, is employed without any lofs in drawing the wheel forward, be- caufe the line of direction A B, wherein the force draws, is parallel to E F, the road whereon the wheel moves ; whereas, in the fecond and third wheels the lines A B, wherein the forces draw, be- ing inclined to E F, whereon the wheels move, fome part of each force muft be loft ; for if each force denoted by A B be refolved into two, to wit C B and A C, whereof C B is parallel to E F, and A C perpendicular thereto ; it is evident, that that force alone which is denoted by C B, acts in moving the wheel forward along E F, whilft the force de- noted by A C does in the fecond wheel prefs it di- rectly downward againft the road, and in the third lifts it directly upward ; whence it follows, that if the force of a horfe be juft fufficient to move the firft wheel, it will not fuffice to ftir the fecond or third. It likewife follows, that if the wheel be fo far inlarged, as that the angle which the line of di- rection A B makes with the plane E F, approaches nearly to a right one, the line C B will bear a very fmall proportion to A B, whilft A C becomes near- ly equal thereto ; fo that almoft the whole of the horfe's force will be fpent in prefling down, and thereby increafing the load ; whence it appears,

that

that notwithstanding the several advantages arising from the largeness of wheels, yet may they be so far increased, as even upon account of their magnitude to render the draft impossible. By the use of small wheels whose axles lie below the level of the horse's chest, provision has been made against the inconvenience last mentioned, and the loss of force (which by reason of the roughness and inequalities of roads cannot wholly be avoided) has been rendered as little as possible, and made to obtain chiefly in level smooth roads, where there is least occasion for the whole force; whereas upon ascents, and in passing over obstacles in rough roads, where the stress is greatest, there little of the force is *lost*; for

Pl. 5.
Fig. 5.

the proof of which, let the wheel be of such a size, that its axle A may be below the horse's breast at B, and let AB, as before, denote the force of the horse; if the wheel be drawn along a smooth level road as EF, CB will express that part of the force which draws the wheel along the road, and AC that part of the force which is employed in lifting up the wheel, which part is lost as to the draft, but however, is not intirely useless; because, by pulling the wheel directly upward, it eases the load, and thereby renders the draft less difficult; tho' at the same time the draft is by no means as easy as it would be, if the force of the horse was applied at G, so as to draw in the direction AG parallel to EF. If the wheel instead of moving along a smooth road, be to pass over the obstacle DH, or which is the same thing, if it be to be drawn up the ascent EHL; and if the force of the horse be applied at G, so as that the direction of the draft AG may be parallel to EF, and consequently, inclined to EHL; it is manifest upon resolving the force AG into two forces, to wit AK and KG, whereof AK is parallel, and KG perpendicular to EHL; that force alone which is expressed by AK, acts in drawing the wheel up EHL; whereas the force expressed by

KG

K G acts in preffing the wheel directly againft EHL, and thereby adds to the weight of the wheel; fo that in this cafe, fome part of the horfe's force is loft, and the load at the fame time increafed, both which inconveniencies are avoided where the breaft of the horfe is fo far elevated above the axle of the wheel, as that the line of direction A B may be parallel to E H L ; for then no part of the horfe's force will be loft, but the whole will be employed in drawing the wheel directly over the obftacle, or up the afcent ; fo that a lefs force will be requifite to draw the wheel over the obftacle D H in the direction A B, than in the direction A G ; and this is fully confirmed by experiments. For whereas the Exp. 9, little carriage with four wheels, each three quarters of an inch in diameter, being fo loaded as that the weight of the carriage and load amounted to 6685 grains, was not drawn over the little obftacle one tenth of an inch in height, by lefs than 2850 grains acting in an horizontal direction, it will be drawn over by 2450 grains, provided the direction be made parallel to the tangents of the wheels in thofe points which touch the obftacle ; and 1950 grains will be fufficient to draw the carriage with the larger wheels over the fame obftacle, if the direction of the draft be made parallel to the forementioned tangents, whereas 2050 grains were neceffary when the direction was parallel to the horizontal plane. And if the direction be ftill farther removed from the parallelifm of the tangents, which may be done by depreffing it below the horizontal plane, the force of 2350 grains will be but juft fufficient to furmount the obftacle, and draw the carriage over.

'Tho' in four wheel carriages, the contrivance of fmall wheels before has its advantages, yet is it not intirely free from inconveniencies ; for by this means the load muft of neceffity be thrown forward, and

a greater

a greater ſtreſs laid on the foremoſt wheels; whereby the reſiſtance that ariſes from the friction of the axle againſt the wheels will become greater in the foremoſt than in the hindmoſt wheels, in proportion to the greater weight which they ſuſtain. Beſides, as the ſpaces deſcribed by wheels in each revolution are nearly equal to the peripheries of the wheels, it is manifeſt that the foremoſt wheels muſt revolve oftener than the hindmoſt, in order to rid the ſame ground. And this frequency of turning requiſite in the foremoſt wheels joined to the greater ſtreſs upon them from the load, as alſo to the greater reſiſtance which they meet with from obſtacles in the road, is the true reaſon why they are more frequently out of order, and ſtand in need of repair much oftener than thoſe behind.

LECTURE X.

MOTION OF BODIES DOWN INCLINED PLANES.

Lᴇᴄᴛ. X. **M**Y deſign in this lecture is to explain the chief properties of the PENDULUM; and in order thereto, I ſhall lay down the following PROPOSITIONS concerning the motion of bodies down inclined planes and curve ſurfaces.

Pl. 5.
Fig. 6.
PROP. I. *The force wherewith a body deſcends upon an inclined plane, as* AC, *is to the abſolute force of gravity wherewith the ſame body falls freely and perpendicularly, as the height of the plane to the length thereof, that is, as* AB *to* AC.

For it has been proved, that the force requiſite to ſuſtain a body upon an inclined plane, is to the abſolute weight of the body, as the height of the

plane

plane to its length ; but the force wherewith a body L ᴇ ᴄ ᴛ.
endeavours to defcend upon an inclined plane, muft   X.
be equal to the force which is neceffary to fupport
it upon that plane ; confequently, the propofition is
true.

Corol. I. Hence it follows, that the motion
of a body defcending on an inclined plane is uniformly accelerated ; for fince the force which carries
a body down an inclined plane, has every where,
and in all parts of the plane, the fame proportion
to the abfolute weight of the body, and fince the
abfolute weight remains unvaried, the other force
muft do fo too ; confequently, as it acts inceffantly
in equal times, it makes equal impreffions on the
defcending body, fo as to generate equal degrees of
velocity in the motion thereof ; that is, in other
words, the motion of a body defcending on an inclined plane is uniformly accelerated.

Corol. II. On account of this uniform acceleration of the motion, the times of defcending, as
alfo the velocities acquired at the end of the defcent,
are as the fquare roots of the fpaces defcribed, as in
the cafe of bodies falling freely ; that is to fay, the
time wherein a body defcends upon the inclined
plane from A to D, is to the time of the defcent
from A to C, as the fquare root of A D, to the
fquare root of A C ; and the velocity of the body
when it has defcended as far as D, is to the velocity
thereof when it arrives at C in the fame proportion
of the root of A D to the root of A C.

Prop. II. *The velocity acquired in any given
time by a body defcending on an inclined plane, is to the
velocity acquired in the fame time by a body falling
freely and perpendicularly, as the height of the plane
to its length, that is, as* A B *to* A C.

For, by the firft *corollary* of the foregoing *propofition,* the motion of a body down an inclined plane

4   is

is uniformly accelerated, in the fame manner as the motion of a body falling freely; confequently, at the end of any given time, the velocities acquired muft be as the accelerating forces; but by the foregoing *propofition*, the accelerating force of a body moving down an inclined plane as AC, is to the accelerating force of a body falling freely and perpendicularly, as the height of the plane to its length; and therefore the velocities acquired in any given time muft be in the fame proportion.

PROP. III. *The fpaces defcribed in a given time by two bodies moving from a ftate of reft, whereof one defcends on an inclined plane, and the other falls freely, are in the fame ratio of the height of the plane to its length; that is, the fpace defcribed by a body moving along* AC, *is to the fpace defcribed by a body falling down the perpendicular* AB, *as* AB *to* AC.

For where the motions are equable, the fpaces defcribed in a given time, are as the velocities wherewith they are defcribed; if therefore the velocities be increafed in a conftant uniform manner, the fpaces defcribed will likewife increafe in the fame manner; but by the fecond *propofition*, the velocities are augmented in fuch a manner as in a given time to bear the fame proportion to one another, as the height of the plane does to its length; confequently, the fpaces defcribed in a given time muft be in that proportion.

COROL. I. If from B, the line BD be drawn perpendicular to AC, AD will be the fpace defcribed by a body moving down the plane AC, in the fame time that a body falls freely down the height of the plane from A to B.

For, from the nature of fimilar triangles, AC is to AB, as AB to AD; but by the *propofition*, as AC is to AB, fo is the fpace defcribed in a given time by a body falling freely, to the fpace defcribed
by

by a body defcending upon the inclined plane A C; confequently, fince A B is the fpace defcribed by the body falling freely, A D muft be the fpace defcribed in the fame time by a body defcending along A C.

Cᴏʀᴏʟ. II. All the chords of a circle are defcribed in the fame time by bodies running down them. For if a circle be defcribed with the diameter A B, which is the height of the inclined plane A C, the point D, which determines the fpace A D thro' which a body defcends upon the inclined plane, whilft another falls freely from A to B, will be in the periphery of the circle, becaufe the angle A D B in the femicircle is always a right one; and for the fame reafon, if the height of the plane continuing the fame, the inclination thereof be varied, fo as that it may become A G, the point E which determines the fpace A E, thro' which a body moves along the plane A G, during the time of a body's fall from A to B, will likewife be in the periphery of the circle; confequently, in the femicircle A D B all the chords as A D and A E will be defcribed in the fame time; and as in the femicircle A F B, whatever chords as B F and B H are drawn thro' the point B, other chords as A D and A E may be drawn in the other femicircle parallel thereto and equal; it follows, that whether a body falls freely down the diameter A B, or whether it defcends along a chord as H B or F B, it will in the fame time arrive at the loweft point of the circle; or in other words, all the chords of a circle will be defcribed in equal times by bodies running along them.

Pl. 5.
Fig. 7.

Pʀᴏᴘ. IV. *The time wherein a body moves down an inclined plane as* A C, *is to the time wherein a body falls freely down* A B *the height of the plane, as the length of the plane to its height, that is, the times are as the fpaces defcribed.*

Pl. 5.
Fig. 6.

L

For

For by the second *Corol.* of the firſt *Prop.* the time of a body's motion along the inclined plane from A to C, is to the time of its motion from A to D, as the ſquare root of A C to the ſquare root of A D; but by the ſecond *Corol.* of the third *Prop.* the time of a body's motion along the inclined plane from A to D, is equal to the time of the fall from A to B; and therefore the time of the motion along the plane from A to C, is to the time of the perpendicular fall from A to B, as the ſquare root of A C, to the ſquare root of A D, that is, becauſe from the ſimilarity of triangles A C, A B, and A D are in continued proportion, as A C to A B, or as the length of the plane to its height.

COROL. Hence it follows, that if ſeveral inclined planes have equal altitudes, the times wherein thoſe planes are deſcribed by bodies running down them, are to one another as the lengths of the planes.

PI. 5.
Fig. 7.

For the time of the deſcent along AC, is to the time of the fall down AB, as AC to AB, and the time of the fall down AB, is to the time of the deſcent along AG, as AB to AG; conſequently, the time of the deſcent from A to C, is to the time of the deſcent from A to G, as AC to AG, that is, the times are as the lengths of the planes.

PI. 5.
Fig. 6.

PROP. V. *The velocity acquired at the end of the fall by a body falling down the perpendicular height of an inclined plane as* AB, *is equal to the velocity acquired at the end of the deſcent by a body moving down the inclined plane, from* A *to* C.

For by the firſt *Prop.* the accelerating force of a body falling freely from A to B, is to the accelerating force of a body moving along the plane AC, as AC to AB; and by the fourth *Prop.* as AB is

to

to AC, fo is the time of the fall from A to B, to the time of the defcent from A to C; fo that the forces which accelerate the bodies during their motions, are to one another, reciprocally as the times that they continue to act; confequently, at the end of thofe times, the velocities generated muft be equal. For inftance, if AB be one half of AC, the force which accelerates the body in its fall from A to B, is to the force which accelerates the body in its defcent from A to C, as two to one; but the time that a body takes to fall from A to B, is but one half of the time that a body takes to defcend from A to C; fo that the accelerating force which acts upon the body during its motion from A to C, tho' it be but one half of the accelerating force which acts upon the body during its fall from A to B, yet does it continue to act twice as long; and therefore muft in the end produce the fame velocity.

COROL. Hence it follows, that the velocities acquired by bodies in falling down inclined planes, are equal where the heights of the planes are equal.

For, the velocity acquired in falling from A to C, is equal to the velocity acquired in falling from A to B, as is alfo the velocity acquired in falling from A to G; confequently, the velocities acquired in falling from A to C, and from A to G, are equal.

Pl. 5. Fig. 7.

PROP. VI. *If a body defcends along feveral contiguous planes as* AB, BC, *and* CD, *the velocity which it acquires in its defcent from A to D, is equal to the velocity acquired by the perpendicular fall from* H *to* D, *on fuppofition that the body is not retarded by the fhocks it fuffers in the angles* B *and* C.

Pl. 5. Fig. 8.

For drawing the horizontal lines HE and DF thro' the points A and D, and producing the planes CB and DC as far as G and E; by the *Corol.* of the

laft

LECT.
X.
laſt *Propoſition*, the ſame velocity is acquired in the point B, by a body in deſcending from A to B, as in deſcending from G to B ; conſequently, the ſame velocity is acquired in the point C, by a body deſcending from A thro' B to C, as in deſcending from G to C ; but by the ſame *Corollary*, the velocity acquired in deſcending from G to C, is equal to the velocity acquired in deſcending from E to C ; wherefore, the velocity in the point D acquired by the deſcent along the three planes AB, BC, and CD, is equal to the velocity acquired by the deſcent from E to D, which velocity by the foregoing *Propoſition*, is equal to the velocity acquired by the perpendicular fall from H to D.

Pl. 5.
Fig. 9.
COROL. Hence it follows, that if a body deſcends along the arch of a circle as AB or of any other curve, the velocity acquired at the end of the deſcent, is equal to the velocity acquired by falling down CB, the perpendicular height of the arch.

For curves may be looked upon as compoſed of an infinite number of right lines inclined one to another.

Pl. 5.
Fig. 10.
PROP. VII. *If two planes as* AB *and* BD *joined together at* B, *have equal degrees of elevation with two other planes as* EF *and* FH *joined together at* F, *and if* AB *be to* EF *as* BD *to* FH ; *the time of a body's fall down the planes* ABD, *will be to the time of the fall down* EFH, *as the ſquare root of* AB *and* BD *taken together, to the ſquare root of* EF *and* FH *taken together.*

Let AB and EF be produced till BC becomes equal to BD, and FG equal to FH. Since AB is to BC, as EF to FG, AB is to AC, as EF to EG ; and ſince thoſe four quantities AB, AC, EF and EG are proportional, their ſquare roots will be ſo too. Again, ſince the planes AC and EG are
equally

equally elevated, they may be looked upon as parts
of one and the fame plane, and therefore, by the
fecond *Corol.* of the firft *Prop.* the time of a body's
fall from A to C, is to the time of the fall from
E to G, as the fquare root of AC to the fquare root
of EG, or as the fquare root of AB to the fquare
root of EF; but the time of a body's fall from A
to B, is to the time of the fall from E to F, as the
fquare root of AB to the fquare root of EF; fo
that the time of the fall from A to C, is to the
time of the fall from E to G, in the fame proportion
of the time of the fall from A to B, to the time
of the fall from E to F; confequently, the time of
the fall from B to C, fuppofing the motion to begin from A, muft be to the time of the fall from
F to G, fuppofing the motion to begin from E, in
the fame proportion of the root of AB to the root of
EF; if the bodies after their fall from A to B, and
from E to F, inftead of moving along BC and FG
continue their motions along BD and FH, fince
thofe two planes are equally inclined to AB and EF,
and fince BD is equal to BC, and FH equal to FG,
whatever proportion the time of the body's motion
along BD bears to the time of its motion along BC,
the fame will the time of the motion along FH
bear to the time of the motion along FG; but it
has been already proved, that the time of the motion along BC, is to the time of the motion along
FG, as the fquare root of AB to the fquare root of
EF; wherefore, the time of the motion along BD,
is to the time along FH, as the fquare root of AB to
the fquare root of EF, that is, in the fame proportion
with the time along AB to the time along EF; and
therefore, the fums of thofe times will be in the
fame proportion; that is to fay, the time of the
motion along AB, added to the time of the motion
along BD, is to the time of the motion along EF,
added to the time of the motion along FH, as the
fquare root of AB to the fquare root of EF; but

L 3   it

it has been proved, that the square root of AB is to the square root of EF, as the square root of AB and BD taken together, to the square root of EF and FH taken together; and therefore, the time of a body's fall from A thro' B to D, is to the time of the fall from E thro' F to H, as the square root of ABD to the square root of EFH, which was to be proved. And what has been thus proved with regard to two planes on each side, is in like manner demonstrable with regard to any number of planes, provided those on one side be proportional to those on the other, and that the corresponding planes have equal degrees of elevation.

Pl. 5.
Fig. 11.

COROL. Hence it follows, that if bodies descend thro' the arches of circles, the times of describing similar arches similarly posited, are as the square roots of the arches. For instance, if bodies move down the similar arches AB and CD, which are similarly posited with regard to the horizontal plane ED, the time of describing AB, is to the time of describing CD, as the square root of AB to the square root of CD.

For all circles whatever, may be considered as similar polygons, consisting of an indefinite number of sides indefinitely small; and therefore, similar arches must consist of an equal number of sides proportional the one to the other; and forasmuch as the angles which those sides contain are equal, if the arches be similarly posited, the corresponding sides in each arch must have equal degrees of elevation; and consequently, the times of describing the arches will be as their square roots.

In my lecture upon gravity, I shewed you, that if a body be thrown directly upward, it will rise to the same height, whence, if it fell from a state of rest, it would by the end of the fall acquire the same velocity wherewith it is thrown up; I likewise shewed you, that the time of the rise is equal to that of the fall. I now say,

PROP.

PROP. VIII. *That the same things do likewise obtain with regard to bodies thrown up obliquely, whether they ascend upon inclined planes or along the arcs of curves.*

Because the same forces which accelerate the motions of bodies descending on such planes or curves, do in the very same manner retard the motions of such bodies as ascend thereon; and therefore, whatever be the time requisite for a body to descend upon an inclined plane or thro' the arc of a curve, in order to acquire any velocity, the same must the time be, wherein that velocity is destroyed in a body ascending upon the same plane or curve, and whatever be the length of the plane or curve, thro' which a body descends in order to acquire any velocity, the same must the length of the plane or curve be, thro' which it must ascend in order to have that velocity destroyed.

COROL. Hence it follows, that if by any contrivance a body be made to descend thro' the arch of a circle as from C to A, and with the velocity acquired by the descent to ascend along the arch AD of the same circle, the arch AD which it describes in its ascent, will be equal to the arch CA described in the descent; and the times in which those arches are described will be equal.

Pl. 5.
Fig. 12.

And this is the case of the PENDULUM; which is a heavy body as A, hanging by a small cord as BA, and moveable therewith about the point B, to which the cord is fixed. If when the cord is stretched the weight be raised as high as C, and thence let fall, it will by its own gravity descend thro' the circular arch CA; and by the *Corol.* of the sixth *Prop.* it will have the same velocity in the point A, that a body would acquire in falling perpendicularly from E to A; and by the first LAW OF NATURE, it will endeavour to go off with that velocity in the tangent AF; but being by the force of

Exp. 1.

L 4   the

the cord made to move in the periphery C A D,
it will rife thro' the arch A D as high as D, where
lofing all its velocity, it will be turned back by its
gravity, and defcending thro' the arch D A, will,
upon its arrival at A, have the fame velocity as before, with which it will afcend to C; and thus it
will continue its motion forward and backward a-
long the curve C A D, which motion is called an *of-
cillatory* or *vibratory* motion; and each fwing from
C to D, as alfo from D to C, is called a *vibration*;
and if the pendulum fuffered no retardation in its
motion from the refiftance of the air, nor from the
friction of the cord againft the center about which
it moves, the arches defcribed in each vibration
would be exactly equal, and the motion of the pendulum would continue for ever; but whereas the
motion of the pendulum is continually retarded by
the forementioned caufes, the arches defcribed in
each vibration muft grow lefs and lefs continually,
and at laft vanifh together with the motion of the
pendulum.

Pl. 5.
Fig. 13.
The vibrations of one and the fame pendulum
vibrating in unequal circular arches, are performed
very nearly in equal times, provided the arches are
but fmall. Thus, in the pendulum A B, the vibration thro' the arch C A D, is performed very nearly
in the fame time wherein the pendulum vibrates
thro' the arch E A F, on fuppofition that the arches
C A and E A are but fmall.

For, drawing the chords C A and A D, as alfo
E A and A F, inafmuch as the arches are fuppofed
to be fmall, they will not differ much either as to
length or declivity from their refpective chords;
confequently the times of defcribing the arches
C A and E A, by a heavy body running along
them, will be nearly equal to the times of defcribing the chords; but by the fecond *Corol.* of the
third *Prop.* the times of defcribing the chords are
equal; wherefore the times of defcribing the
arches

arches C A and E A, muſt be nearly equal; and
ſo likewiſe muſt the double of thoſe times, or the
times wherein the pendulum vibrates thro' the unequal arches C A D and E A F. And this is confirmed by experiment. For if two pendulums of
an equal length, be ſet going at the ſame inſtant of
time, ſo as to vibrate thro' ſmall but unequal arches,
they will for a long time keep pace together; and
continue to begin and end their ſwings without any
ſenſible difference as to point of time, during a
great number of vibrations.

If a pendulum as B A vibrates thro' the circu- Pl. 5.
lar arches C A D and E A F, the velocity which it Fig. 14.
acquires by that time it arrives at the loweſt point
A, is as the chord of the arch which it deſcribes in
its deſcent; that is, the velocity which it acquires
in deſcending from C to A, is to the velocity acquired in its deſcent from E to A, as the chord
C A to the chord E A.

For, drawing the horizontal lines E K and C H,
the velocity acquired in falling from H to A, is to
the velocity acquired in falling from G to A, in
the ſubduplicate *ratio* of H A to G A, as I proved
in my lecture upon gravity; that is, becauſe, from
the nature of the circle H A, C A, and G A, are in
continued proportion, as C A to G A; for the ſame
reaſon the velocity acquired in falling from G to
A, is to the velocity acquired in falling from K to
A, as G A to E A; conſequently, the velocity acquired in falling from H to A, is to the velocity acquired in falling from K to A, as C A to E A; but
by the *Corol.* of the ſixth *Prop.* the velocity acquired
in falling from H to A, is equal to the velocity acquired in the deſcent from C to A, and the velocity
acquired in falling from K to A, is equal to the velocity acquired in the deſcent from E to A; wherefore, the velocity acquired in deſcending thro' the
arch C A, is to the velocity acquired in deſcending
thro' the arch E A, as the chord C A to the chord E A.

Hence

Hence it appears, that if the arch of a circle wherein a pendulum vibrates, be so divided in the points 1, 2, 3, 4, and so on, beginning from the lowest point A, as that the chords drawn from A to the several points of division, may be to one another, as those numbers, the velocities acquired by a pendulum in the lowest point A, when let fall successively from the several points of division, will be as the numbers affixed to the respective points ; and it was upon this account, that in the experiments relating to the collision of bodies, the balls were constantly let fall from such heights, as that the chords of the arches which they described in their descent, might be to one another in the same proportion with the velocities wherewith the balls were supposed to meet at the lowest point.

The times wherein pendulums of unequal lengths vibrating in similar arches, perform their vibrations, are to one another, as the square roots of their lengths ; for instance, the time wherein the pendulum B A vibrates thro' the arch F G, is to the time wherein the pendulum B C vibrates thro' the arch D E similar to F G, as the square root of B A to the square root of B C.

Pl. 5.
Fig. 15.

For, by the *Corol.* of the seventh *Prop.* since the arches F A and D C are similar and similarly posited, the time of the descent thro' F A, is to the time of the descent thro' D C, as the square root of F A, to the square root of D C ; but by the *Corol.* of the eighth *Prop.* the time of the descent thro' F A, is one half of the time of the vibration from F to G, and the time of the descent thro' D C, is one half of the time of the vibration from D to E ; consequently, the time of the vibration thro' F G, is to the time of the vibration thro' D E, as the square root of F A, to the square root of D C ; that is, because the arches F A and D C are similar, as the square root of B A to the square root of B C, that is, the times of the vibrations are

as

as the square roots of the lengths of the pendulums. And forafmuch as the times wherein pendulums perform their vibrations, are to one another inverfly as the number of vibrations performed in a given time; the numbers of vibrations performed by pendulums in a given time, are to one another inverfly as the fquare roots of the lengths of the pendulums. For inftance, if the length of the pendulum B A, be to the length of the pendulum B C, as one to four, the number of vibrations performed in any given time by the fhorter pendulum, is to the number of vibrations performed in the fame time by the longer, as the fquare root of four to the fquare root of one, that is, as two to one; which cafe is experimentally confirmed by two pendulums, whereof the longer being 39.125 inches, vibrates in one fecond of time; and the fhorter being 9.781 inches, vibrates in half a fecond, and performs two vibrations in the fame time that the longer performs one.

Exp. 3.

	Inches.	
Length of a pendulum vibrating.	in a 2d. $\begin{cases} 39.125 & \textit{Halley.} \\ 39.207 & \textit{Newton.} \end{cases}$	
	in $\frac{1}{2}$ a 2d. $\begin{cases} 9.781 & \textit{Halley.} \\ 9.801 & \textit{Newton.} \end{cases}$	

The time of a pendulum's vibration is no way altered by varying the weight thereof; for fince the gravity of every body is proportional to its quantity of matter, as I proved in my lecture upon gravity, all bodies in the fame circumftances are moved by the force of gravity with the fame velocity; and therefore, if the length of a pendulum continues the fame, it will perform its vibrations in the fame time, whatever be the magnitude of the appending weight; which may be confirmed by the following experiment. Let two unequal weights be hung by two threads fo as to conftitute two pendulums equal in length, and let them at the fame inftant of time

Exp. 4.

fall

fall from equal heights, they will keep pace together fo as to perform their vibrations in equal times.

In the foregoing part of this lecture I shewed
you, that the vibrations of one and the same pendulum vibrating thro' unequal but small circular
arches, are performed in times that are very nearly,
but not precisely, equal. Whence it follows, that
however useful such a pendulum may be in mea-
suring time where great exactness is not requisite,
yet can it by no means be admitted as an accurate
measure of time, unless by some contrivance it be
made to perform all its vibrations in equal arches,
which, considering the unavoidable imperfections
of all machines, is extremely difficult, if not impossible; for it has been found by experience, that
the best regulated pendulum clocks, wherein the
greatest care has been taken to make the pendulums
vibrate in equal arches, have notwithstanding varied in a course of time, so as to stand in need of a
new regulation, which they could not possibly do
in case the pendulums, whereon the regularity of
all the other movements depends, continued constantly to vibrate in equal arches.

In order therefore to obtain an exact unerring
measure of time, it is necessary to make a pendulum vibrate in such a manner, as that all its swings,
whether they be thro' larger or smaller arches, may
be performed in times exactly equal; and this may
be done by making a pendulum vibrate in the
curve of a cycloid, as I shall now demonstrate;
but I shall first shew you the manner wherein that
curve is generated, and what its chief properties
are, as also by what contrivance a pendulum is
made to vibrate in such a curve.

Pl. 5.
Fig. 16. If a circle as C E F, which touches the right line
A B in the point C, be moved along that line in
the manner of a wheel from C to D, so as to
perform an intire revolution; the point C will by
virtue of its double motion describe the curve line
CID,

CID, which curve line is called a *cycloid*; and the right line CD is called the *bafe*, the line IK perpendicular to the bafe at its middle point is called the *axis of the cycloid*, and the point I the *vertex*, and the circle CEF or KLI is called the *generating circle*.

From any point in the cycloid as H, let a right line as HL, be drawn parallel to the bafe CD, and continued till it meets the generating circle KLI, defcribed about the axis IK; and let the line HM touch the cycloid in the point H; this being done, the chief properties of the cycloid are thefe three.

Firft, The arch IPL of the generating circle, intercepted between the vertex of the cycloid and the point L, wherein the right line HL meets the generating circle, is equal in length to the right line HL.

Secondly, The chord IL of the circular arch IPL, is parallel to the right line MH, which touches the cycloid in the point H.

Thirdly, The cycloidal arch IH intercepted between the vertex and the point H, is double the chord IL.

The demonftrations of thefe properties may be feen in HUYGENS, WALLIS, COTES, and others who have wrote of the cycloid.

The contrivance whereby a pendulum is made to vibrate in the curve of a cycloid, is thus. A cycloid as AVB, being defcribed on the bafe AB, let the axis VD be produced towards C, till DC becomes equal to VD; thro' the points C and A, and C and B, let two femi-cycloids CA and CB be drawn, each equal to half of AVB, their vertices being at A and B; if then we fuppofe CA and CB to be two plates of fome breadth, and an heavy body to hang from the point C by a ftring equal in length to CV, and to vibrate between the plates CA and CB, the upper part of the ftring will con-

Pl. *c.* Fig. 17.

ftantly

ſtantly apply itſelf to that plate towards which the
body moves, and by ſo doing cauſe it to move in
the cycloid AVB as has been proved by Huygens
the author of this contrivance; and likewiſe by
Cotes in his treatiſe *de motu pendulorum*, where he
has delivered the whole doctrine of pendulums in
four Theorems, which I ſhall here lay down and
explain.

Pl. 5.
Fig. 17.
    Theorem I. *If a pendulum vibrating in a cy-
cloid as* BVA, *begins its motions downward towards*
V, *from any point taken at pleaſure as* L, *and if upon a radius as* VL, *equal in length to the cycloidal
arch* VL, *a circle be deſcribed; the velocities of the
pendulum in the ſeveral points of the cycloidal arch,
will be as the right ſines in the circle which are raiſed
from the correſponding points in the* radius; *for inſtance, if in the* radius LM *be taken equal to* LM *in
the cycloid, and from the point* M *in the* radius correſponding to the point M *in the cycloid, be raiſed the right
ſine* MX, *the velocity of the pendulum in the point* M,
after it has deſcended from L, *will be as the ſine* MX.

    For the proof of which, from the points L and
M in the cycloid, let the right lines LOR and
MQS be drawn perpendicular to the axis, cutting
the generating circle in O and Q, from whence to
the vertex, let the right lines OV and QV be
drawn. By the *Corol.* of the ſixth *Prop.* the velocity which the pendulum acquires in deſcending
along the cycloid from L to M, is equal to the velocity acquired by a body in falling perpendicularly
from R to S; but the velocity which a body acquires in falling perpendicularly, is in the ſubduplicate *ratio* of the ſpace deſcribed, as I proved in
my lecture upon gravity; conſequently, the velocity acquired by the pendulum in its deſcent from
L to M, may be expreſſed by the ſquare root of
RS;

RS; but RS, being equal to the difference between RV and SV, the velocity in the point M
may be expreſſed by the ſquare root of the difference between RV and SV ; or, becauſe, RV multiplied into the axis DV, is to SV multiplied into the
ſame DV, as RV to SV, the velocity may be expreſſed by the ſquare root of the difference between the
product of RV × DV and SV × DV; but from the
nature of the circle, the product of RV × DV is
equal to the ſquare of VO; and the product of
SV × DV is equal to the ſquare of VQ; wherefore,
the velocity at M may be expreſſed by the ſquare
root of the difference between the ſquare of VO and
the ſquare of VQ; but, by the third property of
the cycloid, VO is equal to one half of the cycloidal
arch VL, and VQ to one half of the arch VM ;
wherefore, as VO ſquare, is to VQ ſquare, ſo is
VL ſquare, to VM ſquare; conſequently, the velocity of the pendulum at M may be expreſſed by
the ſquare root of the difference between the ſquare
of VL and the ſquare of VM; but the cycloidal
arches VL and VM are by ſuppoſition equal to
VL and VM in the *radius* of the circle; and, from
the nature of a right-angled triangle, the difference
between the ſquare of VX, which is equal to VL,
and the ſquare of VM, is equal to the ſquare of
MX; wherefore, the velocity of the pendulum at
the point M, is as the ſquare root of MX ſquare,
that is, as MX, as was aſſerted in the *Theorem*.
And what has been thus proved with regard to the
velocity at the point M, is in like manner demonſtrable with regard to the velocity at any other
point as N; namely, that it is as the right ſine
NY raiſed from the point N in the *radius* correſponding to the point N in the cycloid; ſo that the
velocities of a pendulum deſcending in a cycloid,
are in the ſeveral points of the cycloidal arch, as the
right ſines in a circle which are raiſed from the correſponding points of the *radius*, the *radius* being
equal

equal in length to the cycloidal arch intercepted between the vertex and that point from which the
pendulum begins its motion. Thus, VM and VN
in the *radius* of the circle, being taken equal to
VM and VN in the cycloid, so as that the points
M, N, V, in the *radius*, may correspond to the
points M, N, V, in the cycloid, the velocities of
the pendulum in those points are to one another, as
the sines MX, NY, and VZ, the *radius* VZ expressing the greatest velocity at the vertex V.

Pl. 5.
Fig. 17.
THEOREM II. *If a body be supposed to move uniformly in the curve of the circle, with a velocity equal
to the velocity acquired by the pendulum in its descent
from* L *to* V, *which velocity is, as was just now
shewn, expressed by the* radius VZ; *any arch of the
circle as* XY *taken at pleasure, will be described by the
body moving along it in the forementioned manner, in
the same time that the pendulum, which begins its motion from the point* L *in the cycloid, describes the cycloidal arch* MN, *corresponding to and equal in length
to* MN, *that part of the* radius, *which lies between
the sines* MX *and* NY, *which terminate at the extremities of the circular arch* XY.

Let the sine FGH, be drawn indefinitely near
to the sine MX, and let XG be drawn parallel to
MF : and let MF in the cycloid be equal to MF in
the *radius* of the circle. By the foregoing *Theorem*,
the velocity of the pendulum in the point M, is as
MX ; and therefore, since F is supposed to be indefinitely near to M, the little cycloidal arch MF,
equal to MF in the *radius*, is to be looked upon as
described by the pendulum with a velocity which is
as MX ; and the little circular arch XH, is by
supposition described with a velocity which is as
VZ, equal to VX ; and the triangles MXV and
GXH being similar, inasmuch as the angles at M
and G are right ones, and the angle MXV equal to
GXH,

GXH, becaufe GXV is the complement of each of them to a right one; XH is to XG equal to MF, as VX or VZ to MX; that is, XH and MF are to one another, as the velocities wherewith they are defcribed; confequently, they muft be defcribed in the fame time. And what has been thus demonftrated of MF and XH, is in like manner demonftrable of the feveral correfponding parts in the cycloidal arch MN, and circular arch XY; confequently, the whole cycloidal arch MN, will be defcribed by the pendulum in the fame time, that the circular arch XY is defcribed by a body moving along it uniformly with the velocity expreffed by VZ; and by the fame way of reafoning, the time of defcribing any other cycloidal arch as LV, is equal to the time of defcribing the correfponding circular arch LZ.

Corol. As a *Corollary* it follows, that the time wherein a pendulum defcribes any arch of a cycloid as MN, may be expreffed by the correfponding circular arch XY.

For, as the motion along the curve of the circle is fuppofed to be uniform, the time of defcribing any arch as XY, muft be as the length of the arch; but by the *Theorem*, the times of defcribing the circular arch XY, and the cycloidal arch MN, are equal; confequently, the time in which the pendulum defcribes the cycloidal arch MN, is as the circular arch XY.

Theorem III. *The time of one intire vibration of a pendulum moving in a cycloid, is to the time wherein a body falls perpendicularly thro' a fpace equal in length to the axis of the cycloid, as the periphery of a circle to its diameter.*

All things being fuppofed as before, the time of defcribing the femicircular periphery LZP with the

Pl. 5.
Fig. 17.

velocity

M

velocity expreſſed by VZ, is to the time of deſcribing the ſemidiameter LV with the ſame velocity,
as the ſemicircular periphery to the ſemidiameter,
or as the whole periphery to the diameter ; but the
time of deſcribing the ſemicircular periphery LZP
with the velocity VZ, is equal to the time of an intire vibration ; for, by the eighth *Prop.* the time
wherein the pendulum deſcribes the cycloidal arch
LV, is one half of the time wherein it performs an
intire vibration ; and by the ſecond *Theorem,* the
time wherein a pendulum deſcribes the cycloidal
arch LV, is equal to the time wherein the quadrantal arch of the circle, to wit LZ, is deſcribed with
the velocity expreſſed by VZ ; conſequently, the
time of an intire vibration, is equal to the time of
deſcribing the ſemicircular periphery LZP ; and the
time of deſcribing the ſemidiameter LV with the
velocity VZ, is equal to the time of a body's fall
down the height of the axis DV ; for, by the ſecond *Corol.* of the third *Prop.* the fall down the
axis DV, is performed in the ſame time with the deſcent along the chord OV ; and by the eighth *Prop.*
the velocity acquired at the end of the deſcent along
the chord OV, will in the ſame time with that of
the deſcent deſcribe a ſpace equal to twice OV ; but
by the third property of the cycloid, twice VO is
equal to the cycloidal arch LV, which by ſuppoſition is equal to the ſemidiameter VL ; and conſequently, the velocity acquired at the end of the deſcent along the chord OV, is ſuch, as will in a time
equal to that of the fall down the axis DV, deſcribe
the ſemidiameter LV ; but, by the *Corol.* of the
ſixth *Prop.* the velocity acquired at the end of the
deſcent along the chord OV, is equal to the velocity
acquired by the pendulum in its deſcent along the
cycloidal arch from L to V, which by the firſt
Theorem, is as VZ ; wherefore, the time of deſcribing the ſemidiameter LV with the velocity VZ, is
equal

4

equal to the time of the fall down the axis DV; but it has been already proved, that the time of describing the femicircular arch LZP with the velocity VZ, is to the time of describing the femidiameter LV with the fame velocity as the periphery of the circle to its diameter; and it has been likewife proved, that the time of describing the femicircular arch with the velocity VZ, is equal to the time of an intire vibration of the pendulum; confequently, the time of fuch a vibration, is to the time of the fall down the axis, as the periphery of the circle to its diameter.

Corol. From what has been proved it follows, that the time of a vibration of a pendulum moving in a given cycloid is given; or in other words, that all the vibrations of fuch a pendulum, whether they be in larger or fmaller arches, are performed in times exactly equal.

For, as it has been proved, that the time of the vibration which begins from the point L, is to the time of the fall down the axis, as the periphery of the circle defcribed on a *radius* equal to the cycloidal arch VL, to its diameter; it may in like manner be demonftrated, that if the vibration begins from any other point as M, the time thereof will bear the fame proportion to the time of the fall down the axis, that the periphery of a circle defcribed on a *radius* equal in length to the cycloidal arch VM, does to its diameter; but the *ratio* of the periphery to the diameter in any one circle, is the fame with that in any other; wherefore, the times of the vibrations thro' unequal arches, have all the fame *ratio*, to the time of the fall down the axis, and of confequence muft be equal.

From this equality in the times of the fwings it is, that this kind of pendulum is preferable to fuch as vibrate in circular arches, as being a more exact and juft meafure of time; a minute of mean or equal time being precifely meafured by fixty fwings

of

of a pendulum of this kind, whose length is equal to three horary feet, which answers to 39 inches and one eighth of our measure, according to Doctor HALLEY; or to 39 inches and one fifth, according to Sir ISAAC NEWTON; and now that I have mentioned *mean* or *equal*, otherwise called *true time*, it will not be improper in this place, to shew you wherein it differs from that time, which by astronomers is called *unequal* and *apparent time.*

As time in itself does not fall under the notice of our senses, and as the parts thereof go on in a continued succession one after another, no two existing together, it is impossible to discover the equality or inequality of any two portions of time, by an immediate comparison of one with the other; and therefore, it was necessary for those who first thought of distinguishing the parts of time, to have recourse to something sensible, and of a different nature from time, as a measure thereof. And as nothing seems better fitted to serve this purpose, than such natural apppearances as fall under every man's notice, and at the same time have frequent returns, it is highly probable, that in the first ages of the world, men observing the frequent risings and settings of the sun, took the one or the other for their first measure of time, calling that portion of time which passed between two risings or settings, which immediately succeeded each other, by the name of a *day*; in like manner it is rational to suppose, that upon observing the frequent returns of the full and new moons, they made the one or the other their second measure of time, calling that space which passed between two successive new or full moons by the name of a *moon* or *month*. And it is likely, that for some time they contented themselves with these measures, without knowing or considering whether they were exact or not: but in process of time, as men became better acquainted with the motions of the heavenly bodies, they dis-

covered

2

covered some irregularities in the apparent motion of the sun, and of consequence, an inequality in the natural days which depend on that motion; inasmuch as the portion of time, which passes between the sun's departure from the plane of any meridian and its next return thereunto, is not always the same. By considering the causes of this inequality, they were led into a method of making such corrections in the natural days, by adding to some, and taking from others, as reduced them all to a mean equal length; each day being made to consist of 24 equal *hours*, each of which is divided into sixty equal parts called *minutes*, and each of these into sixty others called *seconds*, and these again into *thirds*, and so on in a *sexagesimal* progression, the parts of each denomination being constantly equal among themselves. And these parts of time thus reduced to an equality constitute the mean or equal time, as it stands distinguished by astronomers, from the unequal or apparent time, which is measured by the apparent motion of the sun.

In order to have a constant measure of equal time, HUYGENS contrived a method of adapting pendulums to clocks, whereby their motions are so exactly regulated, as that in a clock whose movements are rightly adjusted, the seconds, minutes, and hours, are for some time pointed out with the greatest exactness; I say, for some time only, because it is not possible that any clock whatever should continue exactly true for a long course of time; for as the pendulums of clocks according to HUYGENS's first contrivance, and by the general practice of clock-makers at this day, are made to vibrate in circular arches, where the times of the vibrations are not precisely equal, unless the arches thro' which the pendulum moves be so too. If the wheels on account of the thickening of the oil by frosty weather, or from any other cause grow more

sluggish,

ſluggiſh, ſo' as to give the weight, which in clocks
is the moving power, greater reſiſtance than according to the firſt adjuſtment, the force of the
crown-wheel upon the palates of the pendulum will
likewiſe be diminiſhed, and of conſequence, the
pendulum being thrown leſs forcibly will move
thro' ſmaller arches than before, and by ſo doing,
will meaſure out ſmaller portions of time, the time
of ſixty ſwings not amounting to a minute, upon
which account the clock muſt gain, and go too faſt.
On the other hand, whenever the parts of the move-
ment which rub one againſt another do, by reaſon
of the thinning of the oil by the heat of the weather, grow more ſlippery, or from their conſtant
friction become more ſmooth, ſo as to give leſs reſiſtance to the moving power than according to the
firſt adjuſtment, the crown-wheel acts more forcibly on the pendulum, and cauſes it to vibrate in
larger arches, by which means the time of each
ſwing is inlarged, and of courſe the clock loſes
and goes too ſlow.   To remedy theſe inconveniences HUYGENS thought of a ſecond method of
adapting pendulums to clocks, ſo as to make them
perform their vibrations in cycloidal arches; by
which means, tho' the force of the crown-wheel
upon the pendulum ſhould vary, ſo as to cauſe it
to vibrate ſometimes in larger and ſometimes in
ſmaller arches, yet will not any variation ariſe from
thence in the times of the vibrations; as is evident from the *Corollary* of the third *Theorem*; ſo
that in clocks whoſe motions are governed by pendulums vibrating in cycloidal arches, the irregularities ariſing from the variation of the force of the
crown-wheel upon the pendulum are wholly avoided; and yet a clock of this kind will not always
go true; for as the pendulum cannot vibrate in the
curve of a cycloid, unleſs the uppermoſt part of
the ſtring does as often as it moves from the perpendicular

pendicular towards either fide, form itfelf into a cycloidal arch ; and as this cannot be done unlefs that part of the ftring be made of filk, or fome other foft and pliable fubftance, which as fuch is apt to imbibe the moifture of the air; whenever the weather becomes remarkably moift, the watery particles which float in the air, will infinuate themfelves into the pores of the ftring, and by fo doing caufe it to contract and fhorten; upon which account, the vibrations of the pendulum will be quickened, as will appear from the next *Theorem*, and the clock will gain. So that neither a clock of this, nor of any other kind, can go exactly true for any long courfe of time, which is a thing well known to clock-makers, who have frequently experienced the beft regulated clocks to vary in the compafs of a few months, fome feconds from the equation table, fo as to ftand in need of a new regulation.

THEOREM IV. *The times wherein pendulums of different lengths as* CV *and* AB *perform their vibrations, are to one another in the fame proportion with the fquare roots of the lengths of the pendulums.* Pl. 5. Fig. 17, 18.

For, by the third *Theorem*, the time wherein the pendulum CV performs its vibrations, is to the time wherein a body falls down the axis DV, as the circumference of a circle to its diameter ; and by the fame *Theorem*, as the circumference of a circle is to the diameter, fo is the time wherein the pendulum AB performs its vibrations, to the time wherein a body falls down the axis EB ; confequently, the time wherein the pendulum CV performs its vibrations, is to the time wherein the pendulum AB performs its vibrations, as the time of the fall down DV, is to the time of the fall down EB ; but, as I proved in my lecture upon gravity, the time of the fall down DV, is to the time of the Pl. 5. Fig. 17.

Pl. 5. Fig. 18.

M 4 fall

fall down EB, as the square root of DV, to the square root of EB; or, because DV is one half of CV, and EB one half of AB, as the square root of CV, to the square root of AB; wherefore, the time wherein the pendulum CV performs its vibrations, is to the time wherein the pendulum AB performs its vibrations, as the square root of CV, to the square root of AB, that is, the times are as the square roots of the lengths of the pendulums.; so that if one pendulum be four times as long as another, the shorter will vibrate in half the time, so as to perform two vibrations in the same time that the longer performs one.

In this *Theorem*, as also in every thing else that has been hitherto said concerning the pendulum, the force of gravity is supposed to be given; whence it follows, that if pendulums of different lengths, as CV and AB, perform their vibrations in equal times, the force of gravity in such pendulums must vary, and that in proportion to the lengths of the pendulums, that is to say, the force of gravity in the pendulum CV, must be to the force of gravity in the pendulum AB, as CV to AB. For, as the times of the vibrations are supposed to be equal, the times of the perpendicular falls down the axes DV, and EB must likewise be equal, inasmuch as they have been proved to be proportional to the times of the vibrations; since therefore, forces which act constantly and uniformly are to one another as the velocities which they generate in any given time, the force of gravity which carries a body down DV, must be to the force of gravity which in the same time carries a body down EB, as the velocity acquired at the end of the fall down DV, to the velocity acquired at the end of the fall down EB; but I proved in my lecture upon gravity, that the velocity acquired in falling down DV, is such as will in a space of time equal to that of the fall, carry a body thro' a space equal to twice DV, that is, thro' a space equal to the

Pl. 5.
Fig. 17,
18.

the length of the pendulum CV; and therefore, the time being given, the velocity may be expressed by the length of the pendulum; and for the same reason, the velocity acquired in falling down E B, may be expressed by the length of the pendulum A B; consequently, the force of gravity which moves the pendulum C V, is to the force of gravity which acts upon the pendulum A B, as the length of the former, to the length of the latter. Since therefore, it has been found by experience, that a pendulum which vibrates in a second of time under the line, must be lengthened as it is removed from the line, and that more and more as its distance therefrom increases; it is manifest, that the force of gravity is less in the æquatorial parts of the earth, than in any other, and that it increases continually as the distance from the line increases, so as to be greatest under the poles; in what proportion this increase of gravity is made, and from what cause it proceeds, I shewed in my lecture upon gravity.

As the several parts of the cycloidal arch LV, have different inclinations to the plane of the horizon, it is evident, from what has been said concerning the motion of bodies upon inclined planes, that the force which accelerates the motion of a pendulum in its descent from L to V, must continually vary; it being greatest in the point L, and thence continually lessening as the cycloidal arch shortens, till at length in the point V it intirely vanishes; and what is particularly remarkable in this case is, that the accelerating forces in the several points of the cycloid, are to one another in the same proportion with the cycloidal arches intercepted between the vertex and the respective points; for instance, the force which accelerates the pendulum in the point L, is to the force which accelerates the same in the point M, as the arch LV, to the arch MY.

Pl. 5. Fig. 17.

For,

For, as the points L and M have the same directions with their tangents, the accelerating forces
in those points must be the same with the forces
which accelerate the motions of bodies descending
along the tangents ; or because the chords O V and
QV in the generating circle are, by the second property of the cycloid, parallel to the tangents at L
and M, as the forces which accelerate . bodies in
their descent upon the chords O V and QV; but
forasmuch as those accelerating forces act constantly
and uniformly, they must be to one another, as the
velocities which they generate in a given time ; and
therefore, since it has been proved, that the chords
O V and QV are described in the same time, the
accelerating forces are as the velocities acquired at
the end of the descent along those chords ; but it
has likewise been proved, that those velocities are
as the lengths of the chords ; consequently, the
force which accelerates a body descending along the
chord O V, is to the force which accelerates a body
descending along the chord QV, as O V to QV ;
but forasmuch as by the third property of the cycloid, O V is one half of LV, and QV one half of
M V, as O V is to QV, so is LV to M V ; and
therefore, the accelerating force along O V, is to the
accelerating force along QV, as the cycloidal arch
LV, to the arch MV ; but it has been proved, that
the accelerating force along O V, is the same with
the accelerating force in the point L, and that the
accelerating force along QV, is the same with the
accelerating force in the point M ; consequently,
the force at L, is to the force at M, as LV to MV ;
as what has been thus demonstrated of the forces
at the points L and M, is in like manner demonstrable of the forces at any other points, so that in
a pendulum descending in the arch of a cycloid, the
accelerating force is in every point as the length of
the cycloidal arch intercepted between the point and
the

the vertex ; or in other words, the force is every where proportional to the fpace to be defcribed.

This then being the law of the accelerating force, and it having been proved, that the pendulum, whether it begins its motion from L or M, or any other point in the cycloid, will arrive in the fame time at the loweft point V ; it follows, that if feveral bodies, placed at different diftances from any point or center, begin to move towards it at the fame inftant of time, with forces that are every where proportional to the diftances from the center, they will all arrive at the center at the fame inftant of time ; which I thought fit to mention in this place, in order to avoid the trouble of demonftrating the fame, when I come to treat of the motions of mufical ftrings, towards the explaining of which this property will be of ufe.

LECTURE XI.

Of the Motion of Projects.

AS the Doctrine of Projects, whereof I intend to treat in this lecture, cannot be rightly apprehended without fome knowledge of the *Parabola* ; I fhall by way of introduction fhew the manner wherein that curve is generated, and point out fuch of its properties as I fhall have occafion to make ufe of in explaining the motion of projects, referring you for their demonftrations to thofe authors who have wrote of the *conick fections*. Lect. XI.

If a cone as A B C, be touched by a plane in the right line A B, and be cut by another plane parallel to the former, the curve which arifes from the interfection of the plane with the furface of the cone is called a *Parabola* ; being fuch as is reprefented in *Fig.* 2, in which the higheft point P is called the *principal* Pl. 6. Fig. 1.

Pl. 6. Fig. 2.

principal vertex; the right line CAP paſſing thro'
the point P, and perpendicular to the tangent at that
point, is called the *axis*; a right line as DA, drawn
from any point in the curve perpendicular to the axis,
is called an *ordinate to the axis*; PA the part of the
axis intercepted between the vertex and the ordinate,
is called the *abſciſſe to that ordinate*; a right line,
being a third proportional to the abſciſſe and its reſpective ordinate, is called the *principal parameter*,
or the *parameter to the axis*; a right line as DEH,
drawn from any point in the curve parallel to the
axis, is called a *diameter*; a right line as PE, intercepted between any point in the curve and the diameter, and parallel to BD which touches the curve
in the point D, is called an *ordinate to that diameter*;
DE the part of the diameter lying between the vertex D and the point E, is called the *abſciſſe to the
ordinate* PE; and a right line, being a third proportional to the abſciſſe DE and the reſpective ordinate EP, is called the *parameter to the diameter*
DH, or *to the vertex* D.

The ſquare of any ordinate divided by the reſpective abſciſſe, is equal to the reſpective parameter; thus the ſquare of DA divided by PA, or the
ſquare of OQ divided by PQ, is equal to the principal parameter; and the ſquare of EP divided by
DE, as alſo the ſquare of LM divided by DL, is
equal to the parameter belonging to the vertex D.
The ſquares of the ordinates to the axis, or to one
and the ſame diameter, are to one another in the
ſame proportion with their reſpective abſciſſa's.
Thus, the ſquare of DA is to the ſquare of OQ, as
PA to PQ; and the ſquare of PE is to the ſquare
of ML, as DE to DL.

In one and the ſame *parabola*, the principal parameter is the leaſt of all the parameters; and the
other parameters increaſe, as the diſtance of their
vertices from the principal vertex increaſes, tho' not
in the ſame proportion.

If

If from any point in a *parabola* as D, an ordinate be drawn to the axis, and if from the same point a tangent be drawn upward, it will meet the axis when produced; and AB, the part of the axis intercepted between the ordinate DA and the tangent DB, will be bisected by P the principal vertex.

These things being premised; if a body be thrown into any direction whatever that is not perpendicular to the plane of the horizon, it will in its motion describe a *parabola*.

For the proof of which, let AE be the direction of the projection, which in the 3d Fig. is parallel to the horizon, and in the 4th and 5th inclined thereto; and let AE be the space which the project would describe in any given time by means of the force impressed, supposing it had no motion downward from the force of gravity; likewise let AB be the space thro' which it would descend in the given time by virtue of its own gravity, supposing it had no other motion; then compleating the parallelogram ABCE, it is manifest from what was formerly said concerning the composition of motion, that at the end of the given time, the project must by virtue of its double motion, be found in the point C; but, forasmuch as the motion impressed in the direction AE is uniform, the space described, that is AE, must be as the time in which it is described; consequently, AE square, or BC square, is as the square of the time; but AB or EC, which is the space described in the same time by the force of gravity, is likewise as the square of the time, as I proved in my lecture upon gravity; consequently, AB is as the square of BC; and therefore, from the nature of the *parabola*, the point C thro' which the project moves, must be in the curve of a *parabola*, whose diameter is AB, the vertex A, the point from whence the project begins its motion, and the parameter

Pl. 6. Fig. 3, 4, 5.

parameter belonging to that vertex, BC fquare divided by AB, or AE fquare divided by EC ; and
what has been thus demonftrated of the point C, is
in like manner demonftrable of all the other points
thro' which the project moves; confequently, the
line which it defcribes is a *parabola*.

The velocity of a project in any point of the *parabola* as A, is fuch as a body acquires in falling
down the fourth part of the parameter belonging to
that point. For the velocity of the project in the
point A is fuch, as would carry it from A to E in
the fame time that a body defcends from E to C;
and the velocity acquired in the defcent from E to C
is fuch, as in the fame fpace of time with that of
the fall, would carry a body thro' a fpace equal to
double EC ; confequently, that velocity is to the
velocity of the project in the point A, as twice EC
to AE, or as EC to $\frac{1}{2}$ AE ; but as EC is to $\frac{1}{2}$ AE,
fo is the velocity acquired in falling from E to C,
to the velocity acquired in falling down the fourth
part of the parameter belonging to the vertex A ;
for, by the nature of the *parabola*, the parameter
belonging to the vertex A, is equal to $\frac{AEq}{EC}$;
wherefore, the velocity acquired in falling from E
to C, is to the velocity acquired in falling down the
fourth part of the parameter, as the fquare root of
EC to the fquare root of $\frac{\frac{1}{4}AEq}{EC}$, which fquare
roots are to one another, as EC to $\frac{1}{2}$ AE, as may
appear by multiplying each into the fquare root of
EC ; fo that the velocity acquired in falling thro'
a fourth part of the parameter belonging to the
vertex A, and the velocity of the project in the
point A, have one and the fame proportion to the
velocity acquired in falling from E to C ; confequently, from the nature of proportionals, thofe
two velocities muft be equal.

Hence

Hence it follows, that if projects move through the fame or different *parabolas*, the fquares of their velocities in the feveral points of the *parabolas*, are to one another, as the parameters belonging to the refpective points; for, fince the velocities in the feveral points are equal to the velocities acquired in falling down the fourth part of the parameters belonging to thofe points, and fince the fquares of the velocities acquired in falling down the fourth part of the parameters, are to one another as the fpaces defcribed, as I proved in my lecture upon gravity, it is evident that the fquares of the velocities wherewith projects move thro' the feveral points of the *parabolas* which they defcribe, are to one another in the fame proportion with the quarter parts of the parameters belonging to thofe points; but the quarter parts of the parameters being to one another as the whole parameters, the fquares of the velocities in the feveral points of the *parabolas* muft bear the fame proportion to one another, that the parameters do which belong to thofe points.

Since this is the cafe, and fince by the nature of the *parabola* the principal parameter is lefs than any other, and that the other parameters grow larger as the points to which they belong are more diftant from the principal vertex; if a project be caft obliquely upward, as in *Fig.* 4. from A towards E, its velocity muft continually decreafe as it rifes and approaches the uppermoft point P, wherein the velocity being leaft muft thence increafe continually as the project defcends and recedes from the point P; and as in one and the fame *parabola*, where the diftances of any two points as A and K, from the principal vertex P, are equal, the parameters belonging to thofe points are likewife equal; it is manifeft, that a project muft have equal velocities in thofe points; and of confequence, fetting afide any difference which may arife from the refiftance of the air,

Pl. 6. Fig. 4.

air, the project will, *cæteris paribus*, ftrike a mark
as forcibly in the point K as it does at its firft fetting out in the point A.

The velocity wherewith a project is thrown being
given, the velocity thereof in any point of the curve
may be thus determined. In the *parabola* of *Fig. 3.*
let the axis BA be continued upward to D, fo as that
DB may equal the height from which a body muft
fall, in order to acquire the fame velocity wherewith the project fets out from G ; then from any
point in the curve taken at pleafure as K, let the
horizontal line KL be drawn, and the velocity of
the project in the point K, will be to the velocity
wherewith it began its motion from G, as the fquare
root of DL, to the fquare root of DB. For, in
my lecture upon gravity, I proved, that if a body
be thrown directly upward from B towards D, with
the fame velocity that it acquires in falling from
D to B, it will in any point of its afcent as L, have
the fame velocity that it would acquire in falling
from D to that point ; but the velocity acquired in
the defcent from D to L, is to the velocity acquired
in the defcent from D to B (which velocity is by
fuppofition equal to the velocity wherewith the body is thrown up) as the fquare root of DL, to the
fquare root of DB ; and by the eighth *Prop.* of my
laft lecture, the velocity of the project at K, is the fame
with the velocity at L ; confequently, the velocity
thereof at K, is to the velocity wherewith it fet out from
G, as the fquare root of DL, to the fquare root of DB.
Whence it follows, that if DB be equal to 1600
feet, and DL to 400, the velocity of the project at K,
is but one half of the velocity which it had at its fetting out from G ; and if DL be equal to 900 feet,
then is the velocity at K, three fourths of the velocity at G ; fo that a project being thrown obliquely upward with fuch a velocity as would carry it to
the height of 1600 feet if thrown directly upward,

will

will lose a fourth part of its velocity by the time it has risen to the perpendicular height of 700 feet, and one half of its velocity when it has risen 500 feet more.

The velocity wherewith a project is thrown from any given place being given, as also the position of a mark, the directions wherein the project must be thrown, in order to hit the mark, may be determined in the following manner.

Pl. 6.
Fig. 6.

Let A be the place from whence the project is thrown, C the mark situated in the line AC whose length is given, as also the angle CAB, which it makes with the horizontal line AB; at A erect the perpendicular AP, equal to the parameter belonging to the point A, which parameter is given, inasmuch as the velocity wherewith the project is cast from the point A is given; for it is equal to four times the height from which a body must fall in order to acquire that velocity. Let AP be bisected by the line KH, cutting it perpendicularly in G; at A erect AK perpendicular to AC, and let it be continued till it meets KH. From the point of concourse K, with the *radius* KA, let the circle AHP be described. This being done, let a right line as BCEI, be erected perpendicular to the horizontal line AB, so as to pass thro' the mark C, and if possible to cut the circle in two points as E and I; AE and AI are the two directions, in either of which, the project being cast with the given velocity, will hit the mark.

For, drawing the lines PE and PI, the angles CAE and APE are equal, from the nature of the circle; and from the nature of parallel lines, the angles CEA and EAP are equal; consequently, the triangle AEC is similar to the triangle PAE; and therefore, PA is to AE, as AE to EC; wherefore, multiplying the extremes and means, and dividing by EC, PA is equal to $\frac{AEq}{EC}$. In like manner

ner

LECT.
XI.
ner, the triangles PAI and CAI being similar, PA is equal to $\frac{\text{AIq}}{\text{IC}}$. Wherefore, since PA is equal to the parameter at the point A, it follows, from the nature of the *parabola*, that those *parabolas* which the project describes when thrown in the directions AE and AI, must pass thro' the point C; consequently, the mark will be hit by a project thrown in either of those directions.

Whenever the mark is placed at such a distance from A on the line ACM, suppose at M, as that the perpendicular NMH which passes thro' the mark, becomes a tangent to the circle at H, the mark is then at the utmost limit on the line AM, to which a project thrown with the given velocity can reach, and there is but one direction, to wit AH, wherewith the mark can be hit; for it is evident, that any other direction must terminate in some point of the circumference above or below the point H; whence if a perpendicular be let fall to the horizontal line AN, it must of necessity fall on this side of HN with respect to A, and of consequence, cut the line AM in a point less distant from A than is the point M.

The line AH, which denotes the direction of the project, when thrown to the greatest distance possible on the line AM, bisects the angle PAM, which measures the visible distance between the zenith or vertical point P and the mark M. For, by the nature of the circle, the angle MAH is equal to the angle HPA; and forasmuch as in the triangles HPG and HAG, the sides PG and AG are equal by construction, and GH common to both, and the angles at G right ones, the angle HPG is equal to HAG, consequently, HAG or HAP is equal to MAH, that is, the angle PAC is bisected by the line AH.

2

If

If the line ACM be fituated below the horizontal
line AN, which is the cafe when the mark is feated
on a defcent, let all things be conftructed as before, Pl. 6.
and the fame things will obtain; to wit, A I and Fig. 7.
AE, will be the directions neceffary to hit the mark
at C; and the line AH will bifect the angle PAM,
which meafures the apparent diftance between the
zenith and the mark; and the point M will be the
utmoft limit on the line A M, of a project thrown
with the given velocity; the demonftrations of
which are exactly the fame as in the foregoing
cafe.

If the mark be placed on a level, the line ACM
will coincide with the horizontal line ABN, and Pl. 6.
the parameter AP, will pafs thro' the center of the Fig. 8.
circle and become a diameter, the points K and G
coinciding.

In this cafe, the horizontal diftance of the mark,
to wit AC or AB, is as the fine of the doubled
angle of elevation; or in other words, the horizontal range, or the diftance to which a project is
thrown on the plane of the horizon with a given
velocity, is as the fine of the doubled angle of elevation.

For AC, the horizontal range of a project thrown
with a given velocity in the direction AE, is equal
to DE, the fine of the angle AKE; but, from
the nature of the circle, the angle AKE is double
the angle APE, which is equal to CAE, the angle
of elevation; confequently, A C, the horizontal
diftance of the mark, or the diftance to which a
project is thrown on the plane of the horizon with
a given velocity, is as the fine of the doubled angle
of elevation.

Hence it follows, that in order to throw a project with a given velocity, to the greateft diftance
poffible on the plane of the horizon, the direction
of the projection muft be elevated in an angle of 45
degrees; for, fince the fine of twice 45 or 90 de-

N 2   grees

grees is equal to the *radius*, and of confequence, the greateſt of all the ſines; and ſince the horizontal ranges at the ſeveral angles of elevation are to one another, as the ſines of the doubled angles of elevation, it is manifeſt, that the greateſt range, or as it is uſually called by gunners, the greateſt random, muſt be when the project is caſt in a direction whoſe elevation is 45 degrees; moreover, the greateſt random is ever equal to one half the parameter at the point from which the projection is made; for the line AM, which expreſſes the greateſt random, is equal to the *radius* KH, or half the diameter AP, which by the conſtruction, is equal to the parameter belonging to the point A; ſo that where the velocity with which a project is thrown is given, the utmoſt diſtance which that project can reach on the horizontal plane, is likewiſe given; for it is equal to twice the height, from which a heavy body muſt fall in order to acquire the velocity wherewith the project is thrown; the parameter belonging to the point A, having been already proved equal to four times that height.

A ſecond conſequence of the horizontal ranges being as the ſines of the doubled angles of elevation is, that if two projects be thrown with equal velocities, in directions whoſe elevations are equally diſtant from 45 degrees above and below, for inſtance, if the elevation of one be 60 degrees, and that of the other 30, whereof the former exceeds 45 degrees, and the latter falls ſhort thereof by 15 degrees, the horizontal ranges will be equal, or, in other words, the two projects will fall on the plane of the horizon, at the ſame diſtance from the place of projection; for as the ſum of any two arches of a quadrant, whereof one exceeds 45 degrees, as much as the other is exceeded thereby, is equal to a quadrant, it is manifeſt, that two ſuch arches are complements to each other; wherefore, ſince by the nature of the circle, the ſine of a doubled arch is equa.

4

to the fine of its doubled complement, the fines of
the doubled angles of two elevations equally diftant
from 45 degrees above and below, muft be equal;
and fo of confequence muft the horizontal ranges
which are proportional to thofe fines.   And thus it
would conftantly be, were it not for two caufes which
do in fome meafure difturb this law of projects, fo
as to make the horizontal ranges of the higher elevations to fall fhort of thofe of the lower.

The firft of thefe difturbing caufes is the *air*,
which as it refifts, and thereby retards the motions
of projects, muft, *cæteris paribus*, caufe a greater
retardation in thofe motions which are of longeft
continuance; confequently, fince the higher the
elevation of the direction is, the longer is the time
of the project's motion, as fhall be fhewn hereafter;
if the directions wherein two projects are caft with
equal velocities, be equally diftant from 45 degrees,
the one above and the other below, the project which
is thrown in the higher direction, will be more retarded than that which is thrown in the lower; and
of courfe, will fall on the plane of the horizon at
a lefs diftance from the place of projection.

The fecond difturbing caufe, obtains with regard
to fuch projects only as are thrown by the force of
gun-powder.   As the force of the powder acts upon
the ball during its continuance in the barrel, fo does
it likewife to fome diftance beyond the muzzle;
and by fo doing makes the ball to move forward in
a right line, which line is commonly called the *line
of impulfe of fire*; at the end of which, the ball
quitting the blaft of the powder, begins to move in
the curve of a *parabola*.

Now, tho' the air gave no refiftance to projects,
yet muft the horizontal ranges of a ball fhot out of
the fame piece with equal charges, in two directions
equally diftant above and below 45 degrees, be different on account of the different directions of the

line

line of impulse of fire; for, let us suppose a gun
at A, to discharge two equal balls with equal quantities of powder, one in the direction AB, and the
other in the direction AC, AB being as far above
45 degrees, as AC is below it; and let AB and AC
denote the lines of impulse of fire, so that at B and
C the balls will begin to move in *parabolick* curves;
from which points let fall the perpendiculars BD
and CE; it is manifest, that AD, which is the
sine of the complement of the higher elevation,
will denote that part of the horizontal range which
is owing to the line of fire, when the project is
thrown according to the higher elevation; and
AE, the sine of the complement of the lower elevation, will be that part of the horizontal distance,
which is owing to the line of impulse when the
project is thrown according to the lower elevation; consequently, since the sine of the complement of a lesser angle is ever greater than that of
a larger angle, the horizontal range of the lower
elevation must exceed that of the higher, so that
where projects are thrown with the same velocity by
the force of powder, in directions equally distant
above and below 45 degrees, those must range far-
thest which are thrown according to the lower elevations, as well on account of the line of fire, as
of the resistance of the air.

The altitude to which a project rises, is as the
versed sine of the doubled angle of elevation; for
Pl. 6.
Fig. 8.
the proof of which, let AE be the direction of the
projection, and let AC or AB be bisected in T,
and from the point of bisection erect the perpendicular TR; since the point T is equally distant
from A, where the project begins its motion, and
from B, where the motion of the project ceases,
TR will be the axis of the *parabola* which the project describes; and, from the nature of the *parabola*,
will be bisected in V by the principal vertex;
where-,

wherefore, TV will be the height to which the project rifes; but, from the nature of fimilar triangles, fince AT is one half of AC, TR is likewife one half of CE, and confequently, TV one fourth of CE; wherefore, from the nature of proportionals, TV is as CE; but CE is equal to AD the verfed fine of the angle AKE, which by the nature of the circle, is double APE; and APE is likewife, by the nature of the circle, equal to BAE the angle of elevation; wherefore, TV, or the height to which the projeft rifes, is as CE the verfed fine of the doubled angle of elevation; hence it follows, that the greater the elevation is, the higher the projeft will rife, inafmuch as the verfed fines of the doubled angles of elevation increafe continually with the elevation, till at length the elevation becoming perpendicular, the verfed fine of the doubled elevation becomes equal to the diameter, which being the greateft of all the verfed fines, the altitude of the perpendicular projeftion muft likewife be greateft; and it is equal to one fourth of the parameter; for, I fhewed you in my lefture upon gravity, that if a body be thrown up with any velocity, it will rife to the fame height, from whence if it fell from a ftate of reft, it would by the end of the fall acquire the fame velocity wherewith it is thrown up; and in this lefture I proved, that the velocity wherewith a projeft moves in any point of the *parabola*, is equal to the velocity acquired by a heavy body in falling down the fourth part of the parameter belonging to that point; confequently, a projeft thrown up with a given velocity from the point A, will rife to a height equal to the fourth part of the parameter belonging to that point. Hence it appears, that the greateft height of the perpendicular projeftion, is equal to half the greateft random, inafmuch as the greateft random has been proved equal to half the parameter belonging to the point A.

The

  The time of the flight of a project thrown with a given velocity, is as the fine of the angle of elevation : for inftance, the time of the flight of a project thrown with a given velocity in the direction AE, is as the fine of CAE, the angle of elevation ; for fince the project moves thro' the curve of a *parabola* from A to C, by virtue of its uniform motion in the direction AE, and of its accelerated motion in the direction EC, it is evident, that the time of its flight thro' the *parabola*, muft be equal to the time of its uniform motion from A to E ; but as the velocity is given, the time of the motion from A to E ; muft be as the fpace defcribed, that is, as AE ; or by the nature of proportionals, as one half of AE ; but AE being the chord of the arch AE, which meafures AKE, the doubled angle of elevation, one half of AE is equal to the fine of half the arch AE, that is, to the fine of the arch which meafures CAE, the angle of elevation ; confequently, the time of the flight is as the fine of the angle of elevation. Hence it follows, that the greater the elevation is, the longer the time of the flight will be ; as alfo that the time of the perpendicular flight is greateft of all, the fine of the perpendicular elevation being equal to *radius*.

  If the velocity wherewith a project is thrown be required, it may be determined from experiments in the following manner ; by the help of a *pendulum* or any other exact *chronometer*, let the time of the perpendicular flight be taken ; then, forafmuch as the times of the afcent and defcent are equal, the time of the defcent muft be equal to one half of the time of the flight, confequently, that time will be known ; and, forafmuch as a heavy body defcends from a ftate of reft at the rate of 16 feet in the firft fecond of time, and that the fpaces thro' which bodies defcend are as the fquares of the times ; if we fay, as one fecond is to fixteen feet, fo is the fquare

of

of the number of feconds which exprefs the time of
the defcent of the projeƈt, to a fourth proportional,
we fhall have the number of feet thro' which the
projeƈt fell, which being doubled, will give us the
number of feet which the projeƈt would defcribe in
the fame time with that of the fall, fuppofing it
moved with an uniform velocity, equal to that
which it acquired by the end of the fall; which laft
found number of feet, being divided by the number of feconds which exprefs the time of the projeƈt's defcent, will give a quotient, expreffing the
number of feet thro' which the projeƈt would move
in one fecond of time with a velocity equal to that
which it acquired in its defcent, which velocity is
equal to the velocity wherewith the projeƈt was
thrown up; confequently, the velocity wherewith
the projeƈt was thrown up is difcovered. To illuf-
trate this by an inftance, let us fuppofe half the time
of the perpendicular flight to be 8 feconds; then,
as one is to 16, fo is 64, the fquare of 8 feconds, to
1024; which being doubled, and then divided by
8, gives 256 in the quotient; which fhews that the
projeƈt was thrown upward with fuch a velocity as
would carry it, fuppofing it moved uniformly, at the
rate of 256 feet in one fecond of time.

Perhaps it may be objeƈted, that the method here
laid down for difcovering the velocities of projeƈts,
is founded on experiments in which projeƈts are fuppofed to move freely without any let or impediment,
whereas the air refifts and retards all projeƈts in their
motions, fo as not to fuffer them to rife to the fame
height, or to return with the fame velocity, that
they would in cafe they moved *in vacuo*; in anfwer
to which it muft be confeffed, that in the experiments here made ufe of, the air does refift and im-
pede the motions of projeƈts, fo as to fhorten their
afcent, and to leffen the velocity of their return;
but then this does very little affeƈt the truth of
the

the conclufions which are gathered from thefe experiments concerning the velocities, wherewith projects begin their motions; for, as in the method
laid down, the only thing neceffary to be known
from experiment, is the true time of the flight of a
project, fuppofing it to move *in vacuo*; if that
time can be had from thefe experiments, the velocity wherewith the project fets out may be rightly
determined, notwithftanding the refiftance of the
air; but the time of the flight of a project thrown
directly upward, is very nearly the fame *in vacuo*, as
in the air; for, as much as the time of a project's
afcent is fhortened by the refiftance of the air, fo
much very nearly is the time of its defcent length-
ned by the fame refiftance, confequently, the whole
time of the flight in air muft be very nearly equal
to the time of the flight *in vacuo*; and therefore,
the time of the flight *in vacuo* is got, by taking
the time of the flight in air.

Degrees.	Sines.	Verfed fines.
30	50	13
45	70	29
60	86	51
90	100	100
120		148

Experiments

Experiments concerning projects made with a small mortar, the length of whose chase was 5½ inches; the diameter of the ball and chase 3½ inches; weight of the hollow ball 23000 grains; length of the chamber 2 inches, and its diameter ¾ inch.

Quantity of powder in grains.	Degrees of elevation.	Horizontal ranges in feet.	Times of the flights in half seconds.
60	30	135	5
60	45	150	6
60	60	120	8
60	90		11
90	30	200	6½
90	45	220	8
90	60	200	9½
90	90		11
120	30	420	9
120	45	450	11
120	60	300	12½
120	90		16
140	45	660	15
140	90		18
180	30	1000	13
180	45	1100	17
180	60	900	21
180	90		26
240	45	1750	20
240	60	1390	25
240	90		32

LECTURE

LECTURE XII.

Of Hydrostaticks.

IN this lecture I shall give you an account of the gravitation and pressure of WATER, and such other FLUIDS, as are commonly called LIQUIDS.

A *fluid* in general is a body, whose parts yield to any force impressed, and in yielding are easily moved one among another.

The minute particles of fluids do not seem to differ from those of solid bodies ; inasmuch as fluids and solids are frequently converted into one another. Thus water and watery fluids are by cold changed into ice ; which by heat is again reduced to its fluid state. Metals of all kinds being melted become fluid, and upon cooling grow solid again. The most solid and ponderous woods, as also the hardest stones, may by the force of fire in a great measure be converted into water, as is well known to the chymists. And there are not wanting instances in nature of the grossest bodies being turned into the subtile fluids of air and light, and these again into gross bodies. Which changes can scarcely be accounted for, unless we suppose the minute particles of fluids to be of the same nature with those of solid bodies. But be this as it will, most certain it is, that fluids as well as solids consist of heavy particles, whose gravity is ever proportional to the quantity of matter which they contain. This having been found as far as experience reaches to be the universal property of matter, whatever be the form under which it appears. Most indeed of the antient naturalists, not being sensible of any weight or pressure from the air about them, or from the incumbent water when immersed therein, were of opinion, that the parts of one and the same element did

did not gravitate one upon another; which opinion L e c t. has been exploded by the moderns as erroneous; XII. and that it is so, will appear from the following experiment.

Let an empty phial close stopped and immersed Exp. 1. in water, be suspended from one end of a balance and poised; then let the stopple be taken out, that the water may run in, the phial upon receiving the water will preponderate, and bear down the arm of the beam from which it hangs; which evidently proves, that the parts of water retain their gravity in water, so as to press and bear down upon the parts beneath them; otherwise the phial would not become heavier upon the admission of the water.

From the gravity of the parts it follows, that setting aside all external impediments, the surface of a liquid contained in a vessel must be smooth and level; for should any part stand higher than the rest, it must descend by the force of its gravity, and in so doing, spread and diffuse itself till it comes to be on a level with the other parts. As the gravity of the parts reduces the upper surface to a level, so does it likewise occasion a pressure on the lower parts, greater or less in proportion to their depth below the surface, each part sustaining a pressure equal to the weight of all those which lie above it; whence it follows, that the parts which are at equal depths below the surface, are equally pressed, and of consequence must be at rest, contrary to the opinion of those, who make the nature of fluidity to consist in the constant actual motion of the parts one among another. Should this equality of pressure at any time be destroyed, then indeed a motion will arise in the parts of the fluid, and continue till the pressure becomes equal again, as may appear from the following experiment; whereby the truth of what has been said concerning the pressure of the superior parts of fluids on those beneath them, will likewise be confirmed.

Take

Take a glaſs tube open at both ends, and ſtop-
ping one end with a finger, immerge the other in
water to any depth whatever; upon the immerſion,
the water will riſe in the tube, but the height to
which it riſes, whilſt the upper orifice continues ſtopped, will be but ſmall; but upon removing the finger, it will riſe to the ſame height with the water
without *.

When the tube is immerſed, that portion of water which lies beneath the orifice ceaſes to be equally preſſed with the other portions that are at the
ſame depth; for that portion bears no other preſſure than what ariſes from the ſpring of air included in the tube, (which preſſure is equal, as ſhall be
ſhewn hereafter, to the preſſure ariſing from the
weight of the external air) whereas, the other portions do not only bear the preſſure of the air, but
likewiſe the weight of the incumbent water; foraſmuch therefore, as the portion of water which lies
beneath the orifice, is preſſed down leſs forcibly than
the adjacent portions, it muſt give way and riſe in
the tube; but the height to which it riſes, whilſt the
upper orifice of the tube continues ſtopped, can be
but ſmall; becauſe, as the water riſes it compreſſes
the air in the tube, and thereby ſtrengthens its ſpring,
ſo as to make it preſs with greater force; and when
the air is ſo far compreſſed by the riſing water, as
that the force of its ſpring, added to the weight of
the elevated water, makes the ſame preſſure on that
portion of water which lies beneath the orifice, as
the joint weight of the atmoſphere and external
water does on the other portions, which are at the
ſame depth with the former, then the water ceaſes
to riſe. Upon opening the upper orifice of the
tube, by the removal of the finger, the compreſſed
air finding a paſſage thro' that orifice, expands and
dilates itſelf till it becomes of an equal denſity with

---

* The water is tinged of a fine blue purple colour with a
few grains of Sal Armoniack and Copper.

the

the external air; by which means, the preſſure ariſing from the condenſation of the air is taken off, and of conſequence, the water which lies beneath the orifice is leſs preſſed than the adjacent portions, and for that reaſon muſt riſe, and continue ſo to do, till the elevated water in the tube gravitates as forcibly on the water beneath the orifice, as the external water does on the neighbouring portions; but this it cannot poſſibly do, till it comes to be of an equal height with the external water.

Should a lighter liquid be poured on the external water, the water within the tube will riſe yet higher than before; and the height to which it riſes above the ſurface of the external water, will be ſo much leſs than the height of the lighter liquor above the ſame ſurface, by how much the ſpecifick gravity of the water exceeds that of the lighter liquor; for inſtance, if the ſpecifick gravity of the water be to the ſpecifick gravity of the lighter liquor, as two to one, the height of the water in the tube above the level of the external water, will be to the height of the lighter liquid, as one to two; becauſe in that caſe, one part of water makes an equal preſſure with two parts of the lighter liquid. To illuſtrate this by an experiment.

Let oil of terpentine, whoſe ſpecifick gravity is to the ſpecifick gravity of water, as 83 to 100, be poured on the external water to the height of eight inches and an half, and the water will riſe in the tube to the height of 7 inches and $\frac{3}{10}$ above the level of the external water, that is, the heights of the water and oil will be in the reciprocal proportion of their ſpecifick gravities; for 7 and $\frac{3}{10}$ is to 8 and $\frac{1}{2}$, or, which is the ſame thing, 73 is to 85, very nearly, as 83 to 100.

The ſame thing is in like manner confirmed by the following experiment.

Let one end of a ſmall tube open at both ends, be immerſed in mercury contained in a larger tube, and

Exp. 3.

Exp. 4.

and let water be poured upon the mercury in the
larger tube to the height of 34 inches; the mercury will rife in the fmaller tube to the height of two
inches and an half above the level of the mercury
in the larger tube; fo that the height of the mercury in the fmaller tube above the level of the mercury in the larger, will be to the height of the water above the fame level in the reciprocal proportion
of their fpecifick gravities; for $2\frac{1}{2}$ is to 34, as 1 to
$13\frac{3}{5}$, which numbers exprefs the proportion of the
fpecifick gravity of water to that of mercury.

The preffure which the lower parts of a liquid
fuftain from the weight of thofe which lie above
them, exerts itfelf every way in all manner of directions, and that equally; or in other words, whatever be the force wherewith a drop of any liquid is
preffed downward by the weight of the incumbent
liquid, with the very fame force is that drop preffed
upward, as alfo laterally and obliquely, and in a
word, in all kind of directions whatever; otherwife
the drop, which from the nature of fluidity, readily
yields and gives way to any impreffion, muft by
reafon of the preffure from above move out of its
place; but this it cannot poffibly do, becaufe the
drops all around it being equally preffed from above,
do on all fides refift the motion of that drop, with
the fame force that it endeavours to move; confequently, the drop muft continue at reft, and be
preffed on all fides with the fame force that it is
from above; and what has been thus proved of one
drop, is in like manner demonftrable of all the reft;
and therefore, the preffure on the lower parts of a
liquid exerts itfelf equally every way, as will appear from the following experiment.

Exp. 5.　　Let four tubes open at both ends be immerfed in
water to the fame depth, their upper orifices being
firft ftopped, and let the lower orifices be fo fituated, as that the water in entring may move directly
upward in one, and directly downward in another,
obliquely

obliquely in a third, and horizontally in the fourth; upon opening the upper orifices, the water will rife in all of them to the fame height with the external water, as being preffed in the feveral directions with a force equal to the weight of the incumbent water.

From the preffure of liquids upwards it is, that folid bodies fpecifically lighter than liquids, are made to afcend when immerfed therein. For when a folid body is immerfed in a liquid, it preffes that part of the liquid whereon it refts, with a force equal to the weight of a column compofed of the body it-felf, and that portion of liquid which lies upon it; and the water preffes upward againft the body, with a force equal to the weight of a like column of the liquid alone; which force, inafmuch as the liquid is heavier than the folid, muft overcome the force wherewith the body preffes downward, and of con-fequence, the body muft rife with the difference of thofe forces; as fhall be fhewn more fully in my next lecture. If by any contrivance the preffure of the liquid from beneath can be taken off, a body tho' fpecifically lighter will not rife in a liquid, but remain immerfed, as in the following experi-ment.

A brafs plate being joined to one end of a cylin- Exp. 6. drical piece of wood, and another plate of the fame fize and fhape being fixed in water; let the cylin-der be totally immerfed, and let its plate be laid upon the other in fuch a manner, as that no water may get between; the cylinder tho' fpecifically lighter will remain beneath the water, being preffed down by its own weight and that of the incumbent water, whilft the contrary preffure of the water from beneath, is kept off by means of the plate whereon the cylinder refts.

As bodies fpecifically lighter than liquids, are forced up, on account of the preffure from below being greater than the force wherewith the bodies

O   prefs

prefs downward ; fo on the other hand, bodies fpecifically heavier muſt ſink, becauſe the force wherewith they prefs downward exceeds the preſſure from
beneath which oppoſes their deſcent ; and the force
wherewith they deſcend is equal to the difference of
thoſe forces ; as ſhall likewiſe be ſhewn in my next
lecture. If by any contrivance thoſe two forces
can be reduced to an equality, then the bodies will
not deſcend, but remain ſuſpended in the liquid ;
as in the following experiment.

Exp. 7. Let a braſs plate, whoſe ſpecifick gravity is to
that of water, as 9 to 1, be adapted to the neck of
a glaſs veſſel in ſuch a manner, as that being immerſed in water no part of the water may get upon its upper ſurface ; let it then be immerſed to the
depth of nine times its own thickneſs, (that is, to
the depth of 2 inches and $\frac{7}{10}$, the thickneſs of the
plate being $\frac{3}{10}$ of an inch) and it will remain ſuſpended ; but upon pouring ever ſo little water upon
its upper ſurface, it will immediately deſcend and
fall to the bottom.

The plate being immerſed to the depth of nine
times its own thickneſs is preſſed upward by a force
equal to the weight of a column of water, whoſe
height is nine times as great as the thickneſs of the
plate ; which weight, inaſmuch as the ſpecifick gravity of the water is to that of the plate, as 1 to 9,
is equal to the weight of the plate, that is, to the
force wherewith the plate preſſes downward ; for as
none of the water lies on its upper ſurface, it can
prefs downward with no other force than what ariſes
from its own gravity ; conſequently, in this caſe,
the force which reſiſts the deſcent, is equal to the
force which promotes it, and of courſe, the plate
muſt remain in its place. When a little water is
poured on the plate, the weight of that added to the
weight of the plate, overcomes the reſiſting force

Exp. 8. of the water, and cauſes the plate to deſcend. Should
the plate be immerſed to twice the former depth,

it·

it will not defcend tho' loaded with water to the height of nine times its own thicknefs; for as in this cafe, the depth to which it is immerfed is double the former, fo likewife is the force wherewith the water preffes upward; confequently, that force is fufficient to fupport twice the weight of the plate, and therefore will fuftain the plate when loaded with water, to the height of nine times its own thicknefs, fuch a quantity of water being juft equal in weight to the plate.

If by pouring on more water, the force wherewith the plate preffes downward be increafed, or by raifing the plate nearer to the furface of the water, the force wherewith the water preffes upward be diminifhed, the plate will fall to the bottom; and on the other hand, if by immerfing the plate to a greater depth, the preffure of the water upward be increafed, the plate will be thruft upward againft the glafs, and would actually afcend were it not hindered by the glafs.

From what has been faid it follows, that if S be put to denote the number expreffing the fpecifick gravity of the plate, that of water being unity; D be the depth to which the plate is immerfed expreffed in the thicknefs of the plate, and H the height of the water upon the plate expreffed likewife in the thicknefs of the plate; D muft be equal to the fum of S and H in all cafes where the plate remains fufpended; and if there be no water upon the plate, then D and S muft be equal; wherefore, if in the former cafe, D be greater than S and H taken together, or in the latter, than S alone, the plate will afcend if not hindered by the glafs; and on the other hand, if D be lefs, the plate will defcend and fall to the bottom.

The preffure which the bottom of a veffel fuftains from a liquid contained in it, whatever be the fhape of the veffel, is equal to the weight of a pillar of the liquid, whofe bafe is equal to the *area* of

the

the bottom, and whose height is the same with the
perpendicular height of the liquor.

That this is the case in vessels that are equally
wide from top to bottom is plain and obvious; inasmuch as the bottom of every such vessel does actually sustain such a pillar of liquor. But that the
case should be the same in irregular vessels, is not
so easy to conceive; for instance, that in a vessel
which from a large bottom grows narrower as it
rises, so as perhaps at length to be contracted into
a tube, the bottom should bear the same pressure
when the vessel is filled, as it would were the vessel
equally wide throughout from bottom to top, seems
strange and surprizing, and yet it is what necessarily follows from the nature of fluidity; for that part
of the bottom which lies directly beneath the tube,
sustains the weight of a pillar of liquor which reaches
to the top of the tube, the vessel being supposed to
be full, and being pressed with the weight of that
pillar, reacts with an equal pressure on that portion
of the liquor which touches it; and that pressure, inasmuch as it exerts itself equally in the liquor every way, is propagated laterally thro' the
several portions of liquor which are contiguous to
the bottom of the vessel; and forasmuch as this lateral pressure does in like manner exert itself equally every way, the bottom of the vessel must be e-
qually pressed in every point; consequently, since
that portion of it which lies beneath the tube, bears
a pressure equal to the weight of a pillar of liquor,
whose height reaches to the top of the vessel, every
other equal portion must bear a pressure equal to
the same weight; and of course, the whole bottom
must be pressed as forcibly, as if the vessel continued of the same wideness to the top, and was filled with the liquor.

Exp. 9.   To confirm this by an experiment. Let there
Pl. 6.   be two glasses open at both ends, and of such shapes
Fig. 10, as are exhibited in the two figures, whose lower parts
11.

4   MN

MN are cylindrical and equal, and of a capacity juſt ſufficient to admit the braſs plate made uſe of in the laſt experiment; which muſt be fitted to each of them ſucceſſively, in order to conſtitute two veſſels of equal bottoms, but of different capacities; and being ſo fitted, let it be immerſed in water, as in the laſt experiment, to ſuch a depth, as that it will be neceſſary to load it with water in order to make it ſink; that is, let the depth be more than nine times the thickneſs of the plate, which depth muſt be the ſame in both caſes; let then water be poured on the plate, and let it be obſerved what height of water is requiſite to force down the plate when the wider veſſel is made uſe of, and it will be found, that the ſame height will ſuffice in the narrower veſſel; conſequently, the ſmall pillar of water in the narrower veſſel, muſt preſs the plate with a force equal to the weight of a pillar of water of the ſame height, and of a baſe equal to the *area* of the plate; for ſuch a pillar does actually preſs the plate in the larger veſſel, as is evident from the bare inſpection of the figures, and the preſſures made on the plate in both veſſels are equal, inaſmuch as they overcome equal reſiſtances.

From what has been ſaid it appears, that where the baſe of a veſſel is given, the preſſures upon it are as the perpendicular heights of the liquid, whatever be the ſhape of the veſſel. And univerſally, the preſſure on any baſe is meaſured by the product of the *area* of that baſe into the perpendicular height of the liquor above it, without any regard to the quantity of liquor contained in the veſſel; ſo that if we ſuppoſe a hogſhead ſet on one end, and filled with a liquor, and a ſmall pipe to iſſue perpendicularly upward from the other end to any height whatever, and to be filled with the ſame liquor, the bottom will be as ſtrongly preſſed, and be in as much danger of burſting out, as if the hogſhead

was

was continued to the fame height with the pipe, and filled with the liquor.

As the bottom of a veffel bears a preffure proportional to the height of the liquor, fo likewife do thofe parts of the fides which are contiguous to the bottom; becaufe the preffure of fluids is equal every way. And as the preffure which the lower parts of a fluid fuftain from the weight of thofe above them exerts itfelf equally every way, and is likewife proportional to the height of the incumbent fluid, the fides of a veffel muft every where fuftain a preffure proportional to their diftance from the upper furface of the liquor. Whence it follows, that in a veffel full of liquor, the fides bear the greateft ftrefs in thofe parts next the bottom; and that the ftrefs upon the fides decreafes with the increafe of the diftance from the bottom, and in the fame proportion; fo that in veffels of a confiderable height, the lower parts ought to be much ftronger than the upper, that they may be able to withftand the greater preffure.

LECTURE XIII.

OF HYDROSTATICKS.

IN this lecture I fhall explain to you, that part of HYDROSTATICKS which is of ufe in difcovering the *denfities* and *fpecifick gravities* of *bodies*.

The DENSITY of any body is meafured by the proportion which its quantity of matter bears to its bulk. For the more numerous the particles of matter are in proportion to the fpace which they poffefs, the greater is the denfity of the body; and the fewer the particles, the lefs the denfity; wherefore, putting D to denote the denfity of a body,

Q its

Q its quantity of matter, and M its magnitude, $D = \frac{Q}{M}$; and forasmuch as the quantity of matter in any body is ever proportional to, and measured by, the weight, as I shewed in my lecture upon gravity; if instead of the quantity of matter the weight of the body be substituted, and if that weight be denoted by W, then $D = \frac{W}{M}$; that is, the density is as the weight of the body directly, and the magnitude inversly.

By the specifick gravity of a body is meant the gravity peculiar to that species of matter, whereof the body is a part; and it is measured by the proportion of the absolute weight to the bulk; which proportion in one and the same kind of matter, remains unvaried; and in different kinds, as this proportion is greater or less, so is the specifick gravity which is measured by it. Let S denote the specifick gravity of a body, its weight and magnitude being denoted by W and M as before; then, from what has been said, $S = \frac{W}{M}$; and by consequence, since D is likewise $= \frac{W}{M}$, $S = D$; that is, the specifick gravity of a body is as its density. And therefore, by finding out the proportion which the specifick gravities of bodies bear to one another, the proportion of their densities is likewise discovered; for which reason I shall take no farther notice of the densities of bodies, but confine myself to the consideration of their specifick gravities alone.

When a solid body is immersed in a liquid, it presses downward, and endeavours to descend by the force of its gravity; but forasmuch as it cannot descend without moving as much of the liquid out of its place, as is equal to it in bulk, it is manifest that it is resisted, and, as I may say, pressed upward

by

by a force equal to the weight of such a portion of the liquid as is equal to it in bulk; consequently, if the specifick gravity of the solid be greater than that of the liquid, that is, if the solid weighs more than an equal bulk of the liquid, the body will descend with a force equal to the excess of its gravity above the gravity of the liquid; on the other hand, if the gravity of the liquid exceeds that of the solid, the body being as it were pressed upward by a force greater than that whereby it endeavours to go down, will ascend with the difference of those forces, that is, with a force equal to the excess of the specifick gravity of the liquid above that of the solid. When the specifick gravities are equal, the body will neither rise nor fall, but remain suspended at any depth; being pressed as strongly upward by the resisting force of the liquid, as it is downward by its own weight. Hence it follows, that if by any contrivance the specifick gravity of a solid can be varied, so as to be one while greater, another less, and then equal to the specifick gravity of a liquid wherein it is immersed, the body will sink, or rise, or remain suspended according to the variation of its specifick gravity. And this is the case in that ludicrous experiment of the little glass images in water, which are made to descend, or rise, or remain suspended at pleasure; the reason of which I shall explain to you, after you have seen the experiment.

Exp. I.   The images being set to float on the water, the top of the vessel must be covered with a bladder closely bound about the neck of the vessel, to the end that the air, which lies upon the surface of the water, may not force its way out when it is condensed by the hand pressing on the bladder. The images themselves tho' lighter, are yet nearly of the same specifick gravity with the water, and being hollow, are full of air, which by means of small

holes

holes in their heels communicates with the air without. When the air which lies beneath the bladder
is preſſed by the hand, it preſſes on the ſurface of
the water ; and foraſmuch as the preſſure is propagated thro' all the water, thoſe portions which are
contiguous to the heels of the images, are thereby
forced into the holes, by which means the air within is condenſed, and at the ſame time, the weight of
the images is increaſed by the additional weight of
the influent water. And when ſo much water is
forced in, as to render the ſpecifick gravity of the
images greater than that of the water, the images
deſcend and fall to the bottom ; where they remain
as long as the preſſure above continues, but when
that is taken off by the removal of the hand, the
condenſed air in the images dilates and expands itſelf, and in ſo doing drives out the water ; upon
which account the images become ſpecifically lighter than the water, and of courſe aſcend. As the
preſſure on the bladder is greater or leſs, ſo muſt the
quantity of water which is forced into the images ;
and therefore, whenever it happens that during the
aſcent or deſcent of an image, ſuch a preſſure is
made as ſuffices to force in juſt as much water as is
requiſite to reduce the image to the ſame ſpecifick
gravity with the water, the image ſtops and remains
ſuſpended, upon increaſing the preſſure it deſcends,
and aſcends if the ſame be leſſened. Some of the
images begin to deſcend ſooner, as alſo to riſe later,
than others, for one or both of theſe reaſons ; firſt,
becauſe ſome are ſpecifically heavier than others ;
and, ſecondly, becauſe the cavities in the legs are
greater in ſome images in proportion to their magnitudes, than they are in others ; upon both which
accounts, a leſs preſſure is requiſite to make ſome
deſcend, and to keep them down, than what is neceſſary to produce the ſame effects in others. For,
firſt, let us ſuppoſe the ſpecifick gravities of two
images

images to be different, but the cavities in their legs, when taken of a given height, to be proportional to their respective magnitudes; since the air is equally denfe in both images, it is manifest, that it gives the fame oppofition in both to the influent water; confequently, the water when forced in by the prefure from above, muft rife to equal heights in the cavities of both; fince therefore the cavities whofe heights are equal, are fuppofed to be proportional to the magnitudes of the images, it is manifest, that the quantities of water contained in thofe cavities muft be fo too; confequently, each image receives an addition of weight from the influent water proportional to its magnitude; or in other words, the fpecifick gravities of the two images are equally augmented; forafmuch therefore as one of the images is fuppofed to be fpecifically heavier than the other, it is evident, that when the fpecifick gravity of the former has received fuch an addition, from the influent water, as makes it a little exceed the fpecifick gravity of the water, the fpecifick gravity of the latter muft fall fhort thereof; confequently, the former muft fink, and leave the other above.

Secondly, Let us fuppofe the fpecifick gravities of the two images to be equal; but let one image be lefs in proportion to the cavity in its legs, than the other is in proportion to its cavity, the height of the cavities being given; since the water does from the fame preffure rife to an equal height in both, it is plain, from what I juft now faid, that the former muft receive a greater quantity of water in proportion to its magnitude, and confequently, a greater addition to its fpecifick gravity than the latter, and of courfe muft defcend fooner.

From what has been faid it follows, that if the proportion which the cavity in the legs bears to the magnitude of the image be given, the greater
the

the fpecifick gravity of the image is, the more apt it will be to defcend; confequently, in this cafe the aptitude or promptnefs of an image to defcend is proportional to, and may be expreffed by, the fpecifick gravity. In like manner, if the fpecifick gravity be given, the greater the proportion is which the cavity in the leg bears to the magnitude of the image, the more apt the image is to defcend; and therefore in this cafe, the aptitude is proportional to, and may be expreffed by, the cavity applied to the magnitude of the image. But if neither the fpecifick gravity of the image, nor the proportion of the cavity to the magnitude of the image, be given, the aptitude of an image to defcend, is as the fpecifick gravity into the cavity applied to the magnitude of the image; that is, putting A to denote the aptitude, S the fpecifick gravity of the image, C the cavity in the leg, (the height whereof is always fuppofed to be given) and M the magnitude of the image; $A = \frac{SC}{M}$; or, fubftituting the abfolute weight of the image applied to its magnitude, in the room of the fpecifick gravity, $A = \frac{WC}{M^2}$; that is, the aptitude an image has to defcend, is as the weight of the image into the cavity of the leg directly, and the fquare of the image's magnitude inverfly.

A folid fpecifically heavier than a liquid, being immerfed therein, lofes as much of its weight as is the weight of a portion of the liquid equal to it in bulk; for it has been already fhewn, that a folid is carried down in a liquid by the excefs only of its gravity, above the gravity of a portion of the liquid equal to it in bulk; confequently, the other part of its gravity is loft, as to any effect it has on the body itfelf; as will appear from the following experiment.

Let

Exp. 2.   Let a fmall cylinder of brafs, fufpended at one end of a balance and counterpoifed, be immerfed in water; upon the immerfion it will become lighter, fuppofe by 200 grains, which is the weight of as much water as is equal in bulk to the cylinder; for a cylindrical veffel, juft large enough to contain the cylinder, being hung at one end of a balance and poifed, and then filled with water, preponderates with the weight of 200 grains.

Since a folid when immerfed in a liquid, lofes as much of its weight, as is equal to the weight of a portion of the liquid of the fame dimenfions with the folid, it follows, that all bodies whatever, whofe magnitudes are equal, however different their fpecifick gravities may be, do fuffer an equal lofs of weight in the fame liquid. Thus a cylinder of Exp. 3. block-tin, equal in dimenfions to the brafs cylinder, but fpecifically lighter, being immerfed in water, lofes 200 grains, as did that of brafs.

Tho' a folid lofes part of its weight when immerfed in a liquid, yet it muft not be imagined that the weight fo loft by the folid, is actually deftroyed, but that it is imparted to the liquid, the liquid conftantly gaining in weight what the folid lofes. For Exp. 4. if the veffel with the water wherein the cylinders were immerfed, be put into a fcale and poifed; upon the immerfion of either cylinder, it will preponderate with the weight of 200 grains, which is what the cylinder lofes.

Solids equal in weight, but of different fpecifick gravities, being immerfed in the fame liquid, fuffer loffes of weight reciprocally proportional to their fpecifick gravities; for as the lofs of weight which any body fuffers in a liquid, is equal to the weight of as much of the liquid as is equal in bulk to the folid, the lofs fuftained is ever proportional to the magnitude of the body; whatever proportion therefore the magnitudes of bodies have to one another, the fame will the loffes of weight have which they

fuffer;

suffer; but the magnitudes of bodies equal in weight,
but of different specifick gravities, are to one another in the reciprocal proportion of their specifick
gravities; consequently, so are the losses of weight
which they suffer. Which is confirmed by the following experiment.

Let two cones, one of lead, the other of tin,
whose specifick gravities are to one another, as 112
to 74, and the weight of each 400 grains, be immersed in water, after the manner of the cylinders;
upon the immersion, the lead will lose $35\frac{1}{4}$ grains,
and the tin 54; but $35\frac{1}{2}$ is to 54, as 74 to 112,
that is, reciprocally as the specifick gravities of the
metals.

From the losses of weight being reicprocally proportional to the specifick gravities, it follows, that
if two bodies of different specifick gravities, which
balance each other in air, be immersed in water or
any other liquor, the *æquilibrium* will be destroyed,
and that which has the greatest specifick gravity will
descend; as will appear, by hanging the cones, one
at each end of a balance, and then immersing them
in water, for the lead will preponderate.

The specifick gravity of a solid specifically heavier than a liquid, is to the specifick gravity of the
liquid, as the absolute weight of the solid, to the
loss of weight which it suffers in the liquid; for the
specifick gravities of bodies being as the absolute
weights applied to the magnitudes, where the
magnitudes are equal, the specifick gravities are directly as the absolute weights; if therefore we compare the solid with a quantity of the liquid equal to
it in magnitude, their specifick gravities must be as
their weights; but the absolute weight of such a
quantity of the liquid, is equal to the loss of weight
sustained by the solid; consequently, the specifick
gravity of the solid, is to that of the liquid, as the
whole weight of the solid, to the loss which it sustains in the liquid.

Hence

Hence we have a method of difcovering the fpecifick gravities of fuch folid bodies as are heavier
than water; I mean, of difcovering the proportions
of their fpecifick gravities to the fpecifick gravity
of water. For if we fuppofe the fpecifick gravity of
water to be unity, and put L to denote the lofs of
weight which any body, whofe fpecifick gravity we
look for, fuftains in water, and W its whole weight,
then $L : W :: 1 : \dfrac{W}{L}$; confequently, $\dfrac{W}{L}$ expreffes the fpecifick gravity of the folid, that of water
being unity; and therefore, in order to know the
fpecifick gravity of any folid heavier than water,
nothing more is requifite, but to difcover the quantities denoted by W and L, and to divide the firft
by the laft; the firft is had, by taking the weight
of the body in air, and the laft, by taking the weight
in water, and fubducting it from the weight in air;
for the remainder is the lofs of weight, which dividing the weight in air, gives a quotient expreffing

Exp. 7.
the fpecifick gravity of the body. To apply this
to a particular cafe, let it be propofed to difcover
the fpecifick gravity of a piece of tin, which being
weighed in air, is found to be 300 grains, and in
water, $259\frac{1}{2}$, which being fubducted from the former, leaves $40\frac{1}{2}$ for the lofs of weight; fo that in
this cafe, W denotes 300, and L $40\frac{1}{2}$; and therefore, dividing 300 by $40\frac{1}{2}$, we fhall have $7\frac{4}{10}$ for
the fpecifick gravity of tin, that of water being
unity. Whence it appears, that tin, bulk for bulk,
is more weighty than water, in the proportion of
74 to 10.

If the body, whofe fpecifick gravity is required,
be lighter than water; then, forafmuch as its gravity is not fufficient to caufe a total immerfion, the
lofs of weight which it fuffers in water cannot be
found out by weighing it alone in that liquid; let
it therefore be joined to fome other body fo weighty,
that the compound may fink; but firft let the lofs

of

of weight which the heavier body alone fuſtains in
water be found out, as before; and then let the loſs
of weight which the compound fuſtains be likewife
difcovered, whence deducting the loſs of weight
fuſtained by the heavier, the remainder will exhibit
the loſs fuſtained by the lighter; confequently, dividing the weight of the lighter by that remainder,
the quotient will expreſs the ſpecifick gravity required; that is, putting W for the weight of the
body whoſe ſpecifick gravity is fought, L for the
loſs of weight fuſtained by the compound, and
l for the loſs fuſtained by the heavier body $\dfrac{W}{L.-l}$
expreſſes the ſpecifick gravity of the body. To
apply this to a particular cafe; let the weight of a Exp. 8.
piece of wood ſpecifically lighter than water be
220 grains, and let it be joined to a piece of tin of
160 grains, whoſe loſs in water is found to be 17
grains; then the compound being weighed in water, will be found to lofe 334 grains; fo that in
this cafe, W is equal to 220 grains, L to 334,
and l to 17; and L lefs l, is equal to 317 grains.
And therefore, dividing 220 by 317, we ſhall
have $\frac{694}{1000}$ for the ſpecifick gravity of the wood,
that of water being unity. So that that kind of
wood is bulk for bulk lighter than water, in the
proportion of 694 to 1000.

If the body whoſe ſpecifick gravity is fought be
diſſolvable in water, then inſtead of water, let fome
other liquor be made ufe of, which will not diffolve
the body; and let the proportion of the ſpecifick
gravity of the body to the ſpecifick gravity of that
liquor, be difcovered by the foregoing method; as
alfo the proportion of the ſpecifick gravity of that
liquor to the ſpecifick gravity of water, by the method which ſhall be ſhewn prefently. Then in whatever proportion the ſpecifick gravity of the liquor exceeds or falls ſhort of the ſpecifick gravity of water,

in

in the same proportion let the specifick gravity of
the body with regard to that of the liquor be increased or diminished, and it will give the specifick
gravity of the body with respect to that of water;
that is, if we put A for unity or the specifick gravity of water, B for the specifick gravity of the
other liquor, and C for the specifick gravity of the
body with regard to that liquor; then by saying, as
A is to B, so C to a fourth proportional, we shall
have $\frac{BC}{A}$ for the specifick gravity of the body with
respect to that of water; or rejecting the divisor as
being equal to unity, and putting S for the specifick
gravity of the body with respect to water, we shall
Exp. 9. have $S = BC$. To apply this, let the specifick gravity of Roman-vitriol be required; let the weight
of a piece in air be 67 grains, and in spirit of wine
41 grains; consequently, its loss of weight in the
spirit is 26 grains, which dividing 67, gives
2.576 for the specifick gravity of the vitriol with
regard to the specifick gravity of the spirit, which
in this case is supposed to be unity; but the specifick gravity of the spirit with regard to that of water is less than unity, being only $\frac{87}{100}$, as shall be
shewn presently; wherefore B is $= 0.87$, and C
is $= 2.576$; consequently 2.24, which is the product arising from the multiplication of those two
numbers, expresses the specifick gravity of Roman-
vitriol with respect to that of water, which is as
unity; and therefore, in whole numbers, the specifick gravity of Roman-vitriol exceeds that of
water, in the proportion of 224 to 100.

The specifick gravities of liquors are discovered
by taking the losses of weight sustained by one and
the same solid in the several liquors; for since the
loss of weight in each liquor, is equal to the weight
of as much of the liquor as is equal in bulk to the
body; by taking the losses of weight sustained by
the

the fame body in the feveral liquors, we get the
abfolute weights of fuch portions of thofe liquors
as are equal in bulk ; and by confequence, the fpecifick gravities of the liquors, the fpecifick gravities of bodies equal in bulk, being to one another
as their abfolute weights ; wherefore, putting L
for the lofs of weight which a body fuftains in water, and little l for the lofs of weight fuftained by
the fame body in any other liquor ; then, by faying, as L to l, fo is unity to a fourth term, we
fhall have $\frac{l}{L}$ for the fpecifick gravity of the other
liquor, that of water being unity ; fo that to difcover the fpecifick gravity of any liquor, we have
nothing more to do, but to weigh one and the
fame folid, both in the liquor whofe quantity is
fought, and in water, and to divide the lofs of
weight which the folid fuffers in the liquor, by the
lofs which it fuftains in water ; for the quotient will
exprefs the fpecifick gravity of the liquor.   Thus,
a glafs bubble, whofe weight in air is 1727 grains, Exp. 10.
being weighed in water, is found to lofe 641 grains,
and 558 in fpirit of wine ; and therefore, dividing
558 by 641, we fhall have a quotient of 0.87 for
the fpecifick gravity of the fpirit, that of water being unity.

When a body fpecifically lighter than a liquid,
is fet to float upon it, the part immerfed is equal in
bulk to a portion of the liquid whofe weight is equal
to the weight of the whole body ; for fince the body finks in part, by moving fome of the liquor
out of its place, and fince the weight of the body
is the power which moves the liquor, the body
muft continue to fink, till it has removed as much
of the liquor as is equal to it in weight ; confequently, the part immerfed muft be equal in magnitude to fuch a portion of the liquor, as is equal
in weight to the whole body ; which is abundantly
confirmed by the following experiment.

A ball

A ball of pear-tree, a wood fpecifically lighter than water, being fet to float on water contained in a glafs veffel, let the veffel be placed in a fcale and counterpoifed ; then, taking out the ball, let the veffel be filled up with water to the fame height at which it ftood when the ball was in it, and the fame weight will counterpoife it as before.

From the veffel's being filled up to the fame height at which the water ftood when the ball was in, it is manifeft, that the quantity poured in is equal in magnitude to that part of the ball which was im‑ merfed ; and, from the fame weight counterpoifing, it is evident, that the water poured in, is equal in weight to the whole ball.

The part immerfed is to the whole body, as the fpecifick gravity of the body to the fpecifick gravi‑ ty of the liquid ; for the fpecifick gravities of two bodies, being to one another as their abfolute weights applied to their magnitudes, if their weights be equal, their magnitudes are in the reciprocal *ratio* of their fpecifick gravities ; fince therefore, fuch a portion of the liquid as is equal in magni‑ tude to the immerfed part of the folid, is likewife equal in weight to the whole folid ; the magnitude of the immerfed part is to the magnitude of the whole body, as the fpecifick gravity of the folid to the fpecifick gravity of the liquid.

When the fame body is fet to float fucceffively in different liquors, the parts immerfed are to one an‑ other in the reciprocal proportion of the fpecifick gravities of the liquors. For the body defcends in each liquor, till the part immerfed takes up the room of as much liquor as is equal in weight to the whole body ; and therefore, fuch portions of the feveral liquors as are equal in magnitude to the immerfed parts of the body have all equal weights ; but the magnitudes of bodies equal in weight, are to one another reciprocally, as their fpecifick gra‑ vities ; confequently, in one and the fame body floating in different liquors, the parts immersed are reciprocally as the specifick gravities of the liquors. On this principle is founded the HYDROMETER; which is an hollow glass ball, with a small hollow stem of about 5 or 6 inches in length, opposite to which, on the other side of the ball, adheres a smaller ball filled in part with mercury, or some other weighty body, to the intent, that when the ball is set to float in water, or any other liquor, the stem may be kept uppermost, and in a position perpendicular to the surface of the liquor; and at the same time, that the machine may be so far immersed, as that the stem only, or some part thereof, may remain above the liquor: the stem being graduated from top to bottom, has numbers annexed to every degree, expressing the magnitudes of the parts which lie below the several degrees.

The use of this little machine is to discover the specifick gravities of liquors, which is done in the following manner. The *hydrometer* being first set to float in water, the degree to which it sinks must be observed, and the number thereto annexed; then being set to float in any other liquor, the degree to which it sinks, with the number annexed, must likewise be noted; for as this number is to the former, so is the specifick gravity of water, to that of the other liquor, as is evident from what was just now said. To illustrate this in the case of water E<small>xp.</small> 12. and spirit of wine. The *hydrometer* being dropt into water, sinks to the degree whose number annexed is 87; and being dropt into spirit of wine, sinks to the degree whose number is 100; whence it appears, that the specifick gravity of water is to that of spirit of wine, as 100 to 87.

Tho' *hydrometers* may be useful in discovering the specifick gravities of liquors for loose and inaccurate computations, yet are they not to be depended on in cases where great exactness is required,

and

and that for two reasons; First, becaufe it is extreamly difficult to graduate the ftems fo exactly,
as that the numbers annexed fhall truly exprefs the
magnitudes of the parts below them. Secondly,
becaufe, partly from the motion of the *hydrometer*
in the liquor, and partly from the rifing of the liquor about the ftem from the attractive force of the
glafs, it is hardly poffible to determine with exactnefs the degree to which the *hydrometer* finks.
Upon both which accounts, as alfo becaufe the
method of determining the fpecifick gravities of
liquors by means of the glafs bubble is much more
eafy and exact, this method by the *hydrometer* is
intirely laid afide.

L E C T U R E XIV.

OF HYDROSTATICKS.

L E C T.
XIV. IN this lecture I fhall give you an account of the
flux of water from RESERVOIRS thro' *orifices*
and *pipes*.

If water, flowing out at an orifice in the bottom
of a veffel, be kept conftantly at the fame height in
the veffel, by being fupplied as faft above as it
runs out below, the velocity wherewith it flows out,
is as the fquare root of the height of the water above
the orifice.

For if we fuppofe the column of water which
ftands directly above the orifice, to be divided into
an indefinite number of plates of an equal but exceedingly fmall thicknefs, it is manifeft, that whatever be the force of gravity, wherewith the uppermoft plate preffes upon the fecond, the fecond preffes upon the third with a double force, and the third
upon the fourth with a triple force, and fo on; fo
that the plate which is next the orifice is preffed
downward

downward by the joint gravities of the feveral plates
which lie above it, and likewife by the force of its
own gravity, inafmuch as there is no other plate
beneath it whereon to reft; confequently, from its
own gravity, and that of the feveral plates above it,
it does all at once receive as many equal impreffions
from gravity, as it would fucceffively in falling
down the height of the water; and of courfe, muft
pafs thro' the orifice, with the fame velocity that it
would acquire in falling down that height; but I
proved in my lecture upon gravity, that the velocity which a body acquires in falling thro' any fpace,
is as the fquare root of the fpace; confequently,
the velocity wherewith the water flows out, is as the
fquare root of the height of the water above the
orifice.

To confirm this by an experiment; let there be
two veffels in all things alike, excepting that one is
four times as tall as the other, the height of one
being 20 inches, and of the other 5; let each of
them have a circular orifice in the bottom, a fifth
part of an inch in diameter; and being both filled
with water, let them be fet a running, and let the
water be fupplied as faft above as it runs out below;
the taller veffel will difcharge about twenty one
ounces in the fpace of a quarter of a minute, and
in the fame time the fhorter will difcharge about
11 ounces. Now, forafmuch as the orifices thro'
which the water flows are equal, and likewife the
times of the flux, the quantities difcharged are as
th evelocities; confequently, the velocity wherewith
the water flows out of the taller veffel, is to the
velocity wherewith it flows out of the fhorter, as
21 to 11, that is, nearly as 2 to 1, which are the
fquare roots of the heights of the water above the
orifices.

As the preffure fuftained by the lower parts of
water from the weight of thofe above, exerts it-

P 3   felf

LECT.
XIV.
self with the same force laterally that it does downward, it matters not whether the orifice through
which the water flows, be at the bottom or side of
a vessel; for the water will flow out of both with
the same velocity, provided they are at equal
depths below the upper surface of the water; and
therefore, the velocity of water flowing out of an
orifice in the side of a vessel, is as the square root
of the height of the water above the orifice; as

Exp. 2. will appear, by repeating the last experiment with
vessels whose orifices are in their sides; for the
quantities discharged will be the same as before.

From what has been said it follows, that if an
orifice in the side of a vessel be situated as far above
an horizontal plane, as it is below the upper surface
of the water, the water will spout from that orifice,
to the distance of twice the height of the orifice

Pl. 6. above the plane. For instance, if AOBC be a ves-
Fig. 12. sel full of water, O an orifice in the side, whose
height OD above the horizontal plane DH, is equal
to OA, the distance of the orifice from the top of the
water; DH the horizontal distance to which the
water spouts, will be double of OD, the height of
the orifice above the plane. For the spouting water has two motions, one uniform from the pressure
of the water in the vessel, in the direction OF perpendicular to the orifice, the other accelerated from
the force of gravity in the direction OD perpendicular to DH; which two motions do by no means
hinder one another, but by their combination cause
the water to spout in the curve of a *parabola*. Now,
the velocity wherewith the water moves in the direction OF, being equal to the velocity acquired by
a body in falling from A to O, or from O to D;
in the same time that it falls from O to D, and by
so doing, reaches the horizontal plane, it will be
carried in the direction OF, thro' a space equal to
twice OD, (inasmuch as all bodies whatever that
move

move uniformly, with a velocity equal to that which
is acquired by a body in falling thro' any height,
do in the fame time with that of the fall, defcribe
a fpace double of that of the fall) ; confequently,
the horizontal diftance to which the water fpouts,
will be equal to twice the height of the orifice above
the plane. Thus, from an orifice in the fide of a
veffel, the depth whereof below the furface of the
water is 20 inches, the water will fpout to the diftance of 38 inches on an horizontal plane, whofe
diftance below the orifice is likewife 20 inches ; and
where the depth of the orifice below the top of the
water is 5 inches, the water will fpout to the diftance of $9\frac{1}{2}$ inches on an horizontàl plane fituated at
the diftance of 5 inches below the orifice ; fo that
in both cafes the diftances to which the water fpouts,
are nearly double the diftances of the planes below
the orifices ; and they would be exactly double,
were it not that the water is retarded a little by the
oppofition it meets with from the air.

The diftances to which water fpouts on an horizontal plane, from orifices in the fides of different
veffels, the orifices being at equal heights above the
plane, are to one another as the fquare roots of the
heights of the water above the orifices.

For fince the orifices are at equal heights above
the plane, the times of the defcent of the water
from the feveral orifices to the plane muft be equal ;
confequently, the horizontal diftances to which the
water fpouts, muft be as the velocities wherewith it
fpouts ; but thofe velocities are as the fquare roots
of the heights of the water above the orifice ; confequently, fo muft the horizontal diftances. Thus,
if two veffels be fo placed, as that the orifices in
their fides fhall be 20 inches above an horizontal
plane, the height of the water in one veffel being
20 inches above the orifice, and in the other 5 ; the
water will fpout from the former, to the diftance

of

of 38 inches, and from the latter, to the diftance of 19 inches.; but 38 is to 19, as 2 to 1; that is, as the fquare roots of the heights of the water above the orifices, for the heights are as 4 and 1.

The diftance to which water fpouts from an orifice in the fide of a veffel, whatever be the height of the orifice above the plane, as alfo of the water above the orifice, may be thus determined; let
Pl. 6.
Fig. 13.
BR reprefent an horizontal plane, F an orifice in the fide of a veffel at any height above the plane, and AB the height of the upper furface of the water above the plane. On AB as a diameter, defcribe the femicircle ADB, and at F fet off FE perpendicular to AB, and meeting the circle in E. The diftance to which the water fpouts on the plane BR from the orifice F, is proportional to the line FE.

For, from the nature of motion, the fpace defcribed, is as a rectangle under the time and velocity; but in this cafe, the time of the motion is as the fquare root of FB, and the velocity wherewith the water fpouts, is as the fquare root of AF; confequently, the fpace thro' which the water runs in the horizontal direction, is as the fquare root of the rectangle AFB; but, by the nature of the circle, the fquare root of the rectangle AFB is equal to FE; confequently, the horizontal diftance to which the water fpouts on the plane BR from the orifice F, is as FE.

Hence it follows, that the diftance to which the water fpouts, is as the fine of the arch AE, whofe verfed fine AF is equal to the height of the water above the orifice. And, forafmuch as any two fines, which are equally diftant from the center, are equal, it follows that the water muft fpout to the fame diftance from two orifices as F and L, whofe diftances from the center are equal; as alfo, that it muft fpout to the greateft diftance from an orifice

in

in the center, the fine C D being in that cafe equal to *radius*, and confequently the greateft.

To confirm what has been faid ; let a veffel whofe height is 16 inches, and which is perforated in the middle, and likewife at the diftance of $5\frac{1}{2}$ inches above and below the middle, be filled with water, and fet upon an horizontal plane ; the water will fpout from the middle orifice to the diftance of above 15 inches, and from each of the other two, to the diftance of about 10 inches.

All things being fuppofed as before, the diftances to which the water fpouts, fetting afide what little difturbance may arife from the refiftance of the air, are equal to twice the fines of the arches, whofe verfed fines are equal to the heights of the water above the orifices. For, the diftance to which the water fpouts from the central orifice C, is to the diftance to which it fpouts from any other orifice as F, as the fine C D is to the fine F E ; but forafmuch as the orifice C is as far diftant above the plane as it is below the furface of the water, the diftance to which the water fpouts from that orifice is equal to twice C B, or twice C D ; confequently, the diftance to which it fpouts from F muft likewife be equal to twice F E, and fo of any other orifice.

Water which fpouts perpendicularly upward fets out with fuch a velocity, as is fufficient to carry it to the fame height with the water in the veffel from which it fpouts. For the velocity wherewith it fets out, is equal to the velocity acquired in falling down the height of the water ; and, in my lecture upon gravity, I fhewed, that a body thrown directly upward rifes to fuch a height, whence if it be let fall, it will by the end of the fall acquire the fame velocity wherewith it was thrown up ; confequently, the water fpouts with a velocity fufficient to carry it to an equal height with the water in the refervoir ; however, it cannot poffibly arrive at that height,

by

by reafon of the refiftance it meets with from the air ; which, as it cannot be taken off, muft leffen the heights of all jets whatever, fo as to make them fall fhort of the heights in the refervoirs ; befides, the water in the uppermoft part of the jet, when it has loft all its motion, refts for fome time on the part next below it, and by its weight obftructs and retards the motion of the whole column, and there-by leffens its height ; and fo great is the refiftance arifing from this caufe, as that the jet is frequently deftroyed by it, the rifing water being by fits and ftarts preffed down to the very orifice from which it fpouts.

Exp. 7. 　 By giving the jet a little inclination, the upper-moft parts, when they have loft their motion up-ward, are made to fall off from the reft, whereby the refiftance which arifes from their weight is taken off. And this is the true reafon why, *cæ-teris paribus*, fuch jets as are a little inclined, rife higher than thofe whofe afcents are perpendi-cular.

The velocity wherewith water flows out of a cy-lindrical pipe inferted horizontally into the fide of a veffel, is as the fquare root of the height of the water in the veffel above the place of the pipe's in-fertion directly; and the fquare root of the length of the pipe inverfly ; for fince the pipe is cylin-drical, the velocity wherewith the water flows out at one end, muft be equal to the velocity where-with it flows in at the other ; but the velocity wherewith it flows in, is in the proportion laid down ; for the preffure of the incumbent water in the veffel, cannot make the water which lies next the orifice flow into the pipe, unlefs at the fame time it drives forward all the water contained in the pipe ; for which reafon, the water in the pipe may be looked upon as an obftacle which refifts and impedes the moving caufe. Now, where a caufe

acts

acts under the difadvantage of a clog or impediment, the potency of fuch a caufe is increafed, either
by diminifhing the impediment, or augmenting the
abfolute ftrength and vigour of the caufe itfelf;
where the ftrength and vigour of the caufe is given,
the potency thereof increafes in proportion as the
impediment leffens, and leffens as that increafes;
and where the impediment is given, the potency of
the caufe increafes, and leffens in proportion to the
increafe and diminution of the abfolute ftrength and
vigour of the caufe; confequently, the potency is
in a *ratio* compounded of the ftrength or magnitude
of the caufe, and of the weaknefs or fmallnefs of
the impediment; that is, it is as the magnitude of
the caufe directly, and as the magnitude of the impediment inverfly; or as the magnitude of the caufe
applied to the magnitude of the impediment. Now,
in the cafe before us, where the preffure of the water in the refervoir is the moving caufe, and the water in the pipe is the impediment, the magnitude
of the former is meafured by a rectangle under the
height of the water, and the orifice of the pipe,
and the magnitude of the latter by a rectangle under the orifice of the pipe, and the length thereof;
or rejecting the orifice as being ever the fame in
both, the magnitude of the moving caufe, is as the
height of the water, and that of the impediment,
as the length of the pipe; and therefore, putting
H for the height of the water in the refervoir above
the place of the pipe's infertion, and L for the length
of the pipe; $\frac{H}{L}$ will denote the preffure of the water in the refervoir, as leffened by the refiftance of
the water in the pipe; and putting O for the orifice of the pipe, $\frac{HO}{L}$ will exprefs the force which
drives the water into the pipe; and forafmuch as
the motion generated in any time by a force acting
conftantly

conſtantly and uniformly, is as a rectangle under the force and time; putting T for the time that the water continues to flow into the pipe, $\dfrac{HOT}{L}$ will be as the motion generated in the water flowing into the pipe; but the motion generated in the influent water, is as the quantity which flows in, multiplied into the velocity wherewith it flows; and therefore, putting Q and V for the quantity and velocity, $\dfrac{HOT}{L}$ is as Q V; or, becauſe the quantity which flows in, is in a *ratio* compounded of the orifice, time, and velocity; by ſubſtituting O, T, V, which denote the orifice, time, and velocity, in the place of Q, we ſhall have $\dfrac{HOT}{L} = OTV^2$; and ſtriking out OT from both ſides, we ſhall have $\dfrac{H}{L} = V^2$; conſequently, V is as $\sqrt{\dfrac{H}{L}}$; that is, the velocity wherewith the water flows out of the reſervoir into the pipe, and conſequently, the velocity wherewith it flows out of the pipe, is as the ſquare root of the height of the water in the reſervoir, applied to the ſquare root of the length of the pipe.

Hence it follows, that if the length of the pipe be varied whilſt the height of the water in the reſervoir continues the ſame, the quantities diſcharged in any given time, will be to one another inverſly as the ſquare roots of the lengths of the pipe; for ſince the diameter of the pipe, and the time of the flux are given, the quantities diſcharged muſt be as the velocities wherewith they run out, that is, in the inverſe *ratio* of the ſquare roots of the lengths of the pipe.

To confirm this by an experiment; let a pipe of 16 feet in length, and half an inch in diameter, be inſerted horizontally into the ſide of a veſſel; and

let

let the water in the veſſel be kept conſtantly at the LECT. height of 3 feet above the place of the pipe's in- XIV. ſertion; the pipe when ſet a running will diſcharge above 161½ ounces in half a minute; let it then be made ſhorter by 12 feet, and ſet a running again, and it will in the ſame ſpace of time diſcharge 321 ounces, that is, near twice as much as before; ſo that the quantities diſcharged, will be to one ano- ther reciprocally as the ſquare roots of the lengths of the pipe, which in this caſe are as 4 and 1.

TABLE I.

L	Q	T
1	436½	
4	321	¾
9	211¾	1¾
16	161½	3
25	132	5
36	87	9
49	72	14
64	65	20½
81	61½	29
100	59	42

TABLE II.

D	Q
1	87¼
4	88½
9	88
16	81
25	74
36	67¼
49	65
64	58¾
81	56
100	54

June the 21ft, 1722, I made feveral experiments concerning the motion and difcharge of water thro' pipes, in the following manner.

There was a refervoir of 3 feet in height, which was kept conftantly full during the flux of the water; at the bottom was inferted horizontally a pipe of half an inch in diameter, whofe length when greateft was 100 feet, but being compofed of feveral pieces, was capable of being made of ten different lengths; which lengths were the fquares of the natural numbers. Into this pipe were inferted horizontally (as occafion was) ten other pipes, each of them 6 inches long, and $\frac{1}{4}$ inch in diameter; the places of their infertion into the main pipe were diftant from the refervoir the fquares of the natural numbers in feet; the axes of the fmall pipes made an angle of 80 degrees, with that of the main pipe; the reafon why they were inferted in fuch an angle was, that it had been obferved that the water flowed out of orifices made in the main pipe nearly in that angle.

In TAB. I. L denotes the length of the main pipe (the fmall pipes not being inferted), Q the quantity in *troy* ounces difcharged in half a minute of time, T the time in feconds which the water took to flow from the refervoir to the extremity of the pipe, the fame having been firft exhaufted.

In TAB. II. D denotes the diftance from the refervoir, at which the fmall pipe was inferted into the main pipe; Q the quantity in *troy* ounces difcharged by the fmall pipe in half a minute of time, the main pipe being ftopped.

TABLE

TABLE III.

1	2	3	4	5	6	7	8	9	10	P	Sum.
84¾										63¼	148
	80									59	139
		74								54½	128½
			65½							49¾	115¼
				56¼						48	104¼
					45¾					45	90¾
						41½				42	83½
							31¼			43	74¼
								26		44½	70½
									6½	59½	66
67¼	57	40	27	19½	9½	5	5¼	4	1¾	8	244¼
69¼	63½	56½	28	17½	7¼	5½	5	5	7		264½
69¾	63½	54½	29½	12½	12¾	10¼	8¾	8			269½
68	62	50	24½	27	17¼	9¾	9				267½
69¼	62½	50¾	24¾	28½	18¾	16½					271
69¼	63½	51½	25	31	27¼						267½
69¼	64	55	32	43							263¼
72¾	69	65	43								249¾
76½	75½	75½									227½
82¾	84										166¾
87¼											87¼
	76	50½	36½	27½	13¼	6	5¼	6½	¾	3½	225¾
		55	42	30½	13¼	6½	5¼	7	2	4½	166
			49	32	14¾	15¼	6¾	9	2¾	5¾	135¼
				40	30¼	17¼	8¾	9¼	3	6¼	114¾
					39	21	10½	10¼	3½	8½	92¾
						37½	16¼	11	6¾	12½	84
							25	16½	7	21	69½
								24	8¾	30½	63¼
									9¼	53	62¼
										59	59

In TAB. III. the numbers at the top denote the ten ſmall pipes, P the main pipe, and the numbers below denote the quantities in *troy* ounces diſcharged in half a minute of time, by the pipes denoted by the numbers directly above them. The blanks denote, that the pipes denoted by the numbers directly above them at the top, were ſtopped at the time that the others diſcharged.

LECTURE

LECTURE XV.

OF PNEUMATICKS.

IN this lecture I shall give an account of the weight and pressure of the air, and of some remarkable effects arising from it.

Tho' the weight of the air which surrounds us is not perceived, by reason of the equal pressure which it makes on all parts of our bodies; yet that it is really heavy appears from hence, that vessels when exhausted are less ponderous than when filled with air. Exp. 1. Thus a glass bottle, whose contents are nearly 40 cubick inches, being exhausted by means of the air pump, will be found to suffer a sensible loss of weight; when I formerly made the experiment, the loss of weight amounted to ten grains, and the magnitude of the exhausted air I found to be 34 cubick inches; for upon immersing the bottle in water, and opening the valve which covered the mouth, the quantity of water which flowed in and possessed the place of the exhausted air, amounted to 8628 grains, which being divided by $253\frac{1}{3}$, the number of grains in a cubick inch of water, give 34 in the quotient; so that from this experiment it is manifest, that 34 cubick inches of that air, which more immediately surrounds us, are equal in weight to ten grains; and that the specifick gravity of the same air is to the specifick gravity of water, as ten to 8628, or, as one to $862\frac{4}{5}$; the specifick gravities of bodies equal in bulk, being to one another as the absolute weights of the bodies.

As the air rises above the surface of the earth, it grows rarer, and consequently lighter; a given bulk of air, being lighter at the distance of a mile than at the earth's surface, and lighter again at the distance

diſtance of two miles, and ſo on continually. And
yet notwithſtanding this diminution of gravity in
the ſuperior parts of air, ſo great is the height of the
atmoſphere, as to render the weight of the whole
very conſiderable ; as will appear from the following experiment.

Let a piece of common glaſs be placed as a cover
on the top of the receiver ; and upon exhauſting the
air, the glaſs will at firſt be preſſed cloſe to the receiver, and at length broken by the weight of the
air, which reſts upon it.

While the air continues undiminiſhed in the receiver, it does by virtue of its elaſticity preſs as
ſtrongly againſt the lower ſurface of the glaſs, as
does the incumbent air by means of its weight upon
the upper ſurface ; as ſhall be ſhewn hereafter ;
conſequently, as long as the air remains undiminiſhed in the receiver, the weight of the incumbent
air can have no ſenſible effect on the glaſs ; but upon leſſening the quantity, and therewith the ſpring
of the included air, the glaſs being no longer ſupported from below, is preſſed down, and broken
by the weight of the air above ; and for the ſame
reaſon, a ſquare glaſs phial when exhauſted cracks
and flies to pieces.

From the weight and preſſure of the air on the
ſurface of liquors it is, that they are made to riſe
in exhauſted tubes open at one end, as will appear
from the following experiments.

Let a glaſs veſſel with mercury be placed under a
receiver, and let a tube open at one end be ſuſpended above the veſſel in ſuch a manner, as that the
open end may at pleaſure be let down into the mercury ; if then, the air being drawn out of the receiver, the tube be let down, the mercury will not
riſe therein as long as the receiver continues empty ;
but upon readmitting the air, it will immediately
aſcend. The reaſon of which is, that upon ex-

Q   hauſting

hausting the receiver, the tube is likewise emptied of air; and therefore, when it is immersed in the mercury, and the air readmitted into the receiver, all parts of the mercury are pressed upon by the air, except that portion which lies beneath the orifice of the tube; consequently, it must rise in the tube, and continue so to do, until the weight of the elevated mercury presses as forcibly on that portion which lies beneath the tube, as the weight of the air does on every other equal portion without the tube. But to proceed to a second experiment of the same kind.

Pl. 6.
Fig. 14.
Exp. 4.
Let two glass tubes as A and B, each above 30 inches long, of which A is open at one end only, but B at both, be so contrived, as by means of screws to be let into the little glass vessel C D, in the manner represented in the figure. A being filled with mercury, and then screwed into the vessel, let mercury be poured into B, till both that and the vessel are full; let then the vessel be inverted; Pl. 6.
Fig. 15. and let the extremity of B be immersed in a vessel of mercury, the mercury will descend thro' B, and continue so to do, till A is emptied, as also so much of the vessel C D as is above the level of the upper orifice of B. This being done, let A be so far unscrewed, as to permit the air to pass between the threads of the screw into the empty part of the vessel; upon the admission of the air, the mercury will rise in the tube A. For, from the circumstances of the experiment it is evident, that the part of A which stands above the level of the mercury remaining in the vessel, is perfectly void of air; consequently, while the mercury all around the tube is pressed by the newly admitted air, that portion which lies beneath the tube suffers no pressure from above; and of course must rise, and continue to rise, until the weight of the elevated mercury becomes a balance to the pressure of the air without.

By

By the weight and preſſure of the air, water is raiſed in common pumps, and fire engines, as will appear by conſidering their ſtructures, and the manner in which they work. AB repreſents the body Pl. 7. Fig. 1. of a pump, which is commonly an hollow cylinder of wood or lead, C a plug fixed near the bottom of the pump, with an hole in the middle, covered by a leathern valve, ſo contrived as to open and give way to the water in paſſing upward, but to ſhut cloſe and obſtruct the paſſage downward; D a ſecond plug of the ſame kind, and perforated in like manner with the former. This plug is commonly called the *ſucker* or *piſton*, and being moveable, is drawn up and thruſt down at pleaſure, by means of the iron rod E to which it is faſtened. The ſides of the ſucker are every where caſed with leather, whereby it is made to fit the cavity of the pump ſo exactly, that neither air nor water can paſs between. At ſome diſtance above the ſucker is an orifice as O in the ſide of the pump, thro' which the water is diſcharged at the time of working, in the following manner. The ſucker being drawn up, the ſpace between that and the lower plug is left void of air; then foraſmuch as the water, which ſtands about the pump, is every where preſſed by the air, except in that part which anſwers to the hole of the plug, it muſt there give way, and paſs up into the cavity of the pump; and upon depreſſing the ſucker again, as it cannot return downward by reaſon of the valve, which ſhuts cloſe upon the hole, and ſtops the paſſage, it riſes up thro' the ſucker, and lodges itſelf thereon; ſo that upon the next elevation of the ſucker, it is carried towards the top of the pump, and thrown out at the orifice O.

If inſtead of an orifice above the ſucker, we ſuppoſe one juſt above the lower plug, with a valve opening outwardly, ſo as to ſuffer the water to flow out, but not to return. And if we ſuppoſe the

Q 2 ſucker

LECT. fucker to be folid without a perforation, the figure
 XV. will reprefent a forcing pump, or fire engine, in
which the water rifes above the lower plug in the
fame manner, and from the fame caufe, that it does
in a common pump; and by the preffure made upon it by the fucker when thruft down, it is forced
out at the orifice, and that fo ftrongly, as by the
help of leathern pipes to be conveyed to the tops
of the higheft houfes.

The air in any particular place does not always
continue of the fame weight, but is fometimes heavier, and fometimes lighter; which plainly argues
a variation in the quantity, inafmuch as the gravity
of any body is proportional to the quantity of matter which it contains. From what caufe this variation arifes, is not eafy to determine. Dr. HALLEY
is of opinion, that the diminution of the quantity of
air in any place, is the effect of two contrary winds
blowing from that place, whereby the air is carried
both ways from it; and of confequence, the incumbent cylinder of air is diminifhed; as for inftance, if in the *German* ocean it fhould blow a gale
of *wefterly* wind, and at the fame time an *eafterly*
wind in the *Irifh* fea; or if in *France* it fhould blow
a *foutherly* wind, and in *Scotland* a *northern*; that part
of the atmofphere which is impendent over *England*
would, he thinks, be thereby carried off and diminifhed. He likewife conceives, that the increafe of
the quantity of air in any place, is occafioned by the
blowing of two contrary winds towards that place,
whereby the air of other places is brought thither
and accumulated. And upon this foot, he endeavours to account for what is commonly obferved in
this part of the world; namely, that the atmofphere,
cæteris paribus, is always heavieft upon an *eafterly*
or *north-eafterly* wind. This happens, fays he, becaufe, that in the great *Atlantick* ocean, on this fide
the thirty fifth degree of *north* latitude, the *wefterly*
and *fouth-wefterly* winds blow almoft always; fo
that

that whenever the wind comes up here at *eaft* or *north-eaft*, it is fure to be checked by a contrary gale, as foon as it reaches the ocean ; for which reafon, the air over us muft needs be heaped up in greater abundance, as often as thofe winds blow. To confirm this hypothefis of contrary winds being the caufe of the variation in the weight of the air, he obferves, that within the *Tropicks*, where there are no contrary currents of air, this variation does not obtain ; but that the atmofphere continues much in the fame ftate of gravity in all kinds of weather. Now, whether this, or whatever elfe be the caufe of it, moft certain it is, that the weight of the air does vary ; and fo confiderable is the variation, that the weight of the air in its heavieft ftate, exceeds the weight thereof when it is lighteft, in the proportion of almoft ten to nine.

The changes which the air undergoes as to its gravity, are obferved by means of the *Barometer* or weather-glafs ; which, as it was the invention of TORRICELLIUS, is known among the naturalifts by the name of the *Torricellian tube or inftrument.* It confifts of a fmall glafs-tube, about three feet long, clofed at one end, which being filled with mercury well purged from air, is inverted into a cylindrical box of timber, wherein fome mercury is lodged ; upon the inverfion fome of the mercury falls out, whereby the upper part of the tube is left empty whilft the lower part continues full. Now, forafmuch as it has appeared from experiments, that the fufpenfion of the mercury in the tube is owing to the preffure of the air on the ftagnant mercury ; the pillar of mercury which is kept up in the tube, muft always be equal in weight to a pillar of the atmofphere of the fame thicknefs ; confequently, as the weight of the atmofphere varies, the height of the mercury in the barometer muft do fo too ; the mercury conftantly rifing as the weight of the air increafes, and finking as that leffens.

leſſens. That the minute variations in the height of the mercury may be obſerved, that part of the tube which lies between the limits of the leaſt and greateſt height, to wit, from 28 to 31 inches, is graduated ; each inch being divided into ten or twelve equal parts by means of a table, whereunto the tube is fixed ; whereon likewiſe are inſcribed in their proper places ſuch conſtitutions of the air and weather, as have been obſerved to accompany different heights of the mercury. In contriving this inſtrument, care muſt be taken to make the box, which contains the ſtagnant mercury, ſo large, as that upon the riſing or falling of the mercury in the tube, the height of that in the box may ſuffer little or no variation ; for ſhould the ſtagnant mercury ſink upon the riſing of the mercury in the tube, or riſe as that ſinks, which muſt be the caſe where the box is ſmall ; the riſe or fall of the mercury in the tube will appear to be leſs than it really is ; as for inſtance, if when the mercury riſes half an inch in the tube, it does at the ſame time fall a quarter in the box, the riſe in the tube, which appears to be only half an inch, is in truth three quarters ; becauſe the height of the mercury is always to be computed from the ſurface of that in the box. So, on the other hand, if the mercury by falling half an inch in the tube riſes a quarter in the box, the true deſcent in the tube is three quarters of an inch ; inaſmuch as the height of the mercury in the tube above the ſurface of the ſtagnant mercury in the box, is leſs after the fall by three quarters of an inch. By making the circular *area* of the box thirty or forty times greater than that of the tube, (which is generally the caſe, the tubes of moſt barometers being but one fifth of an inch wide, and the boxes an inch and a quarter) the ſtagnant mercury in the box may be kept conſtantly at the ſame height very nearly, the greateſt variation of the height not amounting to more than the tenth or

twelfth

twelfth part of an inch, which is inconfiderable.

If the tube inftead of being continued directly
upward, be bent at the height of 28 inches, in the Pl. 7.
manner here reprefented, it is then called an *inflected* Fig. 2.
or *diagonal barometer*; in which the inclined part
A B may conftitute an obtufe angle of any magnitude with the perpendicular part BC ; but the nearer the angle approaches to a right one, the longer
muft the inclined part be ; for it muft be continued
until the perpendicular altitude thereof AH, above
the horizontal line HB, becomes equal to three
inches, which is the difference between the greateft
and leaft height of the mercury in the barometer ;
otherwife, the mercury will not have room to rife to
its utmoft height, at fuch times as the conftitution
of the air requires it. This barometer fhews the
minute variations in the weight of the air much
more accurately than the former ; becaufe the rife
or fall of the mercury in the inclined part AB is
very fenfible, when an alteration in the perpendicular height is fcarcely to be perceived. But then the
box which contains the ftagnant mercury, ought to
be much larger in proportion in this than in the
former ; becaufe in this, a much larger quantity of
mercury rifes into, and falls out of the tube, upon
the changes of the weather.

If the lower part of the tube in the firft barometer, inftead of being inferted into a box, be turned
up in the form of a crook, it is then called a *curved
barometer*, in which the crooked part generally ter-
minates in a large bubble open at top. The bubble Pl. 7.
contains the ftagnant mercury, which, as it is pref- Fig. 4.
fed upon more or lefs by the incumbent air, is forced
up to a greater or fmaller height in the ftrait part
of the tube. In this barometer the bubble ought
to be fo large in proportion to the tube, as that
upon the greateft variation of the height of the
mercury in the tube, the height thereof in the bubble may not vary above one tenth of an inch ; the

Q 4   neceffity

neceſſity there is for this, is evident from what was ſaid concerning the magnitude of the box in the firſt kind of barometer.

Beſides the barometers hitherto mentioned, there is the *wheel*, as alſo the *pendant* or *conical, barometer*, and others of various kinds ; which, however they may differ as to their ſtructures, do all agree in ſhewing the changes in the weight of the air, by the riſing and falling of the mercury in their tubes ; wherein it ſometimes, tho' very rarely, deſcends as low as twenty eight inches ; and at others riſes to thirty one ; the mean height thereof being twenty nine inches and an half. So that a pillar of the atmoſphere, in the mean ſtate of its gravity, is equal in weight to a pillar of mercury of the ſame thickneſs, and whoſe altitude is twenty nine inches and an half. Whence it follows, that an inch ſquare of the earth's ſurface, or of any other body contiguous thereto, ſuſtains a preſſure from the incumbent atmoſphere, when in the mean ſtate of its gravity, equal to ſeventeen pounds, eight ounces, and 374 grains ; that being the weight of a ſquare pillar of mercury one inch thick, and twenty nine and an half high.

From this great preſſure of the air it is, that two brazen hemiſpheres, whoſe diameter is three inches and an half, being laid one upon another, and then exhauſted, cling ſo faſt together, as to require above 150 pounds to ſeparate and draw them aſunder. And it muſt be obſerved, that as the globe in this experiment cannot be perfectly exhauſted, that ſmall portion of air which remains within, by expanding itſelf, contributes to the ſeparation of the hemiſpheres ; for which reaſon, they are drawn aſunder by a leſs weight than that wherewith the air preſſes them together ; for the diameter of the ſphere being three inches and an half, the *area* of its greateſt circle is nine ſquare inches and three fifths nearly ; conſequently, the weight of that pillar of

air

air which preffes the hemifpheres together, is not
lefs than 162 pounds, even in its lighteft ftate,
when the mercury in the barometer ftands at the
height of 28 inches only. If the globe, after it
has been exhaufted, be hung within a receiver, upon drawing the air out of the receiver, the lower
hemifphere will fall off from the other; which
plainly fhews, that their cohefion is owing to nothing elfe but the weight and preffure of the air
upon them.

Since the atmofphere even in its lighteft ftate is
fo ponderous, as that a fquare pillar of it one inch
thick weighs fixteen pounds, nine ounces, and 461
grains; it follows, that a middle fized man, the
furface of whofe body is generally allowed to contain about fifteen fquare feet, fuftains a preffure from
the atmofphere, when in its lighteft ftate, equal to
the weight of 31144 pounds; which preffure on
larger bodies, and in heavier ftates of the air, is ftill
greater; and therefore it may well be afked, how
it comes to pafs, that we are not fenfible of this
preffure, great as it is. In anfwer to which it muft
be obferved, that fuch preffures only are perceived
by us, as do in fome meafure move our fibres, and
put them out of their natural fituation. Now, the
preffure of the air being equal on all parts of the
body, cannot poffibly move the fibres of any one
part, or force them from their fituation; but on
the contrary, muft by reafon of its uniformity
keep all the fibres in their proper places, and as fo
doing, cannot be perceived. And that this is the
cafe is evident from hence, that if the preffure of the
air be taken off from one part of the body, the
preffure on the neighbouring parts immediately becomes fenfible. Thus, if a man covers the top of
an open receiver with his hand, upon exhaufting
the receiver, and thereby taking off the preffure of
the air from the palm of the hand, he will perceive
a weight

a weight on the back of his hand, and that fo great, as to put him to pain, and almoft endanger the breaking of his hand.

LECTURE XVI.

OF PNEUMATICKS.

LECT.
XVI.

BY the elafticity of the air, whereof I intend to treat in this lecture, we are to underftand that force wherewith the particles of air expand themfelves, and recede from each other, whenever the preffure from without, which keeps them together, is taken off. The method which I fhall obferve in treating of this force is, Firft, to fhew from experiments, that the air is really induced with fuch a force ; and, Secondly, to enquire into its nature and laws.

Exp. 1.

As to the firft, if a little warmed ale, or any other liquor fomewhat glutinous, be put into a glafs and included in a receiver, upon exhaufting the receiver the liquor will rife in large frothy bubbles, and run over the glafs.

As the liquor is glutinous, it retains a great number of airy particles, which upon the removal of the outward air, and therewith the preffure which it makes on the liquor, dilate and expand themfelves ; and forafmuch as they cannot readily extricate themfelves from the liquor by reafon of its clamminefs, they raife it up, and carry it over in the form of froth. And for the fame reafon it feems to be, that meath, cyder, and moft other domeftick wines, after they have been bottled a while, upon drawing the cork, fpurt out and fly. For as they are all in fome meafure glutinous, they retain a good quantity of air ; which upon corking the bottle is condenfed

denfed by reafon of the condenfation of the air
which is lodged in the neck of the bottle; befides,
by the flight fermentation which fuch liquors commonly undergo in the bottle, a frefh fupply of air
is generated, equal in denfity to the former. When
therefore upon drawing the cork, the extraordinary
preffure arifing from the condenfed air in the neck
of the bottle is taken off, the air which is difperfed
thro' the liquor expands itfelf with great force,
and not finding a ready paffage between the parts
of the liquor, which by reafon of their clamminefs
do not eafily feparate, drives the liquor before it
in the manner of a fpout. But to proceed;

The expanfive force of the air is likewife evident
from the following experiment, Let a glafs bottle
of a globular form, and containing a fmall quantity of water, have a fmall glafs tube open at both
ends, inferted into it fo far, as that the lower end
may be below the furface of the water; and let the
infertion be made by means of a fcrew and a collar
of leathers, in fuch a manner as that no air may
pafs into or out of the bottle; let then the whole
apparatus be placed under a tall receiver, and upon
exhaufting the air out of the receiver, the water will
rife up thro' the tube in the form of a jet, which
will be higher or lower in proportion as the receiver
is more or lefs exhaufted; the reafon of which is,
that the air included in the bottle, by endeavouring
to expand itfelf, preffes upon the furface of the
water, which therefore muft rife in the tube, as foon
as the preffure of the outward air which keeps it
down is leffened; and the greater the diminution
of that external preffure is, the higher the water
muft be thrown.

Exp. 2.

If a bladder wherein a fmall quantity of air is
included, be placed under a receiver, upon drawing
the air out of the receiver, the bladder will fwell,
and the fwelling will be greater or lefs in proportion as the receiver is more or lefs emptied; which
plainly

Exp. 3.

plainly argues an expansive force in the included air; as does likewise the bursting of a full blown bladder in an exhausted receiver; as also the cracking of a square glass phial when close stopped.

If a small siphon, having a weight fastened from the hand of the piston, and being closed at the end so as that upon drawing up the piston no air can get in, be suspended in an inverted position with the weight downward, and then covered with a receiver; upon drawing part of the air out of the receiver, the weight will descend, and draw down the piston; and upon the readmission of the air it will rise again.

When part of the air is drawn out of the receiver, that portion which remains within expands itself, whereby its spring is so far weakened, as not to be able to stand against and support the weight, for which reason the weight descends; whereas, upon the return of the air which was carried off, the elastick force is so far increased, as to become an overbalance for the weight, and upon that account drives it up.

From this and the foregoing experiments it fully appears, that the air is indued with an expansive force. Whence that force arises, and what the law of its action is, comes now to be considered.

The naturalists were formerly of opinion, that the elasticity of the air was owing to the shape and figure of its parts: for they supposed each particle of air to consist of several branches, which being of a pliable nature, were capable of being compressed and squeezed together by any outward force, and of expanding and spreading themselves abroad upon the removal of the compressing force; and this has been thought by some to be a full and satisfactory account. But that great philosopher Sir ISAAC NEWTON was of opinion, that the expansive force of the air is altogether inexplicable on the foot of

this,

this, or indeed any other hypothefis, except that LECT.
of the air's being indued with a repelling power, XVI.
whereby the particles recede and fly from each
other ; his words are thefe.

" That there is a repulfive virtue, feems alfo to
" follow from the produ&tion of air and vapour.
" The particles when fhaken off from bodies by
" heat or fermentation, fo foon as they are beyond
" the reach of the attra&tion of the body, reced-
" ing from it, and alfo from one another with
" great ftrength, and keeping at a diftance, fo as
" fometimes to take up above a million of times
" more fpace than they did before in the form
" of a denfe body ; which vaft contra&tion and
" expanfion feems unintelligible, by feigning the
" particles of air to be fpringy and ramous, or
" rolled up like hoops, or by any other means
" than a repulfive power."

Now, fuppofing this to be the cafe, and that the
repelling power of each particle exerts itfelf on the
next adjacent particles only, as Sir ISAAC feemed
to imagine, I fhall fhew you what the law of this
repelling power is, or, in other words, how this
power is varied, by varying the diftance of the particles ; and in order thereto, fhall lay down the
following PROPOSITION.

PROP. *If a fluid be compofed of particles endued
with a repulfive power, fo as that each particle repels
thofe, and thofe only, which are next it, and if the
force wherewith two adjacent particles repel each other,
be in a given reciprocal ratio of the interval of their
centers ; that is, putting* I *for the interval of the centers, and* P *for the index of the given power of that
interval ; I fay, if two adjacent particles repel each
other with a force that is as* $\frac{I}{I^P}$, *the force which compreffes the fluid, is as the cubick root of that power of
the*

the denfity of the fluid, whofe index is P *increafed by* 2, *or* P + 2 ; *that is, putting* F *for the compreffing force, and* D *for the denfity of the fluid,* F *is as* D *raifed up to the power whofe index is* $\dfrac{P + 2}{3}$.

Exp. 7.

Pl. 7.
Fig. 5.

For the proof of this, let a portion of the fluid be contained in a given cubick fpace, whofe upper furface is denoted by the fquare ABCG, the compreffing force being applied to that furface.

The elaftick force of the fluid, which withftands the compreffing force, and is exactly equal thereto, is the force of thofe parts only which compofe the upper furface ; becaufe the repelling forces of the particles are fuppofed to exert themfelves on thofe particles only which lie next them, and not to extend to particles more remote. But the force of the fuperficial parts is as the number of particles in the furface, and the force wherewith any two adjacent particles repel each other conjointly. Now, the number of particles in the given fquare furface, is reciprocally as the fquare of the diftance of the centers of two adjacent parts ; that is, as $\dfrac{1}{I^2}$; and by fuppofition, the force wherewith two particles repel each other, is as $\dfrac{1}{I^F}$; and therefore, the elaftick force of the fluid, and of confequence the compreffive force, or F, is as $\dfrac{1}{I^{P+2}}$. The denfity of the fluid contained in the given cubical fpace, is inverfly as the cube of the diftance between the centers of the particles ; that is, D is as, $\dfrac{1}{I^3}$, and I is as $\dfrac{1}{D^{\frac{1}{3}}}$; and therefore, by fubftituting $\dfrac{1}{D^{\frac{1}{3}}}$ in the room of I, F is as D $\dfrac{P+2}{3}$; that is, the compreffive

five force is as the cube root of that power of the
denfity of the fluid, whofe index is P + 2.

COROL. From this propofition it follows, that
if the denfity of an elaftick fluid be as the force
which compreffes it, the particles repel one another
with forces that are inverfly as the diftances of their
centers.

For fince F is as D, $\dfrac{P + 2}{3}$ is equal to unity, and
fo likewife is P; confequently, the P power of I,
whofe reciprocal expreffes the repulfive force of the
particles, is equal to I.

Hence the particles of air muft repel one another
with forces reciprocally proportional to the diftances
of their centers, becaufe the denfity of the air is
proportional to the force which compreffes it; as
will appear from the following experiment.

Let an inflexed tube as AB, open at both ends,
be filled up with mercury to fome fmall height,
fuppofe DC; then ftopping the end B, fo as that
the air may not get out when it is compreffed, and
meafuring the length of BC, that part of the fhorter leg that is filled with air, which air, it is evident,
is compreffed by the weight of the atmofphere;
let mercury be poured in at A, till the height thereof in the longer leg above the height of the fame in
the fhorter, becomes equal to the height at which
it ftands in the barometer, by which means the air
in the fhorter leg will be compreffed by twice the
weight of the atmofphere; let then the length of
that part of the leg which is poffeffed by the air under this double preffure be meafured, and it will be
found to be juft one half of BC; whence it appears,
that the fpaces which a given quantity of air pof-
feffes under different preffures, are reciprocally proportional to the preffures; and confequently, inafmuch as the denfities of bodies where the quantity
of matter is given are reciprocally as their magnitudes,

tudes, the denſity of the air is directly as the compreſſing force. From this property of the air, COTES
has deduced a method for determining the denſity
thereof at any height; what he has delivered concerning this matter, is contained in the 5th chapter
of his *Harmonia Menſurarum*, which I ſhall endeavour to explain to you; and in order thereto, ſhall
lay before you ſuch properties of the logarithmick
curve, as I ſhall have occaſion to make uſe of, referring you for their demonſtrations to the forementioned author, and others who have wrote of that

Pl. 7.
Fig. 7. curve. Let then BDGI be a logarithmick curve,
AH its aſymptot, that is, a right line ſo ſituated
with reſpect to the curve, as not to meet it till it is
drawn to an infinite, or rather indefinite length,
BA, DC, and GF, ordinates, that is, right lines
perpendicular to the aſymptot at the points A, C,
and F, and terminating in the curve. BC a tangent to the curve at the point B. The properties
of this curve, which I ſhall have occaſion to mention, are theſe four.

Firſt, Any portion of the aſymptot intercepted
between two ordinates, is the logarithm or meaſure
of the *ratio* which thoſe ordinates bear one to the
other; thus AC meaſures the *ratio* of BA to DC;
and CF meaſures the *ratio* of DC to GF; and
ſo likewiſe, AF meaſures the *ratio* of BA to GF.
And if AC, AF, and AH be in arithmetick proportion, then DC, GF, and IH are in geometrick
proportion; and if any portion of the aſymptot
be a given quantity, then is the *ratio* of the two
ordinates, which intercept that portion, likewiſe
given.

Secondly, That portion of the aſymptot as AC,
which is intercepted between a tangent and an ordinate, drawn to the ſame point of the curve as B,
is a given quantity, or in other words, to whatever point of the curve the tangent and ordinate are
drawn,

2

drawn, the portion of the afymptot which they intercept is always of one and the fame length. The portion fo intercepted is called the *fubtangent*, and it is the module, or that which regulates the magnitudes of all the logarithms in the fame fyftem ; for they are greater or lefs in proportion to the magnitude of the fubtangent; fo that if in two logarithmick curves, the fubtangent of one be double or triple the fubtangent of the other, the meafures of the fame *ratios* are likewife twice or thrice as great in the former as they are in the latter.

Thirdly, The indefinite *areas* comprehended between the curve and the afymptot, drawn on to an indefinite length beyond HI, are to one another as the ordinates which bound them in their wideft parts ; thus, the indefinite *areas* BAHI, DCHI, and GFHI, are to one another as the ordinates BA, DC, and GF.

Fourthly, The indefinite *area* BAHI, is equal to the parallelogram BACE, comprehended under the ordinate BA, and the fubtangent AC.

Thefe things being premifed, let AB reprefent the earth's furface, and let AH be a line perpendicular thereto ; then, forafmuch as the denfities of the air at different heights, are as the preffures of the incumbent atmofphere, and the ordinates in the curve, as the indefinite *areas* which lie beyond them ; if the indefinite *area* BAHI be made to denote the weight or preffure of all the air, and AB its denfity at the furface of the earth, then, by the third property of the curve, the indefinite *area* DCHI will denote the weight or preffure of all the air which lies above C, and the ordinate DC will denote the denfity of the air at that height ; and thus it is with regard to any other height, fo that at all heights, the denfities of the air will be denoted by the refpective ordinates ; wherefore, by the firft property of the curve, the difference between

R tween

tween any two heights, is the meafure of the *ratio* which the denfities of the air bear to one another at thofe heights; thus C F meafures the proportion which the air's denfity at the height C bears to its denfity at the height F. Let us now fuppofe the force of gravity to ceafe, and that the air is fo comprefled by fome external force, as to be every where from top to bottom of the fame denfity, as it is at the furface of the earth; its weight or pref- fure, which before was denoted by the indefinite *area* BAHI, may now be denoted by the paralle- logram BACE, inafmuch as by the fourth pro- perty of the curve, that *area* and this parallelogram are equal. Since then two fluids which balance each other muft have their heights inverfly as their fpecifick gravities, if we put unity to denote the fpecifick gravity of the air at the furface of the earth, and fay, as unity to 11890, which is the fpecifick gravity of mercury with refpect to that of air, fo is $2\frac{1}{2}$ feet, which is the height of the mercury in the barometer, to a fourth number, we fhall have 29725 feet for the height of the homo- geneal atmofphere; and this height is equal to the fubtangent AC. For fince the preffure of this homogeneal atmofphere is as its denfity into its height, and likewife as the rectangle BACE; and fince the denfity is denoted by BA, the height muft be denoted by AC, the module in this fyftem of logarithms. Hence we have a method for de- termining the denfity of the air at any height; for putting H to denote the height at which the den- fity of the air is required, by the fecond property of the curve, we have this analogy, as the integral number marked A, which is the module of this fyftem, is to the fractional number marked B, which is the module of BRIGGS's fyftem, fo is H expreffed in feet, to a fourth number, which in BRIGGS's tables is the logarithm of the *ratio* of

the

the denfity of the air at the earth's furface, to its L E C T.
denfity at the height H, anfwerable to which in XVI.
the tables is the natural number expreffing that
ratio.

$$\underset{29725}{\overset{A}{\vphantom{.}}} : \underset{0.43429448}{\overset{B}{\vphantom{.}}} :: H : \frac{0.43429448 \times H}{29725}.$$

$$\underset{29725}{\frac{\overset{B}{26400} \times 0.43429448}{A}} = \underset{0.385661.}{\overset{C}{\vphantom{.}}} \qquad \underset{2.4303.}{\overset{D}{\vphantom{.}}}$$

Thus, for inftance, if the denfity of the air at
the height of five miles, or 26400 feet, be required,
by multiplying that number by the fractional number marked B, and dividing the product by the
integral number marked A, we fhall have the logarithm marked C; anfwerable to which in the
tables is the natural number marked D, expreffing
the *ratio* of the air's denfity at the furface of the
earth, to its denfity at the height of five miles;
whence it appears, that at the furface of the earth,
the air is denfer than it is at the height of five miles,
in the proportion of almoft $2\frac{1}{2}$ to one; but then,
this is on fuppofition that the force of gravity continues the fame at all heights, whereas in truth,
that force decreafes in the recefs from the earth's
center in the duplicate *ratio* of the diftance, which
caufes the denfities of the air at different heights to
be fomewhat different from what they would be in
cafe the force of gravity did not vary.

In order therefore to determine the denfities more
accurately, let S be the earth's center, and A B, Pl. 7.
equal to A B in the laft figure, the earth's furface, Fig. 8.
and let F be the height at which the denfity of the
air is required; let S K be a third proportional to
S F and S A, and at the point K, let the ordinate

R 2   K G

KG be drawn, denoting the denſity of the air at F,
then taking the point M at an indefinitely ſmall diſtance above F, let SL be a third proportional to SM
and SA; and at the point L, let the ordinate LN
be drawn, denoting the denſity of the air at M; this
being done, it will appear, that the curve BGN,
which paſſeth thro' the points G and N, is the ſame
logarithmick curve with the former, but in an inverted poſition. For ſince SL is to SA, as SA to
SM, and ſince SK is to the ſame SA, as SA to
SF, then by equality of *ratio*, SL is to SK, as SF
to SM, and by diviſion and permutation, KL is
to FM, as SK to SM; or becauſe FM is indefinitely ſmall, as SK to SF; that is, as SA^q to SF^q;
whence reducing that analogy into an equation, and
dividing by SF^q, we ſhall have $KL = \dfrac{SA^q \times FM}{SF^q}$;
and rejecting SA^q, as being a given quantity, we
ſhall have KL as FM directly, and SF^q inverſly;
but FM is as the quantity of air in the indefinitely
little ſpace FM, and SF^q inverſly is as the gravitation of the ſame air, and KG is as its denſity; conſequently, the rectangle under KL and KG, or the
area KGNL, is as the gravitation, the quantity,
and denſity of that air conjointly, that is, as its preſſure on the air beneath it; and the ſum of all the
ſimilar *areas* below KG, is as the ſum of all the
preſſures above F, that is, as the denſity of the air
at F, or as KG, which denotes that denſity; and
KGNL which is the difference of the two ſums of
all the ſimilar *areas*, one of which ſums begins from
the point K, and the other from the point L, is as
the difference of the air's denſities at F and M, that
is, as KG — LN. Let now KL be given; that is
to ſay, let the ſmall portion intercepted between
KG and LN be always of one and the ſame length,
in whatever parts of the line AS the points K and
L are taken; then KG will be as the *area* KGNL,

and

and confequently, as KG—LN; whence by divifion, KG will be as LN, fo that the *ratio* of KG to LN is given, and of courfe the given line KL will be the meafure of that given *ratio*; whence by the firft property of the logarithmick curve, the curve which paffeth thro' the points G and N is a logarithmick curve; and it is alfo the fame with the former; for taking AO the height above the earth's furface indefinitely fmall, it is evident, that the force of gravity is the fame at O that it is at A, confequently, the denfity of the air at O will come out the fame, whether the law of gravity be taken into the confideration or left out; let then the ordinate OP be drawn in the former curve, and at the fame diftance from A in the latter curve, let the ordinate PQ be drawn. Now, fince one and the fame denfity of the air at the earth's furface is denoted in both curves by the equal ordinates BA, it is evident, that the ordinates OP and PQ, which in the two curves denote one and the fame denfity at O, muft likewife be equal; whence it follows, that both curves have the fame curvature, as alfo the fame inclination of their tangents at the points B, and their fubtangents equal; that is, the latter curve is the fame with the former, but in an inverted pofition. Now, forafmuch as BA in the latter curve denotes the denfity of the air at the furface of the earth, and KG its denfity at F, it is evident by the firft property of the curve, that in this fyftem, AK is the meafure of the *ratio* which the denfity at the furface has to the denfity at F; the firft thing therefore which muft be done, in order to difcover the denfity at F, is to find out the line AK, and this is done by diminifhing AF in the fame proportion that the earth's femidiameter SA is lefs than SF, the diftance of F from the earth's center; for by the conftruction, SF is to SA, as SA to SK; whence by divifion, SF : SA :: AF : AK. AK being thus obtained, let it be called H; then, by the fame

procefs

procefs as before, we may difcover the denfity of the air at the height F.

$$\qquad\qquad\qquad\qquad\qquad\qquad\quad \text{E} \qquad\quad \text{feet.}$$
$$4005 : 4000 :: 5 : \frac{4000 \times 5}{4005} = 4.99375 = 26367.$$

$$\qquad\qquad \text{B} \qquad\qquad\qquad\quad \text{F} \qquad\qquad\qquad \text{G}$$
$$\frac{26367 \times 0.43429448}{\text{A}} = 0.385232. \qquad\qquad 2.4279.$$
$$29725$$

For inftance, if the denfity of the air at the height of five miles be required as before; then by faying, as 4005 miles, that is S F, is to 4000 miles, that is SA, fo is five miles, that is AF, to a fourth, we fhall have the number marked E, expreffing miles, and parts of a mile, equal to 26367 feet, which being multiplied by the fractional number marked B, and the product divided by the integral number marked A, we fhall have the fractional number of BRIGGS's tables marked F, anfwerable to which is the natural number marked G, expreffing the *ratio* of the air's denfity at the furface of the earth, to its denfity at the height of five miles. After the fame manner may the *ratio* of the air's denfity at the furface, to its denfity at any height be computed. The refult of fuch computations I have fet down in the annexed table; the firft column of which contains the heights of the air in *Englifh* miles, whereof 4000 make a femidiameter of the earth. The numbers in the fecond column exprefs the *ratio* of the air's denfity at the furface, to its denfity at the refpective heights, and they likewife denote the rarity or expanfion of the air at thofe heights. The third column contains the denfities and compreffions at the feveral heights. The numbers at the bottom of the fecond column included in crotchets denote, that fo many figures are to be annexed to the five preceding, and thofe included in the crotchets at the bottom of the third column

column denote, that so many decimal cyphers are to be prefixed to the five following figures.

Heights of the air in *English* miles.	Rarity and expansion.	Compression and density.
0	— — — 1 —	1
1/4	— — — 1.0454	0.95676
1/2	— — — 1.0928	0.91509
3/4	— — — 1.1424	0.87535
1	— — — 1.1943	0.85405
1 1/4	— — — 1.2429	0.80456
1 1/2	— — — 1.3052	0.76616
1 3/4	— — — 1.3644	ò.73290
2	— — — 1.4263	0.70118
2 1/4	— — — 1.4871	0.67244
2 1/2	— — — 1.5586	0.64160
2 3/4	— — — 1.6292	0.61379
3	— — — 1.7031	0.58716
3 1/4	— — — 1.7883	0.55919
3 1/2	— — — 1.8596	0.53775
3 3/4	— — — 1.9460	0.51387
4	— — — 2.0336	0.49173
4 1/4	— — — 2.1257	0.47043
4 1/2	— — — 2.2221	0.45002
4 3/4	— — — 2.3226	0.43012
5	— — — 2.4279	0.41187
10	— — — 5.9182	0.16897
20	— — 34.288	0.029164
30	— — 198.34 —	0.0050418
40	— — 1136. —	0.00088028
50	— — 6449.2 —	0.00015505
100	33584[3] — —	0. [7]26798
400	11271[24] —	0. [28]88723
4000	19316[150] —	0.[154]51770
40000	33097[276] —	0.[280]30214
400000	32859[301] —	0.[305]30433
1000000	12002[303] —	0.[307]45450
Infinite.	37311[304] —	0.[308]26802

Cᴏʀᴏʟ.

Corol. Since SF is by conftruction equal to
$\frac{SA}{SK}$, and fince from the nature of mufical proportion, the quotients arifing from the divifion of one
and the fame quantity by quantities in arithmetick
progreffion, conftitute a feries of mufical proportionals, it follows, that if feveral diftances from
the earth's center as SF, be taken, in mufical progreffion, their reciprocals as SK, muft be in arithmetick progreffion; and by the firft property of the
logarithmick curve, the denfities of the air as KG,
muft be in geometrick progreffion.

Since the denfity of the air is proportional to
the compreffing force, and fince the compreffing
force is equal to the elaftick force, it is manifeft,
that if the denfity of the air be increafed, the elafticity will likewife increafe in the fame proportion;
and on this principle are founded artificial fountains,
which play by means of condenfed air; they are of
two kinds, fingle and double. The fingle fountain is made of brafs, and is every where fhut, excepting that thro' the middle of the bafon BB,
there paffes down a pipe PP, whofe lower end
reaches nearly to the bottom of the fountain, and
to the upper end is fitted a ftopcock, by help of
which the pipe may be fhut or opened at pleafure.

Pl. 7.
Fig. 9.

Exp. 8.   Some part of the fountain as ADC, being filled
with water poured in thro' the pipe, a condenfing
or forcing fyphon is fcrewed to the top of the pipe
above the cock, by means whereof a great quantity
of air is thrown into the pipe; which as it cannot
return back, by reafon of a valve which fhuts clofe
upon the hole of the fyphon, forces its way thro'
the water into the upper part of the fountain, and
there remains in a ftate of condenfation, greater than
that of the outward air. When therefore the con-
denfer is taken off, and the cock opened, the included air preffing ftrongly on the water which lies
beneath

beneath it, throws it up thro' the pipe, and thereby makes a jet.

The force wherewith the water is thrown up, is proportional to, and may be expreſſed by the exceſs of the denſity of the included air above that of the external air. For if the included air be equally denſe with that without, its elaſtick force muſt be equal to the compreſſive force of the atmoſphere; conſe-quently thoſe two forces will balance one another, and the water will continue at reſt, being preſſed as ſtrongly downward by the weight of the external air, as it is upward by the expanſive force of the included air; but if the included air be more denſe than the external, its elaſtick force will exceed the compreſſive force of the atmoſphere, in the ſame proportion that its denſity exceeds the denſity of the outward air; conſequently, that part of the expan-ſive force of the included air which raiſes the water, is proportional to, and may be expreſſed by, the exceſs of the denſity of the included air above that of the external air. So that putting F for the force which raiſes the water, D for the denſity of the in-cluded air, and 1 for the denſity of the air without, F is as D — 1.

The height in feet to which the water riſes, ſet-ting aſide all impediments, is equal to the product ariſing from the multiplication of 33 into the ex-ceſs of the denſity of the included air above that of the outward air; that is, putting H for the height of the jet, and x for 33, H = x D — x.

For as water which is driven out of a reſervoir by the preſſure of the incumbent water, if it ſpouts di-rectly upward, riſes to the ſame height with the wa-ter in the reſervoir; ſo if it be driven by any other force, it muſt riſe to an equal height with a pillar of water whoſe preſſure is equal to that of the driv-ing force; foraſmuch therefore as the atmoſphere makes an equal preſſure with a height of water of

33 feet,

33 feet, the water will be thrown to the height of 33 feet by the compreſſive force of the atmoſphere; wherefore if we put 1 for the preſſure of the atmoſphere, and ſay, as one is to 33 or x, ſo is D — 1, which expreſſes that part of the preſſure of the included air which drives out the water, to a fourth proportional, we ſhall have x D — x, or x × $\overline{D — 1}$, for the height to which the water is thrown; whence it appears, that if D — 1 be equal to unity, which is the caſe when the air within is as denſe again as that without, the water will riſe to the height of 33 feet; and if D — 1 be equal to 2, which is the caſe when the included air is thrice as denſe as the external, the height of the jet will be 66 feet, and ſo on.

The double fountain conſiſts of two ſingle fountains, whoſe bottoms are faſtened to an hollow braſs cylinder, one at each end, in the manner repreſent-
Pl. 7.
Fig. 10.
ed in the figure, wherein AA and BB denote the two fountains with their baſons; CC the hollow cylinder, which plays upon the pins DD as upon an axle; each has a pipe as P, whoſe lower end reaches nearly to the bottom of the fountain. From the baſon of the fountain AA, there iſſues another pipe as T, which paſſing thro' AA, and likewiſe the hollow cylinder CC, without communicating with either, opens at E into the fountain BB. And in like manner, ſuch another pipe iſſuing from the baſon of BB, and paſſing thro' that fountain and the cylinder, opens into the fountain AA. The hol-
Exp. 9.
low cylinder being placed in an upright poſture by means of the carriage which ſupports it, and the pipes of the lower fountain being ſtopped, water is conveyed into it thro' the pipe T, which iſſues from the baſon of the upper fountain; by the running in of the water, the air contained in the lower fountain is crowded into a ſmaller ſpace, and thereby condenſed; if then both the pipes of the upper

fountain

fountain be stopped, and the lower fountain be brought into the place of the upper, by turning the cylinder on its pins, the water which it contains will fall to its bottom, and the lower end of the pipe P will be immersed therein, in the manner represented in the upper fountain; so that upon opening that pipe, the water will be driven thro' it by the expansive force of the condensed air; and as it falls into the bason, it will be conveyed thence by the pipe T into the lower fountain; and when the upper is exhausted and ceases to play, then stopping its pipes, and changing the places of the fountains as before, the other may be set a going in the same manner,

LECTURE XVII.

Of Sounds.

IN this lecture I shall first explain to you the Lect. NATURE OF SOUNDS, and then treat of the XVII. VIBRATIONS OF MUSICAL STRINGS.

That SOUNDS have a necessary dependance on the air, will appear from the following experiment.

Let a bell be placed under a receiver in such a Exp. 1. manner as that it may be rung at pleasure; and upon drawing the air out of the receiver, the sound of the bell will grow less and less audible in proportion to the degrees of exhaustion, so as at last almost to die away, and scarcely to be heard at all; and upon re-admitting the air, the sound will revive again, and increase in proportion to the quantity of air that is taken in.

As this experiment proves the air to be necessary to the production of sounds, so the tremblings which great guns, bells, drums, and many other sounding bodies communicate, by means of the interme-

OF SOUNDS.

intermediate air, to such bodies as are near them,
plainly shew, that sounds depend on tremulous motions of the air; which therefore I shall endeavour
to explain to you, together with the cause and manner of their production. When the parts of a bell,
a musical string, or any other elastick body are set
in motion by a stroke, they vibrate, that is, they
go forward and return backward alternately thro'
very short spaces; in going forward they propel,
and thereby compress and condense the air which
lies next them; and in returning backward, they
suffer the compressed air to recede and expand itself, so that the parts of the air which are contiguous to the trembling body, go and return in the
same manner with the parts of the body; and as
they are endued with a repulsive power, they must
by means thereof excite the same vibrations in those
parts which lie next beyond them; and these again
must in like manner agitate the parts beyond them,
and so on continually; so that by one single vibration of an elastick body, a motion is excited in the
air, and propagated directly forward, by which
some parts go forward, whilst others return back,
and that alternately, as far as the motion reaches.

That this motion may more readily be conceived,
Pl. 7.
Fig. 11. let ST represent an elastick string, stretched and
made fast at both ends; and by a force applied to
the middle point H, let it be drawn into the position SET; upon the removal of the force which
inflects it, it will by virtue of its elasticity return to
its former position SHT; and forasmuch as the
restitutive force acts constantly upon it during the
time of its motion from E to H, its motion thro'
that space must be continually accelerated, and the
velocity thereof must be greatest at H. When the
string has recovered the position SHT, it will not
remain therein; but by virtue of the velocity acquired in moving from E to H, it will be carried
forward till it has moved thro' a space as HK, equal

to

to EH, and then its motion forward will ceafe; for as it moves towards K, the elaftick force acts continually upon it in drawing it back; and by fo doing, retards the motion from H to K, in the very fame manner that it accelerated the motion from E to H; confequently, by the time that the ftring has moved from H to K, it will have loft all that velocity which it acquired in moving from E to H; as foon as it ceafes to go forward, it will be brought back again from K to H by the force of elafticity; with an accelerated motion, in the very fame manner as it was at firft from E to H; and when it has arrived at H, it will by virtue of the acquired velocity go on to E, with a retarded motion, in the fame manner as it did from H to K. The motion of the ftring from E to K and back again, is called a *vibration*; and it is evident from what has been faid, that fetting afide all external impediments, a ftring which has made one vibration, muft continue to vibrate for ever thro' the fame fpace; but, whereas it meets with continual refiftance from the air, the fpace thro' which it vibrates muft on that account grow lefs and lefs continually, and at length vanifh; and yet, notwithftanding this variation in the fpace, the times of the vibrations are all equal, as I fhall demonftrate before the clofe of this lecture; but I take notice of it in this place, becaufe one of the chief properties of the pulfes of the air, whereof I fhall have occafion to make mention prefently, has a neceffary dependance thereon.

When the ftring is drawn into the pofition SET, if we fuppofe A, B, C, and fo forth, to be particles of air placed in a right line one beyond another, and that the diftance of the firft particle from the ftring at E, is equal to the interval of any two adjacent particles, as it muft needs be, on fuppofition that the particles of the air fly from other bodies with the fame force that they repel one another; upon
letting

letting the ſtring go, as it cannot move forward
without approaching to the particle A, it muſt in
the very next inſtant after it begins its motion, propel that particle ; which for the ſame reaſon, muſt
in the next inſtant after it begins to move propel
the particle B, and that muſt in the ſame manner
propel C, and C propel D, and ſo on ; ſo that the
ſtring, and the ſeveral particles of air taken in
their order, will begin to move forward ſucceſſively
one after another, at very ſmall intervals of time.
And whereas the ſtring is accelerated in its motion
from E to H, and retarded in its motion from H
to K, the particle A muſt likewiſe be accelerated
in one half of its progreſs, and retarded in the
other ; for ſince A is equally diſtant from the
ſtring, and from B, before the vibration com-
mences, and ſince it begins to move forward a little
later than the ſtring ; it is evident, that upon the
firſt motion of the ſtring, the diſtance between that
and A, muſt become leſs than the diſtance between
A and B ; and foraſmuch as the increments of velocity which are continually generated in the ſtring
by the action of its elaſticity, are not communicated to the particle A, in the inſtant of time wherein
they are generated, but a little later ; it is manifeſt, that the ſtring during its motion from E to
H, muſt continually be nearer to A than A is to
B ; and conſequently, muſt act more forcibly in
driving A forward, than B does in driving it backward, and by ſo doing accelerate its motion. After
the ſtring has arrived at H the middle point of its
progreſs, and ceaſed to be accelerated, in the very
next moment A likewiſe reaches the middle point
of its progreſs, and ceaſes to be accelerated, being
driven as ſtrongly backward by B, as it is forward
by the ſtring. But however, by virtue of the acquired motion it continues to go forward, but with
a retarded motion ; and is at length ſtopped by the
repulſive

repulsive power of B; in the same manner that the L E C T. string in moving from H to K is retarded, and at XVII. last stopped by the action of its elastick force. After the string has reached K, the utmost limit of its progress, in the very next moment does A likewise reach the utmost limit of its progress, and then turning back, pursues the string, which had likewise turned back the moment before. And as the string is accelerated during its return from K to H, and retarded from H to E; so the particle A, during the first half of its return, being nearer to B than it is to the string, must be accelerated; and during the latter half, being nearer to the string, is thereby retarded, and at length stopped upon its arrival at the place from whence it set out, which happens immediately after the string has returned to E; and there it continues at rest, unless by a second vibration of the string it be again driven forward in the same manner as before. As this particle is made to go and return thro' a very short space, by the impulse of the string, so likewise are the several succeeding particles, by the impulses of the foregoing; and as the string and the several particles taken in their order, begin their motions forward, successively one after another at very small intervals of time, so likewise do they begin to return in their order at the same intervals of time; whence it follows, that some of them must go forward, at the same time that others return back. As the particles which go forward begin their motions successively one after another, they must necessarily come nearer together; that is, they must be condensed. And it must be observed, that the condensation goes forward continually; for in the very next instant after any particle as D, has made its nearest approach to E, E must make its nearest approach to F; and in the next instant F must make its nearest approach to G, and so on continually; so that the conden-

fation muſt paſs forward ſucceſſively in a regular
manner thro' the ſeveral particles of air.

But that I may explain this vibratory motion of
the air more particularly, it muſt be obſerved, that
as the ſtring during the firſt half of its progreſs from
E to H is continually accelerated, its diſtance from
the particle A, muſt conſtantly grow leſs; and foraſmuch as during the latter half of its progreſs from
H to K, it is continually retarded, and that in the
ſame uniform manner that it was accelerated from
E to H, its diſtance from A muſt conſtantly be inlarged, and that in the ſame regular manner that
it was diminiſhed during the progreſs of the ſtring
from E to H; ſo that by the time it has arrived at
K, the utmoſt limit of its progreſs, it is juſt as far
diſtant from the particle A, as it was when it firſt
ſet out. Upon the return of the ſtring, inaſmuch
as it is continually accelerated from K to H; its
diſtance from the particle A muſt ſtill be inlarged;
and foraſmuch at it is retarded in its motion from
H to E, in the very ſame manner as it was accelerated from K to H, its diſtance from A muſt conſtantly grow leſs in the ſame regular manner that it
was inlarged during its motion from K to H, ſo
that upon its return to E, it is again juſt as far diſtant from the particle A, as it was at its firſt ſet-
ting out. From what has been ſaid, it is evident,
that the ſtring during the time of its progreſs is
always nearer to the particle A, than it was before
its motion began, and that its leaſt diſtance from
the particle is at H, the middle point of its progreſs; it is likewiſe manifeſt, that during the time
of its return, it is always more diſtant from the particle than it was before its motion began; and that
its greateſt diſtance from the particle is at H, the
middle point of its return. And what has been
thus ſhewn of the ſtring with reſpect to the particle
A, is in like manner true of that particle with reſpect

fpect to the particle B, and of B with refpect to C, and fo on of every particle, with refpect to that which lies immediately beyond it, as far as the motion reaches; fo that each particle with regard to that which lies immediately beyond it, is in a ftate of condenfation during its progrefs, and of rarefaction during its return, its greateft condenfation being at the midft of its progrefs, and its greateft rarefaction at the midft of its return. What proportion thefe rarefactions and condenfations bear to the denfity of the air in its natural ftate, in every point of that fmall fpace thro' which a particle of air vibrates, fhall be fhewn in my next lecture, as alfo the law of this vibratory motion.

As the parts which go forward, do in their progreffive motion ftrike fuch obftacles as they meet in their way, they are for that reafon called *pulfes*; and the fenfations which are excited in the mind by the ftrokes of thefe pulfes on the drum of the ear are called *founds*; fo that founds as confidered in their phyfical caufes, are nothing elfe but the pulfes of the air. In order therefore to explain the nature of founds, I fhall lay before you the chief properties of thefe pulfes.

The firft of which is, that they are propagated from the trembling body all around in a fphærical manner. For tho' the parts of the body, by whofe vibrations the pulfes are generated, do go and return according to certain directions, yet forafmuch as every impreffion which is made on a fluid is propagated every way throughout the fluid, whatever be the direction wherein it is made, the pulfes muft fpread and dilate, fo as to form themfelves into concentrick fphærical furfaces, or rather thin fhells, whofe common center is the place of the founding body. And hence appears the reafon why one and the fame found may be heard by feveral perfons, tho' differently fituated with refpect to the founding body.

S   A fecond

A second property of the pulses is, that they
grow less and less dense as they recede from the
founding body, and that in the same proportion
with the squares of their distances from the body.
For whatever be the force wherewith the founding
body acts on the first sphærical shell of air, with the
very same force does that shell act upon the second,
and that again upon the third, and so on continually; so that the force which condenses the air in
the several shells is given; consequently, the con-
densations which it produces in those shells must be
inversly as the resistances it meets with; but the resistances are as the shells; and therefore, since those
increase continually in the same proportion with the
squares of their distances from the center, their
densities must decrease in the same manner.

By reason of this diminution in the densities of
the pulses, those which are farther removed from
the founding body, make slighter impressions on
the drum of the ear, than those which are less distant; and hence it is, that sounds grow less and
less audible, the farther they go from the founding
body; and at certain distances become so weak as
not to be heard at all.

A third property of the pulses is, that all of them,
whether denser or rarer, move equally swift, so as
to be carried thro' equal spaces in equal times, as I
shall demonstrate in my next lecture.

From this property it follows, that all sounds,
whether they be loud or low, grave or acute, move
equally swift, the softest whisper making equal speed
with the noise of a cannon, or the loudest thunder-
clap; and it has been found by experiment, and I
shall likewise demonstrate in my next lecture, that
sounds move at the rate of 1142 feet in a second of
time or thereabouts; for the velocity is not precisely the same in all seasons of the year, but is
somewhat greater in *summer* than in *winter*, on account of the heat which renders the air more elastick

tick in proportion to its denſity, than it is in the L e c t. XVII.
cold *winter* ſeaſon.

A fourth property of the pulſes is, that all thoſe which are excited by the vibrations of one and the ſame body, are at equal diſtances from one another. For ſince each pulſe is excited by one ſingle vibration of the ſounding body, and ſince all the pulſes move with equal and uniform velocities, it is manifeſt, that they muſt ſucceed one another at diſtances proportional to the times of the vibrations; but the times of the vibrations of one and the ſame body are all equal; conſequently, the intervals of the pulſes are ſo too. And it muſt be obſerved, that the interval between two pulſes, which is by ſome called the *length*, and by others the *breadth* of a pulſe, is that ſpace thro' which the motion of the air is carried, during the time wherein any one particle performs its vibratory motion in going forward and returning back.

On the intervals of the pulſes depend the tones of ſounds; and here I muſt obſerve to you, that all the variety there is in ſounds, reſpects either their *ſtrength* or their *tone*; with regard to their ſtrength, they are diſtinguiſhed into *loud* and *low*; and with reſpect to their tone, into *grave* and *acute*, otherwiſe called *flat* and *ſharp*. The ſtrength of any ſound depends on the magnitude of the ſtroke, which is made by a pulſe on the drum of the ear; the greater the ſtroke is, the louder is the ſound which it excites, and the weaker the ſtroke, the lower the ſound; and whereas all the pulſes move with equal velocities, the magnitude of the ſtroke, and conſequently the ſtrength of the ſound, muſt be as the quantity of matter in the pulſe; that is, as a rectangle under the denſity and breadth of the pulſe; and ſuppoſing the breadth of the pulſe to be given, it muſt be as the denſity.

The tone of a ſound depends on the duration of a ſtroke; the longer a ſtroke is which a pulſe makes

on

on the drum of the ear, the more grave is the found which it produces ; and the ſhorter the ſtroke, the more acute is the found ; but ſince all the pulſes move equally ſwift, the duration of a ſtroke muſt be proportional to the interval between two ſucceſſive pulſes ; and of conſequence, a found is more or leſs grave or acute in proportion to the length of that interval. Hence it follows, that all the founds from the loudeſt to the loweſt, which are excited by the vibrations of one and the ſame body, are of one tone. It likewiſe follows, that all thoſe founding bodies, whoſe parts perform their vibrations in equal times, have the ſame tone ; as alſo, that thoſe bodies which vibrate ſloweſt, have the graveſt or deepeſt tone ; and on the contrary, thoſe which vibrate quickeſt have the ſharpeſt or ſhrilleſt tone.

As there may be an infinite variety in the times wherein founding bodies perform their vibrations, ſo may there likewiſe be in the tones of the founds which depend thereon ; and yet amidſt this great variety, muſicians acknowledge but ſeven principal notes in an *octave* ; for tho' the eighth be requiſite to compleat the ſeven intervals in an octave, yet are there in truth but ſeven notes ; for that which is called the *eighth*, becomes the baſe or ground note in the next octave aſcending ; and as it ſtands in the limits of the two octaves, it is called the *eighth* with reſpect to the baſe note below it, and the ground or baſe note with reſpect to the 15th which is above it ; which 15th is likewiſe the baſe in the next aſcending octave ; for by a repetition of notes, wherein the proportions of the times of the notes in the firſt octave are preſerved, the octaves may be continued on both ways, aſcending and deſcending, and that in *infinitum* ; and yet, notwithſtanding this infinite progreſſion in the octaves, the number of harmonick founds is limited. Mr. SAUVEUR is of opinion, that all the harmonick founds, that is, ſuch founds as can be heard diſtinctly and with pleaſure,

and

and in whofe tones a difference can be clearly perceived by the ear, lie within the compafs of ten octaves; as alfo, that all founds whatever, from the
loweft harmonick found, to the higheft that the human ear can well bear, are contained within the limits of two octaves more. And if this be the cafe,
it follows, that that body which gives the fhrilleft
found that the ear can bear, makes 4096 vibrations
in the fame time that one vibration is performed
by that body which gives the graveft harmonick
found; for fince in every octave, the time of the
eighth is $\frac{1}{2}$ of the time of the bafe note, if $\frac{1}{2}$ be raifed
up to the 12th power, it will exhibit the time of the
fhrilleft found, that of the graveft being unity; but
the 12th power of $\frac{1}{2}$ is the 4069 part of an unite;
confequently, the time of the fhrilleft found that
the ear can well bear, and likewife of the vibration
which produces it, is to the time of the graveft harmonick found, and of the vibration whereby it is
produced, as 1 to 4096; but the times of the vibrations of two bodies are inverfly, as the numbers
of vibrations which they perform in a given time;
confequently, the body which gives the fhrilleft
found performs 4096 vibrations in the fame time
that the body which gives the graveft harmonick
found performs one; and forafmuch as Mr. S<small>AU</small>-
<small>VEUR</small> has found by fome experiments which he
made on organ pipes, of which I fhall give you an
account in my next lecture, that a body which
gives the graveft harmonick found, vibrates twelve
times and an half in a fecond, the fhrilleft founding body muft perform 51100 vibrations in the fame
time; which argues great fwiftnefs in the vibrating
parts; and yet, great as it is, it has nothing extraordinary or furprifing in it, if compared with the
velocity of fome other motions; for if we fuppofe
the parts in each vibration to run thro' a fpace equal
to the 10th part of an inch, tho' it is highly probable, that the lengths they run are much fhorter;

and

and if we fuppofe them to move with the fame velocity during the whole time of their motion; it follows, that they are carried at the rate of 425 feet and
10 inches in a fecond; confequently, they do not move
with much more than two third parts of the velocity
wherewith a ball flies from the mouth of a cannon.

The fifth and laft property of the pulfes is, that
they may be propagated together in great numbers
from different bodies, without difturbance or confufion; as is evident from conforts, wherein the
founds of the feveral inftruments are conveyed dif-
tinctly to the ears of the audience; as they move
along, fome of them coincide and ftrike the drum
of the ear at one and the fame time, and thereby
excite a fmooth regular motion, that is pleafing
and agreeable; whilft others which do not mix
and unite, at leaft not frequently, ftrike the ear at
different inftants of time, and thereby difturb each
other's motion, fo as to render them harfh, grating, and offenfive. And hereon depends almoft
the whole of concords and difcords in mufick;
fuch founds, generally fpeaking, being deemed
concords, as are excited by pulfes which have frequent coincidences; and on the other hand, fuch
founds being called *difcords*, as arife from pulfes
which coincide but rarely.

The frequency or infrequency of the coincidences,
depends on the proportions which the intervals of
the pulfes bear one to another; as I fhall fhew you
in relation to the feveral notes in an octave; in doing of which, inftead of the pulfes and their intervals, I fhall confider the vibrations of the bodies
which excite the pulfes, and the times of thofe vibrations; becaufe the number of pulfes is always
equal to the number of vibrations in the founding
bodies, and the intervals of the pulfes proportional
to the times of the vibrations.

If two vibrating bodies begin their motions together, and vibrate in equal times, it is manifeft,
that

that their vibrations muſt keep pace together, and conſtantly coincide. But if the vibrations be performed in unequal times, it is plain, that they cannot conſtantly keep pace together; for which reaſon ſome of them only will coincide; and which thoſe are may be determined from the times of the vibrations; for ſince the numbers of vibrations, which are performed in a given time, are inverſly as the times of the vibrations, if the numbers which expreſs the times of the vibrations of two bodies be taken reciprocally, they will exhibit the coincident vibrations of the reſpective bodies. For inſtance, if the time of the vibrations of one body, be to the time of the vibrations of another, as 8 to 9, which is the caſe of two bodies, whereof one ſounds a ſecond or tone major to the other, every ninth vibration of the former coincides with every eighth of the latter. So again, if the times of the vibrations be to one another, as 5 to 6, which is the caſe, where one body ſounds a leſſer third to the other, every ſixth vibration of the former falls in with every fifth of the latter.

In this ſcheme, I have ſet down thoſe fractional numbers which expreſs the proportions that the times of the vibrations of thoſe bodies, which found the ſeveral notes in an octave, bear to the time of the vibration of that body which ſounds the baſe note; by the help of which numbers the co-

$\frac{1}{2}$	*Eight.*
$\frac{8}{15}$	*Greater ſeventh.*
$\frac{5}{9}$	*Leſſer ſeventh.*
$\frac{3}{5}$	*Greater ſixth.*
$\frac{5}{8}$	*Leſſer ſixth.*
$\frac{2}{3}$	*Fifth.*
$\frac{3}{4}$	*Fourth.*
$\frac{4}{5}$	*Greater third.*
$\frac{5}{6}$	*Leſſer third.*
$\frac{8}{9}$	*Second or tone major.*
1	*Baſe note.*

incident vibrations may be readily diſcovered. For in each fraction, the denominator exhibits the coinciding vibration of that body which ſounds the note, and the numerator the coinciding vibration of the body which ſounds the baſe note.

S 4

Having

LECT.
XVII.   Having thus explained the nature and properties of found, I come now to give you an account of the VIBRATIONS of MUSICAL STRINGS, and to ſhew you in what proportions the times of the vibrations are varied, by varying the length, thickneſs, or tenſion of the ſtrings; and in order thereto, I ſhall lay down the following PROPOSITION.

Pl. 7.
Fig. 13.   *Let an elaſtick ſtring as* AB, *faſtened at* A, *and paſſing over a ſmall pin or pulley at* B, *be ſtretched by an appending weight as* P, *(which I ſhall call the tending force)*; *and by a force applied at the middle point* C, *(which I ſhall call the* inflecting force), *let it be drawn into the poſition* ADB; *if the diſtance between* C *and* D *be exceedingly ſmall in proportion to the length of the ſtring, or, to ſpeak in the mathematical phraſe, if* CD *be a naſcent quantity, the inflecting force will be meaſured by a rectangle under the ſpace* CD, *and the tending force applied to the length of the ſtring.* For ſince the tending force acts upon the ſtring in the direction DB, it may be denoted by that line, and being ſo denoted, it may be reſolved into two forces, whereof one acts in pulling the ſtring horizontally in the direction CB, and is therefore to be expreſſed by CB; whilſt the other acts in drawing the ſtring perpendicularly upward from D towards C, and is therefore to be expreſſed by the line DC; ſo that that portion of the tending force which acts in moving the ſtring upward, is to the whole force, as DC to DB; or, becauſe D and C are ſuppoſed to be indefinitely near, as DC to CB; but the force which acts in drawing the ſtring upward, is equal to the inflecting force, becauſe they balance each other; conſequently, the inflecting force is to the tending force, as CD to CB; and turning this analogy into an equation, by multiplying the extreams and means, and then dividing by CB, we ſhall have the inflecting force equal to a rectangle under the tending force, and the line CD, applied

to

to half the length of the ſtring; and therefore, foraſmuch as whole quantities are in proportion as their halves, the inflecting force will be as a rectangle under the tending force and the line C D, applied to the length of the ſtring; ſo that putting F for the inflecting force, P for the tending force, S for the line CD, and L for the length of the ſtring, F is as $\frac{SP}{L}$. Hence it follows, that if P and L, that is, if the tending force and length of the ſtring be given, the inflecting force is as the line C D, as will appear from the following experiment.

Let a ſmall braſs wire three feet long, faſtened Exp. 2. at one end, and paſſing over a pin ſo as that when ſtretched it may be in an horizontal poſition, be tended by a weight of three pounds; and let half an ounce, and an ounce, be appended ſucceſſively to the middle of the wire; in the former caſe, the point of ſuſpenſion will be drawn down $\frac{4}{10}$th parts of an inch, and in the latter $\frac{8}{10}$.

Since the force which inflects a ſtring of a given length, and tended by a given force, is as the ſpace C D, thro' which the ſtring is bent; the force wherewith the ſtring reſtores itſelf, muſt likewiſe be as C D, becauſe the reſtitutive force is in all caſes equal to the inflecting force; conſequently, the point D is carried towards C, by a force that varies with the diſtance; and therefore, whatever be the diſtance at which it begins its motion, the time wherein it arrives at C will ſtill be the ſame; as I proved in my lecture on the pendulum. Whence it follows, that the vibrations of one and the ſame ſtring, whether they be through larger or ſmaller ſpaces, are all performed in equal times.

If L and S be given, F is as P; that is, if the length of the ſtring, and the ſpace thro' which it is bent be given, the inflecting force is as the tending force; or, in other words, one and the ſame ſtring, being

being tended by different forces will upon the in-
flexion be drawn down equal fpaces by inflecting
forces, which are to one another in the fame proportion with the tending forces.

Exp. 3.   Let the fame wire as before be tended by a
weight of fix pounds, and it will require one ounce
to draw it down $\frac{4}{10}$ths of an inch, and two ounces
to draw it down $\frac{8}{10}$ths; whereas, when it was
tended by a weight of three pounds only, it was
drawn down the fame fpaces by half an ounce, and
an ounce.

If P and S be given, F is as $\frac{1}{L}$; that is if the
force which tends the ftring, and the fpace thro'
which it is bent be given, the inflecting force is inverfly as the length of the ftring; or, in other
words, if ftrings of different lengths be tended by
equal forces, they will be drawn thro' equal fpaces
by inflecting forces, which are to one another inverfly as the lengths of the ftrings.

Exp. 4.   Let a fmall brafs wire a foot and an half long,
be tended by a weight of three pounds, and it will
require an ounce to bend it down $\frac{4}{10}$ths of an inch,
whereas half an ounce was fufficient to give the
fame bent to the wire which was of a double length,
and under the fame tenfion.

*The time of a vibration of an elaftick ftring is mea-
fured by a rectangle, under the length and diameter of
the ftring, applied to the fquare root of the tending
force.* For if, as in the cafe of gravity, we fuppofe the force wherewith the inflected ftring reftores
itfelf to act uniformly, as we fafely may, becaufe
the fpace thro' which it acts is exceedingly fmall;
then the motion generated will be as a rectangle
under the force and the time of its acting; fo that
putting M for the motion, F for the reftitutive
force, and T for the time of its acting, M is as
                                    FT;

FT; but the motion is as the quantity of matter moved into the velocity wherewith it moves; and in this cafe, the quantity of matter is as a product under the length of the ftring, and the fquare of its diameter; wherefore, putting D, L, and V, to denote the diameter, length, and velocity, FT is as D^2LV; and dividing both fides by F, T is as $\frac{D^2LV}{F}$; but the reftitutive force of the ftring being equal to the force which inflects it, and that having been proved to be as $\frac{SP}{L}$, wherein S denotes the fpace thro' which the ftring is bent, P the tending force, and L the length of the ftring; if inftead of F we fubftitute $\frac{SP}{L}$, T will be as $\frac{D^2L^2V}{PS}$; but the velocity applied to the fpace is inverfly as the time, that is, $\frac{V}{S}$ is as $\frac{1}{T}$; and therefore, inftead of that, fubftituting this, and multiplying both fides by T, we fhall have T^2, as $\frac{D^2L^2}{P}$; and therefore extracting the root, T is as $\frac{DL}{P^{\frac{1}{2}}}$;

that is, the time of a vibration, is as a rectangle under the diameter and length of the ftring, applied to the fquare root of the tending force.

Hence it follows, that if D and P be given, T is as L; that is, if the diameter of the ftring and the tending force be given, the time of the vibrations varies with the length of the ftring; as is manifeft from the divifion of the *monochord*, wherein the parts of the chord which found the feveral notes in an octave, have the fame proportions to the whole chord, that the times of the refpective notes have to the time of the bafe note; as for inftance, one half of the chord founds an octave to the whole, whofe time is one half of the time of the bafe note;

Exp. 5.

and

LECT.
XVII.
and $\frac{2}{3}$ of the chord found a fifth, the time whereof is $\frac{2}{3}$ of the bafe time of the note, and fo of all the reft.

If P and L be given, then T is as D; that is, if the tending force and length of the ftring be

Exp. 6. given, the time of the vibration is as the diameter of the ftring; as will appear, if two wires of equal lengths be tended by equal weights, the diameter of one being the 90th part of an inch, and that of the other the 45th part; for the former will found an octave to the latter.

If D and L be given, then T is inverfly as the fquare root of P; that is, if the diameter and length of the ftring be given, the time of the vibration is inverfly as the fquare root of the tending force; as

Exp. 7. will appear, if eleven wires equal as to length and thicknefs, be tended by weights, whofe fquare roots are to one another inverfly as the times of the notes in an octave; for the wires fo tended will found the refpective notes.

Eighth	$\frac{1}{2}$	240
Greater feventh	$\frac{8}{15}$	$210\frac{15}{16}$
Leffer feventh	$\frac{5}{9}$	$194\frac{2}{5}$
Greater fixth	$\frac{3}{5}$	$166\frac{2}{3}$
Leffer fixth	$\frac{5}{8}$	$153\frac{3}{5}$
Fifth	$\frac{2}{3}$	135
Fourth	$\frac{3}{4}$	$106\frac{2}{3}$
Greater third	$\frac{4}{5}$	$93\frac{3}{4}$
Leffer third	$\frac{5}{6}$	$86\frac{2}{3}$
Tone major, or fecond	$\frac{8}{9}$	$75\frac{15}{16}$
Bafe note	1	60

In the left hand column of this table, the numbers exprefs the times of the feveral notes; and the numbers in the right hand column, exprefs the weights in ounces, whereby the wires which found the refpective notes are tended; the fquare roots of
which

which weights, are to one another inverfly as the times of the refpective notes; as for inftance, the weight which tends the ftring that founds the octave, is to the weight whereby the ftring that founds the bafe note is tended, as 4 to 1, whofe fquare roots are as 2 to 1, that is, inverfly as the time of the octave, to the time of the bafe note; and fo of all the reft.

L E C T U R E XVIII.

Of the Motion of Sound.

IN my laft lecture, wherein I treated of that motion of the air, which is productive of founds, I fhewed you, that each particle of air in going forward and returning back, is twice accelerated, and as often retarded; but I did not then enquire into the law of that acceleration and retardation. I likewife told you, that all the pulfes of the air move equally fwift, the demonftration of which, I promifed to give you in this lecture. Lect. XVIII.

Now Sir ISAAC NEWTON, having in a moft elegant manner, in the 47th *Propofition* of the *Second Book* of his *Principles*, demonftrated that each particle of air, during its vibratory motion, is accelerated and retarded, in the very fame manner as a pendulum vibrating in a cycloid; and having likewife, in the 49th and 50th *Propofitions* of the fame book, determined the velocity of found, I fhall in this lecture lay before you what he has faid, in relation both to the one and the other, in the cleareft light that I am able.

As to the firft, let the line AB denote the length of a pulfe; or that fpace thro' which the motion of the air is propagated, during the time that a particle performs its vibration, by going forward and returning Pl. 8. Fig. 1.

returning back; and let E, F and G, be three particles, or phyſical points of air ſituated in the right
line at equal diſtances, and at reſt; and let EQ,
FR, and GT, be three equal, but exceedingly ſhort
ſpaces, thro' which theſe particles go and return in
their vibrations; which ſpaces tho' they be here
taken of ſome length, to avoid confuſion in the
ſcheme, are in reality ſo exceedingly ſmall, as to
bear no proportion to AB, the length of a pulſe.
Let x, y, and z denote any intermediate points,
in which the particles are found during their motion
forward or backward. Let EF, and FG be ſmall
phyſical lines, or little portions of air, ſituated in
ſtrait lines between thoſe phyſical points; which

Pl. 8.
Fig. 2.
Fig. 1. lines are ſucceſſively moved into the places xy, yz,
and QR, RT. Let the right line PS, be drawn
equal to EQ, and on that line as a diameter, let
the circle SIPi be deſcribed; and let the circumference of that circle denote the time of the vibration
of a particle, and the parts of the circumference,
the proportional parts of the time; ſo as that after
any time as PH, or PHSh, if right lines as HL
and hl be drawn from the points H and h perpendicular to SP, and Ex be taken equal to PL, or

Pl. 8.
Fig. 1. Pl, the particle E may be found at x. By this
means the particle or phyſical point E, in moving
forward thro' x to Q, and thence back again thro'
x to E, will be accelerated and retarded, in the
ſame manner with a pendulum vibrating in a cycloid; inaſmuch as in my lecture on the pendulum,
I ſhewed you, that the ſpaces deſcribed by ſuch a
pendulum, and the times of deſcribing thoſe ſpaces,
are (as we have now ſuppoſed them to be in the
caſe of the air's motion) as the verſed ſines and
arches of a circle, whoſe diameter is equal in length
to the whole cycloid.

     Now, in order to prove that the ſeveral little por-
tions of air are agitated in the forementioned manner

by

by their elasticity, which in this case is the true
moving cause, let us suppose them to be so moved
by some cause or other, be that cause what it will;
and their elasticity will be found to be such in every point of their progress and return, as must of
necessity produce in them the same degrees of acceleration and retardation, that gravity does in a
pendulum vibrating in a cycloid.

In the circumference of the circle, let the equal
arches HI and IK, or hi and ik, be taken, bear-
ing the same proportion to the whole circumference,
that the little right lines EF and FG do to AB the
length of a pulse; and drawing the lines IM and
KN, or im and kn perpendicular to PS, inasmuch
as the points or particles E, F and G, are moved
in the same manner successively one after another,
the motion beginning with E, and each of them
performs its intire vibration, in going forward and
returning back, in the same time that the motion
is propagated thro' a space equal to AB, the length
of a pulse; if PH or PHSh denotes the time from
the beginning of E's motion, PI, or PHSi, will
denote the time from the beginning of F's motion;
and in line manner PK, or PHSk, will denote
the time from the beginning of G's motion. And
if the points E, F and G be found at x, y and z;
the lines Ex, Fy, and Gz, in the first figure, will
be respectively equal in the second, to PL, PM,
and PN, in the progress of the points; and in
their return, equal to Pl, Pm, Pn, those being the
versed sines of the arches which denote the times.
Whence it follows, that xz, which is equal to the
difference between Ex, and the sum of EG and Gz,
is in the progress of the points, equal to EG —
LN, and to EG + ln in the return; but xz is as
the expansion of the little portion of air EG, when
it is in the place xz; consequently, that expansion
is to the mean ordinary expansion, or that expan-

Pl. 8.
Fig. 1.
Fig. 2.

Fig. 1.

4   sion

Pl. 8.
Fig. 2.

fion which it has when at reft before it is put into its vibratory motion, as EG—LN, to EG, when that portion of air in going forward is found in the place xz; and it is as EG + ln; or, becaufe LN and ln are equal, as EG + LN, to EG, when the portion of air in returning back, is found in the fame place. Let now ID be drawn from the point I, perpendicular to HL, and the nafcent triangle HID, will be fimilar to the triangle OIM, becaufe the angles at D and M are right ones, and the angles at I are equal, as being each of them the complement of one and the fame angle DIO, to a right one; confequently, DI, or its equal LM, is to HI, as IM to the *radius* OI, equal to OP; and double LM equal to LN, is to double HI equal to HK, as IM to OP; and by the conftruction, HK is to EG, as the circumference of the circle to AB; or putting R for the *radius* of a circle, whofe circumference is equal to AB, as OP to R; whence reducing thefe two analogies into equations, we fhall have $\frac{LN}{HK} = \frac{IM}{OP}$, and $\frac{HK}{EG} = \frac{OP}{R}$; wherefore, multiplying thefe equations together, we fhall have $\frac{LN}{EG} = \frac{IM}{R}$; and refolving this into an analogy, we fhall have LN : EG :: IM : R; and of courfe, by fubftituting IM and R, in the places of LN and EG, the expanfion of the fmall portion of air EG, or of the phyfical point F, when in the place xz or y, is to its mean ordinary expanfion, as R—IM to R, in its going forward, and as R + im to R, in its returning; and forafmuch as its elafticity is inverfly as its expanfion, its elafticity, when at the point y, is to its ordinary elafticity, as $\frac{1}{R-IM}$ to $\frac{1}{R}$ in its progrefs, and in its regrefs in the fame point, as $\frac{1}{R + im}$ to $\frac{1}{R}$; and

and by the fame way of arguing, the elaftick forces of the phyfical points E and G, when in going forward they are found at x and z, will be to their ordinary elafticity, as $\dfrac{1}{R-HL}$, and $\dfrac{1}{R-KN}$ to $\dfrac{1}{R}$; and by fubducting the latter of thefe quantities from the former, the difference of thofe forces will be as

$$\frac{HL-KN}{R^2-R+HL-R+KN+HL\times KN} \text{ to } \frac{1}{R} \text{ ; or,}$$

rejecting all the terms of the divifor except the firft, as being indefinitely fmall with refpect to that, as $\dfrac{HL-KN}{R^2}$ to $\dfrac{1}{R}$; or, multiplying both fides by R^2, as HL—KN to R ; but forafmuch as R is a given quantity, HL—KN is as unity ; confequently, the difference of the forces is as HL—KN. But from the fimilarity of triangles, HL—KN is to HK, as OM to OI or OP; confequently, fince HK and OP are given, HL—KN is as OM ; or becaufe SP and EQ are equal, if EQ be bifected in C, as cy. And by the fame way of reafoning, the difference of the elaftick forces of the fame points, when in their return they are found at x and z, is as the fame cy ; but that difference, or the excefs of the elaftick force of the point x above the elaftick force of the point z, is the force by which the little line or portion of air xz, which lies between thofe points is accelerated in its progrefs ; and on the other hand, the excefs of the elaftick force of the point z above that of the point x, is the force by which the fame little line or portion of air is accelerated in its return ; fo that the force by which that little portion is accelerated, is every where as its diftance from C, the middle point of its vibration ; confequently, during its vibratory motion, it muft be accelerated and retarded in the fame manner with a pendulum vibrating in a cycloid ; inafmuch as I proved in my lecture on the pendulum, that the

T   force

force which agitates the pendulum in the forementioned manner, is every where as its diſtance from
the middle or loweſt point of the vibration. And
what has been thus proved of the little portion EG,
is in like manner demonſtrable of every other little
portion of air, thro' which the motion is propagated.

As to the velocity of ſound, or what amounts to
the ſame thing, of the pulſes of the air, if a pendulum
be made equal in length to the height of an homogeneal atmoſphere, whoſe weight is equal to that
of our atmoſphere, and its denſity the ſame with
that of the air at the ſurface of the earth; which
height is, as I ſhewed you in a former lecture, equal
to 29725 feet, and which I ſhall now denote by
the letter H; in the ſame time that ſuch a pendulum performs an intire vibration by going forward
and returning back, a pulſe of the air will move
thro' a ſpace equal to the circumference of a circle
PI. 8. deſcribed with the *radius* H. For if the little por-
Fig. 2. tion of air EG, vibrating thro' a ſmall ſpace as
PS, be acted upon at P and S, the extremities of
the ſpace thro' which it vibrates by an elaſtick force
equal to its gravity, it will perform its vibrations
in the ſame time that it would in a cycloid whoſe
length is equal to PS; becauſe equal forces muſt of
neceſſity move equal bodies thro' equal ſpaces in e-
qual times. Since then, the times of vibrations are
in the ſubduplicate *ratio* of the lengths of the pendulums, and the length of any pendulum is equal to
half of the cycloid, wherein it vibrates; the time in
which the ſmall portion of air would vibrate by the
force of its gravity in a cycloid equal in length to
PS, muſt be to the time of the vibration of a pendulum whoſe length is H, in the ſubduplicate *ratio* of
PO to H. But the elaſtick force which acts upon
the little portion of air in the extream points P and
S, was proved to be to its whole or ordinary elaſtick

5   force,

force, as HL—KN to R; that is, in the cafe before us, where the point K coincides with P, as
HK to R; for upon the coincidence of K and P,
KN vaniſhes, and HL, which in this cafe is their
difference, becomes the ſine of HK, and equal to
it, inaſmuch as HK is a naſcent arch. And the
whole elaſtick force of that little portion of air, or,
which is the ſame thing, the weight which compreſſes it, is to its own weight, as the height of the
homogeneal atmoſphere or H, to the ſmall line
EG; whence putting e to denote the elaſtick force,
which agitates the ſmall portion of air in the extream
points of its vibration P and S, and w for its weight,
W for the whole elaſtick force, or the weight
of the compreſſing atmoſphere, and reducing the
two laſt analogies into equations, we ſhall have
$\frac{HK}{R} = \frac{e}{W}$, and $\frac{W}{w} = \frac{H}{EG}$; whence multiplying
the two middle terms together, and likewiſe the
extreams, we ſhall have $\frac{e}{w} = \frac{HK \times H}{R \times EG}$; and by
ſubſtituting PO and R for HK and EG, to which
they are proportional, $\frac{e}{w}$ is equal to $\frac{PO \times H}{R^2}$; that
is, by reſolving this equation into an analogy, the
elaſtick force which agitates the little portion of air
in the extream points of the ſpace thro' which it
vibrates, is to its weight, as PO × H to R²; ſince
then, from the nature of motion, the times wherein equal bodies are moved thro' equal ſpaces, are
reciprocally in the ſubduplicate *ratio* of the moving
forces, it follows, that the time wherein the little
portion of air performs its vibration by virtue of
the elaſtick force denoted by e, is to the time
wherein it can vibrate thro' an equal ſpace by the
force of its gravity, in the ſubduplicate *ratio* of R²
to PO × H, and of courſe, to the time of the vibration of a pendulum whoſe length is H, in a *ratio*

T 2   com-

compounded of the laſt mentioned *ratio*, and of the
ſubduplicate *ratio* of PO to H ; that is, as R² × PO
to H² × PO; that is, by dividing by PO, and extracting the ſquare roots in the ſimple *ratio* of R to H.
But in the time that the little portion of air performs one vibration by going forward and returning back, the pulſe is carried thro' a ſpace equal to
AB ; conſequently, the time in which a pulſe moves
from A to B, is to the time in which a pendulum
whoſe length is H, ſwings forward and backward,
as R to H, or as BC, the circumference of a circle whoſe *radius* is R, to the circumference of a
circle whoſe *radius* is H ; but the time of the pulſe's
motion from A to B, is to the time in which it
moves thro' a ſpace equal to the circumference of
a circle whoſe *radius* is H, in the ſame proportion ;
wherefore, in the ſame time that a pendulum whoſe
length is H, ſwings forward and backward, a pulſe
will move thro' a ſpace equal to the circumference
of a circle whoſe *radius* is H, which was the thing
to be proved.

As a *Corollary* it follows, that the pulſes move
with ſuch a velocity as a heavy body acquires in
falling down half the height denoted by H ; for in
the ſame time with the fall they will, with a velocity equal to that acquired by the fall, deſcribe a
ſpace double that of the fall, that is, a ſpace equal
to H ; and of conſequence, in the time that the
pendulum vibrates forward and backward, they
will run thro' a ſpace equal to the circumference of
a circle whoſe *radius* is H. For, in my lecture on
the pendulum, I ſhewed you, that the time of the
fall thro' half the length of the pendulum, is to the
time of one vibration, as the diameter of a circle
to its circumference ; and of courſe, to the time of
a double vibration, as the *radius* to the circumference. Since then it has been proved, that the
pulſes move with ſuch a velocity as carries them
thro' a ſpace equal to the circumference of a circle
whoſe

whofe *radius* is H, in the fame time that a pendulum whofe length is H, performs a double fwing; and fince it appears that the velocity acquired by a heavy body in falling down half the height H, will carry the pulfes thro' the fame fpace in the fame time, it is manifeft, that they move with that velocity.

As a fecond *Corollary* it follows, that the velocity of the pulfes is in a *ratio* compounded of the fubduplicate *ratio* of the air's elafticity directly, and of the fubduplicate *ratio* of its denfity inverfly; for fince the velocity wherewith they move, is fuch as a body acquires in falling down half the height H, and fince the velocities acquired by falling bodies, are in the fubduplicate *ratios* of the heights from which they fall, it is manifeft, that the velocity of the pulfes is as the fquare root of H, but the height H, is directly as the air's elafticity, and inverfly as its denfity; confequently, the velocity of the pulfes is in the fubduplicate *ratio* of the air's elafticity directly, and the fubduplicate *ratio* of its denfity inverfly. Whence it appears, that the velocity of the pulfes is given, forafmuch as, *cæteris paribus*, the elafticity is as the denfity. In the *winter* time indeed, the motion of the pulfes is fomewhat flower than in *fummer*, becaufe the coldnefs of that feafon does in fome meafure weaken the elafticity, and at the fame time increafe the denfity. From what has been faid, the fpace thro' which found moves in any given time, may readily be determined; for fince it is known by experience, that a pendulum 39⅕ inches long, performs a double vibration by going forward and returning back in two feconds of time, a pendulum whofe length is H, that is 29725 feet long, will perform a like double vibration in 190¾ feconds; confequently, in that time found will move thro' a fpace equal to the circumference of a circle whofe *radius* is 29725 feet; that is, it will move thro' 186768 feet, which being

T 3
di-

divided by 190¾, gives a quotient of 979 feet, for
the space thro' which found moves in one second of
time.   But it muft be obferved, that in this computation no regard has been had to the thicknefs of
the folid particles of air, thro' which found is propagated in an inftant; if that therefore be allowed
for, the velocity of found will come out greater in
the proportion of about ten to nine; for fince the
fpecifick gravity of air is to that of water, as 1 to
870, if we fuppofe the particles of air to be equally
denfe with thofe of water, and that the greater rarity of air is owing to the greater interval between
its particles, it follows, that that interval is about
nine times as great as the diameter of a particle;
confequently, a tenth part of the fpace thro' which
found is propagated is poffeffed by the particles of
air; if therefore to 979 feet, which is the fpace
thro' which found would move in a fecond, in cafe
the particles of air had no magnitude, we add a
ninth part, or 109 feet more on account of the
thicknefs of the particles, we fhall have 1088 feet
for the fpace thro' which found is carried in a fecond
of time.   Befides, as there are vapours difperfed
thro' the air, which being of a different tone and
elafticity, do not partake of that motion of the
true air by virtue whereof found is propagated,
the moving caufe having on that account fewer particles of matter to agitate, muft of neceffity give
them a greater velocity; and from the nature of
motion it is evident, that the velocity will be greater
in the inverfe fubduplicate *ratio* of the quantity
of matter to be moved; that is to fay, if we fuppofe the atmofphere to confift of ten parts of true
air, and one part of vapours, the motion of found
will be quicker in fuch an atmofphere, than in an
atmofphere confifting intirely of true air, in the fubduplicate *ratio* of 11 to 10, or in the fimple *ratio* of
about 21 to 20. If therefore the velocity laft found
be augmented in that proportion, we fhall have

1142 feet for the space thro' which sound moves in one second of time; and this agrees with the most accurate experiments that have been made, for discovering the velocity of sound.

The space thro' which sound moves in a second of time being thus discovered, the length of the pulses excited by the vibrations of a sounding body may likewise be found, provided the number of vibrations performed by the sounding body in a given time, can by any method be determined; for since each vibration excites a new pulse, all that is requisite to be done, is to divide 1142 by the number of vibrations which the sounding body performs in a second, and the quotient will express the length of a pulse in feet. Now, the number of vibrations which a sounding body performs in a given time, has been determined by Mr. SAUVEUR, in the following manner; " Musicians having frequently " observed, that if two organ pipes which are near- " ly unisons, be made to sound together, there are " certain instants of time, and those, as well as " they can be judged of by the ear, at equal inter- " vals, wherein their joint sound is stronger, than " in the intermediate times." This, Mr. SAUVEUR with great appearance of reason, thinks is owing to the coincidence of their vibrations at those instants; for when by the coincidence of their vibrations, they strike the ear at one and the same instant, they must needs make a stronger impression upon it, than when they strike it separately one after another. Taking this for granted, he by the help of a pendulum, took the time between two successive coincidences in the vibrations of two pipes of considerable lengths, and nearly of the same tone; he made choice of long pipes, because the coincidences of their vibrations are rarer, and consequently, the intervals between the coincidences are more easily measured, in long pipes than in short ones. Having thus found the time which passed between two

suc-

succeſſive coincidences, he readily found the number of vibrations performed by each pipe in the
ſame time, they being inverſly as the numbers ex-
preſſing the proportion of the tones of the pipes ; as
for inſtance, if the time between two ſucceſſive coincidences was found to be the ſixth part of a ſecond, and the numbers which expreſſed the proportion of the tones of the pipes were 45 and 46, the
longer pipe performed 45 vibrations, and the ſhorter 46, in the ſixth part of a ſecond. From theſe
experiments he found, that a pipe, whoſe length
was about five *Pariſian* feet, had the ſame tone
with a ſtring that vibrates an hundred times in a ſecond ; conſequently, of the pulſes excited by the
ſounding of ſuch a pipe, there are about one hundred in the ſpace of 1142 *Engliſh*, or, 1070 *Pariſian*
feet ; and of courſe, the length of one pulſe is about
10 *Pariſian* feet and $\frac{7}{10}$ths, that is about twice the
length of the pipe ; whence it is probable, that the
lengths of the pulſes excited by the ſoundings of
open pipes, are in all caſes equal to twice the length
of the pipes.

In a former lecture, ſpeaking of the increaſe
which motion received by being communicated
from a ſmaller elaſtick body to a larger, I took occaſion to give a reaſon for the augmentation of
ſound in ſpeaking trumpets ; I ſhall cloſe this lecture, by accounting for it from the nature of the
pulſes of the air. From what has been ſaid in relation to the properties of thoſe pulſes, it is manifeſt,
that the greater their condenſation is, the ſtronger
is the ſound which they excite ; now, when the
voice acts upon a portion of air confined within a
trumpet, it muſt neceſſarily make a ſtronger im-
preſſion upon it, and of courſe condenſe it more,
than when it acts upon it in an unconfined ſtate ; inaſmuch as in the former caſe, the force of the voice is
wholly imployed in giving motion to that ſmall portion of air which lies within the trumpet, whereas

in

in the latter cafe, not only that portion of air is put
in motion by the force of the voice, but likewife all
that body of air which immediately furrounds it;
the air then in the trumpet being by reafon of its
confinement, more ftrongly agitated and more
clofely condenfed, than it would otherwife be,
muft at the *exit* of the trumpet, communicate to
the air without greater degrees of condenfation; and
of confequence, produce a louder found, than could
poffibly be excited by the fame force of the voice,
were it immediately impreffed on the unconfined
air.

L E C T U R E XIX.

O f L i g h t.

L IG H T, whereof I intend to treat in this
lecture, is a moft fubtile fluid, confifting of
particles exceedingly fmall, but of different magnitudes, as fhall be fhewn hereafter, which are thrown
off from luminous bodies by the vibrating motions
of their parts, with a velocity furprizingly great;
for they do not fpend above feven or eight minutes
of an hour in paffing from the fun to the earth, as
was obferved firft by Mr. ROMER, Profeffor of
Aftronomy to the late King of *France*; and after
him by others, by means of the *eclipfes* of the *fatellites* of JUPITER; for thefe eclipfes, when the earth
is between the fun and JUPITER, are obferved to
happen about feven or eight minutes fooner than
they ought to do by the aftronomical tables; and
on the contrary, when the earth is beyond the fun
with refpect to JUPITER, they happen about feven
or eight minutes later than they ought to do; fo
that in the latter fituation of the earth, they are obferved to happen fourteen or fixteen minutes later
than in the former; forafmuch therefore as the fatellites

L e c t.
XIX.

lites cannot difappear, but muft continue vifible to the eye of an obferver, till all that light which they reflect before their immerfions has paffed by the place of obfervation, it follows, that the reflected light of the fatellites fpends fourteen or fixteen minutes in paffing from one end of the diameter of the earth's orbit to the other; and confequently, half that time in moving from the fun to the earth. Hence, if the diftance of the fun from the earth be 70 millions of miles, as it muft be on fuppofition that its horizontal parallax is twelve feconds of a degree, and fuch the moft accurate obfervations of the lateft aftronomers make it; then light moves at the rate of about 150 thoufand miles in a fecond of time, and its velocity exceeds the velocity of found, in the proportion of above feven hundred thoufand to one.

The motion of light is in its own nature rectilineal, as is evident from the fhadows which all opaque bodies caft when placed in the light of the fun, or of any other luminous body; and yet the beams or rays of light in paffing out of one tranfparent body or medium into another of a different denfity, are bent and turned out of their way; or to fpeak more properly, they are made to change the direction of their motion; and this bending or change of direction is commonly called *refraction*; and it has been found by experience, that the rays in paffing out of a rarer *medium* into a denfer, are bent in fuch a manner as to be brought nearer to a line drawn perpendicular to the refracting furface at the point of incidence; and on the contrary, in their paffage out of a denfer medium into a rarer, they decline from the perpendicular.

Pl. 8.
Fig. 3.
For the illuftration of which, let AB reprefent a ray of light moving in air from A to B, and paffing into water at B, and let HK be perpendicular to the furface of the water at the point B; when the ray goes into the water, it does not continue its

motion

motion ftrait forward in the line BC, but in fome
other line as BD, which is more inclined to the perpendicular BK. And on the other hand, if the
line DB be fuppofed to be a ray of light moving
in water from D to B, and there paffing into air, inftead of continuing its motion in the direction BE,
it goes on in fome other direction as BA, which
being lefs inclined to, is more diftant from, the perpendicular BH ; as will appear from the following experiment. Let an empty veffel as BCDE,
have a fmall object as A, placed at its bottom ; and
let it be fo fituated as that the fight of the object
may be intercepted by the fide of the veffel from
an eye placed at Q; let then the veffel be filled
with water, and the ray AB, which before the pouring in of the water, moved in a right line from A
to K, and by fo doing paffed above the eye, will
upon its emerfion out of the water be bent downward, fo as to ftrike upon the eye, and thereby
render the object vifible.

Lect.
XIX.

Exp. 1.
Pl. 8.
Fig. 4.

This bending of the rays in their paffage out of
one medium into another, feems to be owing to the
attractive force of the denfer medium acting upon
the rays at right angles to the furface, as may appear by confidering the confequences of fuch an
attraction.

Let then AC be a ray of light moving from A
to C, and there entring into a denfer medium, the
furface which feparates the two mediums being denoted by the line HK. The motion of the ray in
the direction AC, being refolved according to the
known method into two, one in the direction AD,
and the other in the direction AB or DC, whereof
the former is parallel, and the latter perpendicular
to HK ; it is manifeft, that as the ray enters into
the denfer medium at C, its perpendicular motion
muft be accelerated by the attraction, whilft its parallel motion continues the fame ; let then the line
CG be taken in the fame proportion to CD, that
the

Pl. 8.
Fig. 5.

LECT.
XIX.
the velocity of the perpendicular motion after refraction has to the velocity thereof before the refraction; and forasmuch as the parallel motion is
the same before and after refraction, let CE be taken equal to AD or BC, and letting fall EF equal
and parallel to CG, and drawing the diagonal CF,
the ray after refraction will describe the line CF in
the same time that it moved from A to C before
the refraction; and forasmuch as GF is equal to
AD, LM, that is, the sine of the angle MCL,
must be less than AD, the sine of ACD; confequently, by the attraction of the denser medium,
the ray in passing into that medium is brought
nearer to the perpendicular.

Again, let FC denote the motion of a ray in the
denfer medium from F to C, and let this motion
be refolved into two others, one in the direction
FG or EC, and the other in the direction FE or
GC, the former being parallel, and the latter perpendicular to HK; when the ray paffes into the rarer *medium* at C, the parallel motion does not fuffer any change from the attraction; but the perpendicular motion is retarded by the attractive
force, which in this cafe acts in direct oppofition to
it; let then CD be to GC, as the perpendicular velocity of the ray in the rarer *medium*, to the perpendicular velocity thereof in the denser; and let
DA be drawn equal and parallel to FG, in order to
denote the parallel motion of the ray after refraction; and the diagonal CA will be the line defcribed by the ray after refraction, in a fpace of time equal
to that wherein it defcribed the line FC before refraction; and forafmuch as AD is equal to GF, it
muft be greater than LM; confequently, the angle
ACD is greater than FCG; and therefore, the ray
in paffing out of a denfer *medium* into a rarer, is by
the attraction of the denfer *medium* bent from the
perpendicular; fo that in both cafes, the refraction
feems to be owing to the attractive force of the
denfer

denfer medium, acting upon the rays at right angles
to its furface; and what farther confirms this opinion is, that the denfer any medium is, and confequently the ftronger its attraction, the greater,
cæteris paribus, is its refractive power; thus oil of
vitriol, whofe denfity exceeds the denfity of water
in the proportion nearly of three to two, acts more
forcibly than water on the rays of light, in bending
and turning them out of their way; as will appear
from the following experiment; let the fixth figure
reprefent a quadrant, whofe *radius* A B is parallel
to the horizon; and let A be a fmall coloured object, placed on the limb of the quadrant at the extremity of the horizontal *radius*; this being viewed
thro' an empty glafs veffel as C, of a prifmatick
form, placed at the center of the quadrant, with
its refracting angle downwards, will appear in its
real place at A. Let then the veffel be filled with
water, and let the object be raifed on the limb of
the quadrant as high as D, that is to fay, to the
height of fifteen degrees and twenty minutes, and
the rays as D B, which go from it towards the prifm,
will be fo bent in paffing thro' the water as to enter the eye in a direction parallel to the horizon,
and reprefent the object as if placed at A. And
the fame thing will happen when the veffel is filled
with oil of vitriol, excepting only that the object
muft be raifed to a greater height, fuppofe to E, fo
as to have an elevation of twenty degrees and eight
minutes; which plainly fhews, that the rays are
more bent, and fuffer a greater refraction under
the fame circumftances from oil of vitriol, than
they do from water.

Exp. 2.
Pl. 8.
Fig. 6.

Refractions taken by the Quadrant and prismatick veſſel.

	Denſity.	Degrees and min.	Sines.
Water	I	15.20	2644342
Oil of vitriol	1.497	20.8	3442060
Salt water	1.2	17.52	3068029
Spirit of hartſhorn	1.011	16.	2756374
Spirit of wine	0.835	17.	2923717
Oil of turpentine	0.869	22.34	3837582
Oil of linſeed	0.939	22.57	3889277

The denſer medium begins to attract the rays at
ſome diſtance from its ſurface, and it acts upon
them more and more forcibly in proportion as their
diſtance from its ſurface leſſens; but however, in
what follows I ſhall ſuppoſe the attractive force to
act with the ſame vigour in all parts of the ſpace
thro' which it extends itſelf; becauſe, as that ſpace
is indefinitely ſmall, no ſenſible error will ariſe from
ſuch a ſuppoſition. If then CD be the ſurface of
the denſer medium, and AB the ſpace thro' which
the attractive force extends itſelf from A to B; a
ray of light in paſſing from B to A will be accelerated in ſuch a manner, as that the perpendicular velocity thereof at the point A will be equal to the
ſquare root of the ſum of the ſquare of the perpendicular velocity of the ray at its incidence on the
point B, and of the ſquare of the perpendicular velocity

Pl. 8.
Fig. 7.

locity which it would have at A, suppofing it began its motion at B, from a ftate of reft. For fince
the attractive force is fuppofed to act uniformly thro'
the fpace BA, the motion which it generates will
as to its properties correfpond with the motion arifing from gravity; if therefore the triangle EGH Pl. 8.
be taken to denote the fpace BA, GH will exprefs Fig. 8.
the velocity of a ray at A, on fuppofition that from
a ftate of reft it begins its motion at B; but if at B
it has a velocity expreffed by any right line as IK,
parallel to GH, let the triangle be continued on till
the portion IFLK becomes equal to EGH, and
FL will exprefs the velocity of the ray at the point
A; and forafmuch as the triangle EFL, is equal
to the fum of the two triangles EGH and EIK, FL
is equal to the fquare root of the fum of the fquares
of GH and IK; that is, the perpendicular velocity
of the ray at A, is equal to the fquare root of the
fum of the fquare of the perpendicular velocity of
the ray at its incidence on the point B, and of
the fquare of the perpendicular velocity which it
would have at A, on fuppofition that it began its
motion at B from a ftate of reft. And this being
fo, the courfe and velocity of a ray of light after
refraction, in paffing out of a rarer *medium* into a
denfer, may be determined in the following man- Pl. 8.
ner. Let Z be a rarer *medium*, and X a denfer, fe- Fig. 9.
parated by the common furface EF, on which let
a ray of light as AC, fall obliquely, and let AC
meafure the velocity of the ray in the rarer *medium*;
which velocity is the fame, whatever be the inclination of the ray. From the center C with the
radius CA, let a circle be defcribed, in which let
NM be drawn thro' the center perpendicular to
EF, and from A let fall AQ perpendicular to EF,
as alfo AO perpendicular to NC. The motion of
the ray in the direction AC being refolved into two
others, one in the direction AO or QC, and the
other in the direction AQ or OC; the line OC
will

will meafure the velocity of the perpendicular motion ; and therefore, if C P be taken to denote the perpendicular velocity generated by the attraction of the denfer medium, the line P O will meafure the perpendicular velocity of the ray in the denfer medium ; and forafmuch as the velocity of the parallel motion is no way altered by the attraction, if C V be taken equal to Q C, and V B be drawn parallel to C M, and equal to P O, it is evident, that the ray after refraction, will defcribe the line C B, and that the velocity of its motion will be meafured by that line.

As a *Corollary*, from what has been proved it follows, that the velocity of the refracted ray in the denfer medium is no way varied by varying the inclination of the incident ray ; for the fquare of B C being equal to the fum of the fquares of B V and C V, or of P O and A O, and the fquare of P O being equal to the fum of the fquares of C O and P C, the fquare of C B is equal to the fum of the fquares of A O, C O, and P C ; but the fquares of A O and C O are equal to the fquare of C A or C N ; confequently, the fquare of C B is equal to the fum of the fquares of P C and C N, which quantities continue unvaried, whatever be the inclination of the incident ray ; and therefore P N or C B is a given quantity ; that is, the meafure of the velocity, and of confequence, the velocity wherewith the rays move after refraction in the denfer medium, is always the fame, however differently inclined the rays may be to the furface of the denfer medium at their incidence thereon.

The angle A C N, which the line defcribed by the incident ray contains, with the perpendicular to the refracting furface at the point of incidence, is called the *angle of incidence* ; and the angle B C M, which the line defcribed by the refracted contains, with the perpendicular to the refracting furface at the point of incidence, is called the *angle of refraction*.

As

As a second *Corollary*, from what has been proved it follows, that the sines of these angles are to
one another in a given *ratio*; or, in other words,
that whatever proportion the sine of any one angle
of incidence bears to the sine of the corresponding
angle of refraction, the same does the sine of any
other angle of incidence bear to the sine of the respective angle of refraction. For since CB is cut
by the circle in the point T, if from B and T, BS
and TR be drawn perpendicular to the *radius*, BS
will be equal to AO, which is the sine of the angle
of incidence, and TR will be the sine of the angle
of refraction; and from the nature of similar triangles, BS is to TR, as CB to CT; that is, the
sine of incidence is to the sine of refraction in the
same proportion with two standing quantities; consequently, that proportion is given, whatever be the
inclination of the incident ray. And what has been
thus proved, with respect to the sines of incidence and refraction, when rays pass out of a rarer
medium into a denser, is in like manner demonstrable of those lines, when the rays move out of a
denser medium into a rarer, with this difference only, that whereas in the former case the angle of incidence exceeds the angle of refraction, in the latter
it is exceeded by it; for as the attraction of the denser medium by accelerating the perpendicular velocity of the rays in their passage from a rarer medium turns them out of their way, so as to bring
them nearer the perpendicular, so on the other hand,
by retarding their perpendicular velocity in their
passage into the rarer medium, it turns them out of
their way so as to remove them farther from the
perpendicular, as has been already shewn; and forasmuch as the rays are turned out of their way in
both cases by one and the same cause acting in the
same uniform manner, it is manifest, that in both
cases, they must be equally bent; consequently, as

U   much

much as the angle of incidence exceeds the angle of
refraction when a ray paffes out of the rarer medium
into the denfer, fo much muft it be exceeded by
it, when the paffage of the ray is made the contrary
way.

Now that the fine of the angle of incidence is to
the fine of the angle of refraction in a given *ratio*,
whatever be the inclination of the incident ray, may
be proved experimentally in the following manner.
Let a brafs quadrant graduated on both fides, and
fixed at its center to a perpendicular pillar in the
manner reprefented, have two indices as A and B,
one on each fide, moveable on the center C; and
let the index A, whereof the ftem D is a continuation, be made to point to the 15th degree, and
the index B to the 15th minute of the 20th degree; let then the pillar be immerfed in water, fo
far as that C E the horizontal edge of the quadrant
may touch the furface of the water, and upon
viewing the ftem D which lies within the water, it
will by reafon of the refraction feem to have changed its fituation, and appear to lie in the fame plane
with the index B.   And the fame thing will likewife obtain, if the index A be fet at the 30th degree, and B at the 30th minute of the 42d degree.
Now in both thefe cafes, the angle of incidence is
equal to the angle contained between F C, the
perpendicular edge of the quadrant, and the index A; and the angle of refraction is the angle
made by the perpendicular edge of the quadrant,
and the index B; fo that one of the angles of incidence is 15 degrees, and the other 30, and the
correfponding angles of refraction are nineteen degrees fifteen minutes, and 41 degrees 30 minutes;
and 25, which is the fine of the leffer angle of incidence, is to 33, the fine of the correfponding
angle of refraction, as 50, the fine of the greater
angle of incidence, to 66, the fine of the angle

Pl. 8.
Fig. 10.

2

of

of refraction, which correfponds thereto; as in the following TABLE.

	Angles of incidence.	Sines.	Angles of refraction.	Sines.
Out of water into air.	d. 15. d. 30.	2588 5000	d. m. 19. 15 d. m. 41. 30	3296 6626
Out of oil of tur-pentine into air.	d. 15. d. 30.	2588 5000	d. 22. d. 47.	3746 7313

LECTURE XX.

OF COLOURS.

NATURALISTS were formerly of opinion, that LIGHT was in its own nature fimple and uniform, without any difference or variety in its parts. And that COLOURS, which are to be the fubject of this lecture, were nothing elfe than certain changes or modifications of light caufed by *refractions*, *reflections*, and *fhadows*. But Sir ISAAC NEWTON, to whom we are indebted for almoft every thing that we know with certainty concerning the nature of light, has fhewn from experiments, that notwithftanding the uniform appearance of light, the particles whereof it is compofed are of different colours; and that the colour of each particle is lafting and permanent, fo as not to be changed either by refraction or reflexion. He has likewife fhewn, that thofe particles which differ

as

as to colour, differ alſo in degrees of refrangibility; by means whereof, the rays of different colours may be ſeparated from each other, and exhibited apart.

Let a beam of the ſun's light paſs into a darkened chamber thro' a round hole as H, about the ſixteenth or twentieth part of an inch wide, ſo as to fall directly on the middle of a double *convex lens* as L, ground to a *radius* of five or ſix feet, and placed at the diſtance of ten or twelve feet from the hole; by which means the image of the hole will be projected to I, on the other ſide of the *lens*, at the diſtance of ten or twelve feet more, and there appear white and round. Let then a priſm of ſolid greeniſh glaſs as P, be placed cloſe behind the *lens*, and in ſuch a poſture as that the beam of light may fall upon it perpendicular to its axis, which is an imaginary ſtrait line, running thro' the middle from one end to the other parallel to its edges; this being done, the image of the hole, inſtead of being round and white, and projected to I, will be long and coloured, and caſt ſidewiſe from I; and the colours of the image taken in their order from that which lies neareſt to I, will be *red, orange, yellow, green, blue, purple,* and *violet*; as in the image MN, where the ſeveral colours are denoted by their initial letters.

From the lengthening of the round image by the refraction of the priſm, it is evident, that of the particles of light which form the image, ſome are more refrangible than others; for were they all alike refrangible, the diſtances to which they are thrown ſidewiſe from their firſt ſituation at I, would be all equal, and of conſequence, the ſecond image would be round as the firſt.

As in the coloured *ſpectrum* the *red* lies neareſt to, and the *violet* fartheſt from I; it is manifeſt, that the red particles in their paſſage thro' the priſm, are puſhed out of their way leſs, and the violet more,

more than any other; and consequently, that the
red particles have the smallest degree of refrangibility, and the violet the greatest; and that the particles of intermediate colours have intermediate degrees of refrangibility, greater or lefs in proportion
as they lie nearer to the one or the other of the two
extreams.

This difference of refrangibility in the particles
of light, argues a difference likewife in their magnitudes; for fince one and the fame caufe, to wit,
the attraction of the glafs, acting upon them all
with equal force, and under like circumftances,
produces unequal changes in the directions of their
motions, it muft needs be that they move with unequal forces, and confequently, that their quantities
of motion are unequal, which inequality of motion
can arife from nothing elfe but the different fize
of the particles, in cafe they all move equally fwift,
as they are generally fuppofed to do; and that they
are all perfectly folid, as their power of penetrating
and diffolving the denfeft bodies, without fuffering
any change themfelves, feems to require; confequently, the particles of light which differ as to
colour, differ alfo in magnitude; thofe of violet being fmalleft, and the particles of other colours increafing continually one above another, as they are
more and more removed from the violet, and approach nearer to the red, whofe particles are largeft
of all; and here it will not be improper to obferve,
that as the red particles are of all others the largeft,
they muft on that account act with the greateft
force, and excite the ftrongeft vibrations in the nervous coat of the eye; which may be one reafon
why reds are found to be more offenfive to the
eyes, than any other colour whatever.

The feven colours whereof the long image is
compofed are permanent and lafting, and cannot
poffibly be changed, either by refraction or reflexion, as will appear from the following experiments.

U 3 Let

Let a small hole be made in the paper whereon the coloured image is formed, thro' which, let each of the seven colours pass succeffively, and falling upon a prism, be again refracted, and they will be found to continue the fame, without the leaft change or alteration; thus, the *red* when refracted, will continue totally of the fame red colour as before; neither *orange*, *yellow*, *green*, *blue*, nor any other new colour, will arife from the refraction; and the like conftancy and immutability will be found in the other fix colours, when refracted fingly and apart from the reft. And as thefe colours are not changeable by refraction, fo neither are they by reflexion; for if bodies of different colours be placed in the red light, they will all appear red, and in the blue light, they will appear blue, in the green light, green, and fo of the other colours; in the light of any one colour, they will all appear totally of that fame colour, with this difference only, that in fome the colour will be more ftrong and full, in others more faint and dilute, every body appearing moft fplendid and luminous in the light of its own colour.

Thus for inftance, if a deep red as *carmine*, and a full blue as *ultramarine*, be held together in the red light, they will both appear red; but the *carmine* will appear of a ftrongly luminous and refplendent red, and the *ultramarine* of a faint obfcure and dark red; and on the other hand, if they be held together in the blue light, they will both appear blue; but the *ultramarine* will appear of a ftrongly luminous and refplendent blue, and the *carmine* of a faint dark blue.

Since the colours of the rays are not capable of being changed either by refraction or reflexion, it is manifeft, that if the fun's light confifted of but one fort of rays, there would be but one colour in the world; and by confequence, that the variety of colours depends upon the compofition of light. It is likewife manifeft, that the permanent colours of

natural

natural bodies arise from hence, that some bodies reflect some sort of rays, and others other sorts more copiously than the rest, and upon that account appear of this or that colour. Thus *minium*, and other red bodies, reflect the red rays most copiously, and thence appear red; *violets*, and all other bodies of the like colour, reflect the violet rays in greater abundance than the rest, and thence have their colour; and so of other bodies, every body reflecting the rays of its own colour more copiously than the rest, and deriving its colour from the excess and predominancy of those rays in the reflected light; for tho' all bodies appear of the same colour, when placed together in the light of any one colour, yet every body looks more splendid and luminous in the light of its own colour than in that of any other, which puts it past dispute, that every body reflects the rays of its own colour in greater abundance, than it does the rest, and thence has its colour.

As natural bodies appear of divers colours, accordingly as they are disposed to reflect most copiously the rays originally indued with those colours, so from the different proportions which the predominant rays bear to the rest of the reflected light, arise different shades or degrees in those colours. Where the predominant rays are very numerous in proportion to the rest, the colour appears strong and full; but as the excess of the predominant rays lessens, the colour, from the mixture of the other rays, abates of its liveliness, and becomes more faint and dilute; and when all the rays are equally reflected, so as that no one kind predominates, the colour becomes white; for whiteness is a mixture of all the colours, and it is more or less intense in proportion as the reflected rays are more or fewer in number; all *grays*, *duns*, *ruffets*, *browns*, and other dark and dirty colours, down to the deepest *black*,

being

being but fo many leffer degrees of *white*, and differ-
ing from perfect whitenefs on no other account but
that they confift of a leffer quantity of light, and
confequently appear lefs glaring and luminous.

The reafon why bodies reflect this or that kind
of ray more copioufly than the reft, and confequently appear of this or that colour, depends altogether on the fize and denfity of the particles
whereof the bodies are compofed. Particles of
coloured bodies reflecting rays of different colours according to their different magnitudes and
denfities, as has been fully proved by Sir ISAAC
NEWTON, from experiments and obfervations made
on the colours of thined bodies of *air, water*, and
glafs; by the help of which he has in the fecond
book of his *Opticks*, given us a table containing
feven orders or *feries* of colours, together with the
thickneffes of the particles of air, water, and glafs,
which exhibit the feveral colours in each order;
which thickneffes are expreffed in parts, whereof
ten hundred thoufand make an inch. The firft
part of that table is here laid before you; and by
infpection thereof it will be found, that in each
order of colours, the *red* is reflected by particles of
the greateft thicknefs, and that the thickneffes of
the particles which reflect the other colours, grow
lefs and lefs, as the colours which they reflect are
more and more removed from the *red*. It is likewife manifeft from the fame table, that among the
particles which reflect one and the fame colour,
thofe which have the greateft denfity, have the
leaft thicknefs; thus for inftance, the thicknefs of
a particle of glafs which reflects the *fcarlet* of the
fecond order, is but $12\frac{2}{7}$; whereas the thicknefs of
water which reflects the fame colour, is $14\frac{3}{4}$, and
that of air ftill greater, to wit $19\frac{2}{7}$; fo that the
thickneffes of the particles which reflect any colour,
increafe as their denfities leffen; for which reafon,

particles

particles of the fame thicknefs may reflect different colours, provided their denfities be unequal; thus the particles of air which reflect the *violet* of the *fecond order*, have very nearly the fame thicknefs with particles of water which reflect the *green*, as alfo with the particles of glafs which reflect the *orange* of the fame *order*.

		Thickneffes of		
		Air.	*Water.*	*Glafs.*
The colours of the *firft order*.	Very Black	$\frac{1}{2}$	$\frac{3}{8}$	$\frac{10}{31}$
	Black	1	$\frac{3}{4}$	$\frac{20}{31}$
	Beginning of black	2	$1\frac{1}{2}$	$1\frac{2}{7}$
	Blue	$2\frac{2}{3}$	$1\frac{4}{5}$	$1\frac{11}{20}$
	White	$5\frac{1}{4}$	$3\frac{7}{8}$	$3\frac{2}{5}$
	Yellow	$7\frac{1}{9}$	$5\frac{1}{3}$	$4\frac{3}{5}$
	Orange	8	6	$5\frac{1}{8}$
	Red	9	$6\frac{3}{4}$	$5\frac{4}{5}$
Of the *fecond order*.	Violet	$11\frac{1}{6}$	$8\frac{3}{8}$	$7\frac{1}{3}$
	Indico	$12\frac{5}{6}$	$9\frac{5}{8}$	$8\frac{2}{11}$
	Blue	14	$10\frac{1}{2}$	9
	Green	$15\frac{1}{8}$	$11\frac{1}{5}$	$9\frac{5}{7}$
	Yellow	$16\frac{2}{7}$	$12\frac{1}{3}$	$10\frac{2}{5}$
	Orange	$17\frac{2}{9}$	13	$11\frac{1}{9}$
	Bright red	$18\frac{1}{3}$	$13\frac{3}{4}$	$11\frac{5}{6}$
	Scarlet	$19\frac{2}{3}$	$14\frac{3}{4}$	$12\frac{2}{3}$

From what has been faid concerning the colours of natural bodies, it follows, that if any change be made in the fize or denfity of the particles whereof a body is compofed, the colour of the body will likewife be changed; for which reafon, if two colourlefs liquors be mixed together, they may in the mixing fuffer fuch changes in the fize and denfity of their parts from their mutual actions one upon another, as to become opaque and coloured; and
fuch

LECT.
XX.
such liquors as are coloured, may for the same reason, when mixed together, either become transparent and colourless, or of such a colour as is different from the colour of either, before the mixture;
as will appear from the experiments now to be
made.

*Colours produced by the mixture of liquors void of
colour.*

1. Rosated spirit of wine, and spirit of vitriol, a *Red.*
2. Solution of mercury, and oil of tartar, *Orange.*
3. Solution of sublimate, and lime water, *Yellow.*
4. Tincture of roses, and oil of tartar,   *Green.*
5. Tincture of roses, and spirit of urine,   *Blue.*
6. Solution of copper, and spirit of sal armoniack,   *Purple.*
7. Solution of sublimate, and spirit of sal
   armoniack,   *White.*
8. Solution of sugar of lead, and the so-
   lution of vitriol,   *Black.*

*Colours arising from the mixture of such liquors as are
coloured.*

1.	*Yellow.* Tincture of saffron	}	*Green.*
	Red. Tincture of red roses		
2.	*Blue.* Tincture of violets	}	*Crimson.*
	Brown. Spirit of sulphur		
3.	*Red.* Tincture of red roses	}	*Blue.*
	Brown. Spirit of hartshorn		
4.	*Blue.* Tincture of violets	}	*Violet.*
	Blue. Solution of copper		
5.	*Blue.* Tincture of violets	}	*Purple.*
	Blue. Solution of Hungarian vitriol		
6.	*Blue.* Tincture of cyanus	}	*Green.*
	Blue. Spirit of sal armon. coloured		

7. *Blue.*

7. { *Blue.* Solution of Hungarian vitriol } *Yellow.*
   { *Brown.* Lixivium

8. { *Blue.* Solution of Hungarian vitriol } *Black.*
   { *Red.* Tincture of red roses

9. { *Blue.* Tincture of cyanus } *Red.*
   { *Green.* Solution of copper

Colours changed and restored.

1. A solution of copper, which is *green*, by spirit of nitre is made *colourless*, and is again restored by oil of tartar.

2. A limpid infusion of galls, is made *black* by a solution of vitriol, and *transparent* again by oil of vitriol, and then *black* again by oil of tartar.

3. Tincture of red roses, is made *black* by a solution of vitriol, and becomes *red* again by oil of tartar.

4. A slight tincture of roses, by spirit of vitriol becomes a fine *red*, then by spirit of sal armoniack turns *green*, and then by oil of vitriol becomes *red* again.

5. Solution of verdegrease, from a *green* by spirit of vitriol becomes *colourless*, then by spirit of sal armoniack turns a *purple*, and then by oil of vitriol becomes *transparent* again.

Among the various *Phænomena* of colours, there is none more remarkable than that of the *rainbow*, which is an appearance observable in those places only where it rains in the sunshine, and where the spectator is placed in a due position between the sun and the rain, with his back to the former; for which reason it is generally allowed, that the bow is made by the refraction of the sun's light in drops of falling rain; the manner wherein it is formed, has in some measure been explained by ANTONIUS DE DOMINIS, archbishop of *Spalato*, and after him by DES CARTES; but as neither of them understood

the

the true origin of colours, it was impoſſible for them not to be defective in their accounts; and therefore Sir ISAAC NEWTON, after he had diſco-vered the true nature and riſe of colours, ſet him-ſelf to the conſideration of this ſubject, and towards the latter end of the firſt book of his *Opticks*, has given a full and ſatisfactory account of the whole matter; the ſubſtance of what he has there deliver-ed concerning the rainbow is as follows.

Pl. 9.
Fig. 1.
Let a drop of rain, or any other ſpherical tran-ſparent body, be repreſented by the ſphere BNFG, and let AN be one of the ſun's rays, incident upon it at N and thence refracted to F, where let it either go out of the ſphere by refraction towards V, or be reflected to G; and there let it either go out by refraction to R, or be reflected to H, where let it go out by refraction towards S, cutting the in-cident ray in Y; let AN and RG be produced till they meet in X. Parallel to the incident ray AN, let the diameter BQ be drawn, and let BL be a quadrant, on every point of which let us ſuppoſe a ray to fall parallel to BQ; as the point of incidence removes from B towards L, the angle AXR which the rays AN and RG contain, will firſt increaſe, and then decreaſe; and on the other hand, the angle AYS, contained between the rays AN and YS, will firſt decreaſe and then increaſe. This be-ing ſo, if we ſuppoſe N to be that point of the quadrant BL, whereon if the incident ray AN falls, it makes the greateſt angle with the ray GR, which emerges after one reflexion; then all the rays which fall on each ſide at a very little diſtance from N, and go out after one reflexion, will emerge pa-rallel or very nearly parallel to GR; whereas thoſe which fall on the quadrant at greater diſtances from N, will notwithſtanding their paralleliſm before their incidence be ſcattered, and diverge from one another after their emergence. If therefore an eye be ſituated in the direction of the former rays

2

which

which go out parallel, they will enter it so copiously as to exhibit the image of the sun in the drop of rain which reflects them; but if the eye be so placed as to receive the latter rays which go out diverging, those which enter the eye will be too few to excite any sensation; and of consequence, the image of the sun will not appear in the drop to an eye so situated.

If N be the point, whereon if the incident ray A N falls it makes the smallest angle with the ray H S, which emerges after two reflexions; then, as before, all the rays which are incident near N, and which emerge after two reflexions, will go out parallel, and for that reason will exhibit the sun's image to an eye situated in their direction; but those rays which are incident at any sensible distance from N, and which emerge after two reflexions, will be scattered as they go out, and upon that account will be too few, and consequently too feeble to excite any sensation in the eye of the spectator.

Now, forasmuch as the rays which are of different colours have likewise different degrees of refrangibility, the greatest angle AXR which can be made by the incident rays, and those which go out after one reflexion, will be of different magnitudes in rays of different colours; so likewise will the smallest angle AYS, that can be made by the incident rays, and those which go out after two reflexions; and it has been found by computation, that in the least refrangible or red rays, the greatest angle AXR, is 42 degrees and two minutes; and the least angle AYS, 50 degrees and 57 minutes; and in the most refrangible or violet rays, the greatest angle AXR, has been found to be 40 degrees and 17 minutes; and the least angle AYS, 54 degrees and 7 minutes.

Suppose now that O is the spectator's eye, and O P a line drawn parallel to the sun's rays; and let

Pl. 9.
Fig. 2.

POE be an angle of 40 degrees and 17 minutes, POF of 42 degrees 2 minutes, POG of 50 degrees 57 minutes, and POH an angle of 54 degrees 7 minutes; and thefe angles turned about their common fide, fhall with their other fides OE, OF, OG, and OH, defcribe verges of two rainbows AFBE and CHDG. For if E, F, G, and H, be drops of rain placed any where in the conical furfaces defcribed by OE, OF, OG, and OH, and be illuminated by the fun's rays SE, SF, SG, and SH, the angle SEO being equal to the angle POE, or 40 degrees and 17 minutes, fhall be the greateft angle in which the moft refrangible rays can after one reflexion be refracted to the eye; and therefore, all the drops in the line OE fhall fend the moft refrangible rays moft copioufly to the eye, and thereby ftrike the fenfes with the deepeft *violet* colour in that region. And in like manner, the angle SFO being equal to the angle POF, or 42 degrees 2 minutes, fhall be the greateft in which the leaft refrangible rays after one reflexion can emerge out of the drops; and therefore, thofe rays fhall come moft copioufly to the eye from the drops in the line OF, and ftrike the fenfes with the deepeft *red* colour in that region. And by the fame argument, the rays which have intermediate degrees of refrangibility, fhall come moft copioufly from drops between E and F, and exhibit the intermediate colours in the order which their degrees of refrangibility require, that is, in the progrefs from E to F, or from the infide of the bow to the outfide in this order, *violet, indigo, blue, green, yellow, orange*, and *red*.

Again, the angle SGO being equal to the angle POG, or 50 degrees and 57 minutes, fhall be the leaft angle in which the leaft refrangible rays can after two reflexions emerge out of the drops, and therefore the leaft refrangible rays fhall come moft copioufly to the eye from the drops in the line OG,

and

and ſtrike the ſenſe with the deepeſt *red* in that region. And the angle SHO being equal to the
angle POH, or 54 degrees and 7 minutes, ſhall
be the leaſt angle, in which the moſt refrangible
rays, after two reflexions, can emerge out of the
drops ; and therefore, thoſe rays ſhall come moſt
copiouſly to the eye from the drops in the line OH,
and ſtrike the ſenſes with the deepeſt *violet* in that
region. And by the ſame argument, the drops in
the regions between G and H, ſhall ſtrike the ſenſes
with the intermediate colours, in the order which
their degrees of refrangibility require, that is, in
the progreſs from G to H, or from the inſide of the
bow to the outſide in this order, *red, orange, yellow,
green, blue, indigo,* and *violet.* And ſince theſe four
lines OE, OF, OG, and OH, may be ſituated any
where in the abovementioned conical ſurfaces, what
is ſaid of the drops and colours in theſe lines, is to
be underſtood of the drops and colours every where
in thoſe ſurfaces. Thus then ſhall there be made
two bows of colours, an interior and ſtronger by
one reflexion in the drops, and an exterior and
fainter by two (for the light becomes fainter by
every reflexion), and their colours ſhall be in a contrary order to one another, the *red* of both bows
bordering upon the ſpace GF, which is between the
bows. The breadth of the interior bow meaſured
croſs the colours, ſhall be one degree and 45 minutes, and the breadth of the exterior, ſhall be three
degrees 10 minutes, and the diſtance between them,
ſhall be 8 degrees 55 minutes ; the greateſt ſemidiameter of the innermoſt, or the angle POF, being 42 degrees and 2 minutes, and the leaſt ſemidiameter of the outermoſt, or the angle POG, being 50 degrees and 57 minutes. And theſe are
the meaſures of the bows as they would be were the
ſun but a point ; for by the breadth of his body,
the breadth of the bows will be increaſed, and their
diſtance leſſened by half a degree ; and ſo the
breadth

Pl 9.
Fig. 3.

breadth of the interior will be 2 degrees 15 minutes, and that of the exterior 3 degrees 40 minutes, and their diftance 8 degrees 25 minutes; the greateft femidiameter of the interior bow 42 degrees 17 minutes, and the leaft of the exterior 50 degrees 42 minutes; and fuch Sir ISAAC NEWTON fays he has found the dimenfions of the bows in the Heavens, when he meafured the fame. This explication of the rainbow is confirmed by the following experiment; let a glafs globe filled with water, as AB, be hung up in the fun-fhine, with a black cloth placed behind it, and let IS be one of the fun's rays incident thereon; let the eye of a fpectator whofe back is to the fun, be placed at O, and let it be directed to fuch a point in the lower part of the globe fuppofe C, as that a ftrait line drawn from the eye thro' that point, and continued on till it meets the incident ray likewife produced, may therewith make an angle OXI, of 42 degrees 2 minutes; and the fpectator fhall then fee a full *red* colour in that fide of the globe oppofed to the fun as at F; let then the eye be raifed up gradually to P, till the angle PZI becomes equal to 40 degrees and 17 minutes, and as the eye rifes, it will perceive other colours, to wit, *yellow*, *green*, and *blue*, fucceffively in the fame fide of the globe.

Again let the eye be placed at Q, and let it be directed to fuch a point in the upper part of the globe fuppofe D, as that a ftrait line, drawn from the eye thro' that point and meeting the incident ray protracted, may therewith make an angle QSI of 50 degrees and 57 minutes, and there will appear a faint red colour in that fide of the globe towards the fun; let then the eye be gradually depreffed to R, till the angle RTI is 54 degrees 7 minutes, as the eye finks, the *red* will turn fucceffively to the other colours, *yellow*, *green*, and *blue*, as in the former cafe upon the raifing of the eye.

LECTURE XXI.

Of Dioptricks.

INTENDING in my next lecture to enquire into the Nature of Vision, where I shall have occasion to take notice of *Defective Eyes*, I shall in this lecture, by way of preparation, lay before you some of the chief properties of such *lenses* or glasses as are most commonly in use for assisting defective eyes; and they are of two sorts, First, such as are equally convex on both sides, and secondly, such as are on both sides equally concave. The former sort is represented in the fourth figure, and the latter in the fifth.

Let ABC be an object placed before the double Pl. 9. *convex lens* HK at any distance greater than the Fig. 6. *radius* of the sphere, whereof the *lens* is a segment; the rays, which issue from the several points of the object, and fall upon the *lens*, will in their passage thro' it be so bent by the refractive power of the glass, as to be made to convene at so many other points behind the *lens*, and at the place of their concourse they will form an image or representation of the object; and this image will be inverted, because the rays which flow from A, the uppermost point of the object, are united at F, the lowermost point of the image, whilst those which flow from C, the lowerest point of the object, are brought together again at D, the highest point of the image. So likewise those rays which issue from the right side of the object, are united in the left side of the image, whilst those which proceed from the left side of the object, concur in the right side of the image; as will appear by placing a lighted candle before a double *convex lens*, at such a distance

X   tance

LECT.
XXI.

tance as that the image thereof may be formed on a piece of white paper placed at a due diſtance behind the *lens*; for the flame will appear inverted with its point downward; and if either ſide of the flame be intercepted by the interpoſition of a dark body, the contrary ſide of the image will be obſcured.

With regard to this experiment, I muſt obſerve to you, that tho' there is one certain diſtance, at which the paper muſt be placed, in order to exhibit the image with the greateſt diſtinctneſs, yet may the diſtance be a little varied without rendring the image confuſed; and it is remarkable, that when the image is projected on the paper at the neareſt diſtance that it can with any degree of diſtinctneſs, it appears bordered all around with red; which redneſs continually decreaſes, as the paper is more and more removed from the *lens*; and when it is removed to ſuch a diſtance as is requiſite to give the image the greateſt advantage in point of diſtinctneſs, the redneſs intirely vaniſhes, and leaves the image equally white all over; but upon a farther removal of the paper, the edges of the image which at the neareſt diſtance were tinged with red, do now appear tinged with blue. If a candle, which is placed at A before the *convex lens* CD, has its image projected on a paper at EF, ſuppoſing that to be the leaſt diſtance at which it can be projected diſtinctly, its edges will appear red, but upon the removal of the paper to GH, they will become white; and when the paper is removed to IK, they will appear blue; the reaſon of theſe different appearances is this, the rays of light as AC and AD, which flow from the candle, being compounded of particles of different colours, whereof the red are leaſt refrangible, and the blue moſt ſo, upon paſſing thro' the *lens*, the blue rays are made to convene ſooneſt, and the red lateſt; as in the figure where the blue are denoted by the pricked lines, and the

Pl. 9.
Fig. 7.

red

red by the continued; so that an image is formed at EF, by the concurrence of some of the more refrangible rays, and it is tinged around its edges by the red rays, which converging more slowly than the rest lie outermost.

After the blue rays have concurred, they cross one another, and go on diverging towards GH, where meeting with the red rays which have not yet concurred, and there mixing with them and the rays of other colours, they produce a white image, whiteness resulting from a due mixture of all the colours; as they proceed forward toward IK, they, by reason of their greater divergence, spread themselves on all sides beyond the other rays, and by so doing, tinge the outlines of the image blue.

On the formation of pictures by means of a double *convex lens*, depend the appearances of the *camera obscura*, which is a small square box with a tube issuing horizontally from one side, at the extremity whereof is fixed a double *convex lens*; within the box is placed a looking-glass in a slanting position, so as to be at half right angles with the bottom of the box, which is parallel to the horizon. On the top of the box is placed horizontally a plate of glass rough on one side, whereon the pictures of objects are represented in the following manner.

Let AB be an object placed before CD, the *lens* fixed in the tube which issues from the box; GH the looking-glass inclined to the bottom of the box, in an angle of 45 degrees, LM the plate of rough glass covering the top of the box horizontally. The rays which flow from A, the uppermost point of the object, after they have passed the *lens*, converge towards F, and would actually meet at that point, but that they are intercepted by the looking-glass GH, which reflects them, and throws them upward; and forasmuch as the inclination of the rays towards one another is no way altered by the reflexi-

Pl. 9.
Fig. 8.
Exp. 2.

X 2

on, they muſt meet at ſome point as K, as far diſtant above the *ſpeculum*, as the point F is behind it.
In like manner, the rays which flow from B, the
loweſt point of the object, and which after they have
paſſed the glaſs are tending towards E, being reflected upward by the *ſpeculum*, are made to convene at I, whoſe diſtance above the *ſpeculum* is
equal to the diſtance of E behind the *ſpeculum* ; and
as the rays from the extream points A and B, are
made to convene at K and I, ſo thoſe which flow
from the intermediate points of the object, are
brought together at correſponding points between
K and I, whereby the image is projected horizontally, but with its right and left ſides correſponding to the contrary ſides of the object ; as may appear by placing a man before the *lens*, and cauſing
him to ſtir one of his hands ; for in the image the
other hand will appear to move.

The diſtance of the image behind the glaſs is always varied by varying the diſtance of the object
before the glaſs ; the image approaching as the object recedes, and receding as that approaches. For
Pl. 9.
Fig. 9. if we ſuppoſe A and C to be two radiating points,
from which the rays AH, AK, and CH, CK fall
upon the *lens* HK, it is manifeſt, that the rays from
the more diſtant point diverge leſs than thoſe from
the nearer point, the angle at A being leſs than that
at C ; conſequently, when they paſs thro’ the glaſs
they muſt be brought together ſooner, and muſt
convene at ſome point as B, leſs diſtant from the
lens, than is the point D, whereat the more diverging rays from the point C are made to convene.

Where the diſtance of the object, and the *radius*
of the *lens*’s convexity are given, and where the
thickneſs of the *lens* is but ſmall, as is commonly
the caſe ; the diſtance of the image from the *lens* is
determined very nearly, by ſaying, as the diſtance
of the object from the *lens*, leſſened by the *radius*

　　　　　　　　　　　　　　　　　of

of the *lens*'s convexity, is to the *radius*, fo is the diftance of the object from the *lens*, to the diftance of the image from the *lens*; that is, putting D for the diftance of the object, R for the *radius* of the convexity, and F for the diftance of the image,

$$D-R : R :: D : F; \text{ confequently, } F = \frac{RD}{D-R}.$$

The truth of this rule is demonftrated by the writers of DIOPTRICKS; but as all the demonftrations which I have hitherto met with are tedious and intricate, I fhall not at prefent trouble you with them, but fhall proceed to confirm the rule by experiments.

Let then the flame of a candle be placed at the diftance of twelve feet and an half from a double *convex lens*, the *radius* of whofe convexity is four feet two inches; that is, let the diftance of the flame from the glafs be equal to thrice the *radius*, and the image will be projected behind the *lens* at the diftance of fix feet three inches, that is, at the diftance of a *radius* and an half; for in this cafe, R being put equal to unity, RD is three, which being divided by D — R, that is, by two, gives one and an half in the quotient.

If the flame be brought nearer to the *lens*, the image will move farther from it, and when the diftance of the flame becomes equal to twice the *radius* of the *lens*'s convexity, the diftance of the image will be equal to that of the flame, the *lens* ftanding in the midway between them; for in this cafe D—R is equal to R, and of confequence, F is equal to D.

The flame being placed at the diftance of the *radius*, the diftance of the image becomes infinite. For in this cafe D—R is nothing, and F is equal to $\frac{DR}{0}$, which expreffion denotes an infinite quantity; fo that in this cafe, there will not be any image of

the

the flame; but the rays of light which flow from
the candle, after they have paſſed thro' the *lens*,
will go on parallel to one another; and by ſo doing, form a bright circular image, equal in ſize to
the *lens*, and the magnitude thereof will remain the
ſame at all diſtances from the glaſs.

Where the diſtance of the flame is leſs than the
radius of the convexity, D—R becomes a negative
quantity, and ſo of conſequence does the quotient
ariſing from the diviſion of DR by D—R; which
ſhews, that the place at which the rays meet, lies on
the ſame ſide of the *lens* with the flame; or to ſpeak
more properly, that the rays after they have paſſed
the *lens*, proceed diverging from one another in
ſuch a manner, as if they had flowed from a point
before the *lens*, more diſtant than the place of the
flame. For the eaſier underſtanding of which, let

Pl. 9.
Fig. 10.
the rays AB and AC flow from the point A, whoſe
diſtance from the *lens* BC, is leſs than the *radius*
of the *lens*'s convexity; after they have paſſed the
glaſs, they will not continue to go on in the directions BD and CE, but in the directions BF and CG,
as if they had proceeded from ſome point as H,
more diſtant from the *lens* than is the point A, from
which they really flow; ſo that in this caſe, the rays
after they paſs the glaſs, go on diverging from one
another, but however they do not diverge as much
as they did before they paſſed the glaſs.

When the diſtance of the flame from the glaſs is
ſo great as that neither the breadth of the *lens*,
nor the *radius* of its convexity bears any ſenſible
proportion to it, then D—R is equal to D; and of
conſequence, F is equal to R; that is, the diſtance
of the image is equal to the *radius* of the glaſs's
convexity, and this is the leaſt diſtance at which an
image can be projected by ſuch a *lens*; and foraſmuch as the rays of the ſun, which by reaſon of
the immenſe diſtance of his body are always united

at

at the fmalleft diftance, are apt to burn at the place of their union; that place is ufually called the *focus* or *burning point*, and fometimes the *abfolute focus*, in contradiftinction to thofe places whereat the images of lefs remote objects are formed, and which are frequently called the *refpective foci*.

The length or breadth of an object, is to the length or breadth of its image, as the diftance of the object from the *lens*, to the diftance of the image from the *lens*. For if AC be the length or breadth of an object, and D F the length or breadth of its image; AB, which is one half of AC, is to F E, which is one half of FD, as BL to EL, the triangles ABL and FEL being fimilar. Hence it follows, that the nearer an object approaches the *lens*, the larger is its image, the image receding, and confequently inlarging, as the object approaches; and thus it appears to be from experiments; for the flame of a candle being placed at a diftance greater than the diameter of the *lens*'s convexity, in which cafe the diftance of the image is lefs, appears larger than the image, but being brought within the diftance of the diameter, the image, which in that cafe is at the fame diftance, becomes equal to it; and upon bringing the flame ftill nigher, the image becomes larger in proportion to the fquare of its greater diftance.

The fame thing is likewife evident from the magick lantern; which is a lantern out of which iffues an horizontal arm, capable of being lengthened or fhortened at pleafure, by means of one part fliding in and out of the other; to the extremity of the moveable part is fitted a double *convex lens*; and to that part of the arm which joins the lantern is adapted a glafs, plane on one fide, and *convex* on the other, the plane fide looking towards the lantern; in the body of the lantern there is placed a candle, whofe diftance from the *plano-convex* glafs

Pl. 9. Fig. 6.

X 4

is

is somewhat less than the focal distance; so that the
light which passes thro' that glass, is thrown very
strongly upon little images painted in dilute colours
on pieces of plane thin glass; which being fixed in
a slider that moves to and fro across the arm, are
placed at a small distance behind the *plano-convex*
glass in an inverted position, and by means of the
lens in the moveable part of the arm, are projected
in an erect position, on a paper or white cloth placed
at a proper distance; if by drawing out the moveable part of the arm, the pictures be removed to a
greater distance from the *lens*, the lantern must be
brought nearer to the cloth, in order to a distinct
representation; because, as the object recedes from
the *lens*, the image approaches, and at the same
time the images will be diminished. But on the
other hand, if by thrusting in the arm the pictures
be brought near the *lens*, the lantern must be removed farther from the cloth, and in this case the
images will appear larger.

As convex glasses cause the rays of light to converge and unite, so those which are concave make
them separate and diverge; for which reason, if
diverging rays fall upon a concave *lens*, they will
diverge more after they have passed thro' it, than
they did before; and such rays as converge before
their incidence, will after their passage converge
less; for instance, if the rays A B and A C, which

Pl. 9.
Fig. 11.
diverge from A, pass thro' the concave *lens* B C,
they will not go on in the directions B D and C E,
but in some other directions as B H and C G, so as
to widen faster than before. On the other hand, if
H B and G C be two rays converging towards K,
after they have passed thro' the glass, they will not
go on towards K, but towards a more distant point
as A, so as to converge more slowly than before.
All which is fully confirmed by experiments. For

Exp. 8. a candle being placed before a *convex lens*, so as to
have

have its image projected on a white paper, placed at a due diftance behind the *lens*, if a concave glafs be placed between the *convex* and the image, fo as that the rays which are converging towards the image may pafs thro' it, the image will thereby be thrown to a greater diftance behind, the rays being made to converge more flowly, and of confequence, to meet at a greater diftance than they did before the concave was interpofed; and it muft be obferved, that as the image is thrown to a greater diftance, it muft for that very reafon be inlarged; and forafmuch as the larger image is compofed of the fame number of rays, or rather fewer, fome of the rays being reflected by the concave *lens*, it muft on that account appear lefs bright and luminous than the fmaller. If by the removal of the *convex-lens*, the rays which flow from the candle be fuffered to fall diverging on the concave, and a white paper be placed clofe behind the glafs, there will appear thereon a dark circle of fome breadth, occafioned by the fhadow of the hoop which contains the glafs; and the circular *area* contained within the fhadow will be inlightened by the rays which pafs thro' the glafs; and becaufe all the rays which fall upon the glafs do not pafs thro' it, fome of them being reflected, the circular *area* will appear fomewhat darker than the other parts of the paper, which are expofed to the light of the candle, without the interpofition of the glafs; upon removing the paper gradually from the glafs, the circular *area* will gradually inlarge, and as that inlarges, the fhadow which environs it will grow narrower, and at length vanifh; and upon the vanifhing of the fhadow, if the paper be removed a little farther, there will arife a bright circle all around the circular *area*, which will grow broader, but lefs bright, as the paper is more and more removed from the glafs; and at the fame time, the circular *area* will continue to widen, and grow darker. All which

appear-

appearances are the natural and neceſſary conſe-
quences of the divergency or ſpreading of the rays,
occaſioned by their paſſage thro' the glaſs; for the
farther they go from the glaſs, the more they muſt
diverge, and by ſo doing, muſt on all ſides ſpread
themſelves into the place of the ſhadow, and render
it equally luminous with the reſt of the *area*; and
when they have ſpread themſelves a little beyond
the limits of the ſhadow, they fall upon ſuch parts
of the paper as were before inlightened, and there,
by their additional light, exhibit that bright circle
which ſurrounds the darker *area*; and the bright
circle, by the farther ſpreading of the rays, as the
paper is more and more removed from the glaſs,
grows broader and leſs luminous; as does likewiſe
the circular *area*, from the ſpreading of the rays
wherewith it is inlightened.

Tho' concave glaſſes do not collect the rays of
light, and conſequently, have not a real *focus*; yet
inaſmuch as the rays after they have paſſed thro'
ſuch glaſſes, do flow in ſuch a manner as that they
either tend to ſome point behind the glaſs, or appear to flow from ſome point before it, thoſe points
are uſually called the *foci*; and in double concaves
of equal concavities, the *foci* for converging rays are
found, by ſaying, as the *radius* of the glaſs's concavity leſſened by the diſtance of the point of convergence from the glaſs, is to the *radius*, ſo is the diſtance of the point of convergence to the *focus*. And
the *foci* for diverging rays are found, by ſaying, as
the ſum of the *radius* and the diſtance of the point of
divergence from the glaſs, is to the *radius*, ſo is the
diſtance of the point of divergence to the *focus*. So
that putting F for the *focus*, R for the *radius*, and
D for the diſtance of the point of convergence, or
divergence, $F = \dfrac{RD}{R \times D}$; the negative ſign being to
be prefixed to D when the rays converge, and the
affirmative when they diverge.

The

The demonſtration of this *Theorem* I ſhall for the preſent omit, on account of its tedioufneſs and intricacy, and ſhall cloſe the lecture with this ob-ſervation ; that if rays which are converging to-wards a *focus* be intercepted by a concave *lens*, whoſe diſtance from the *focus* is equal to the *radius* of its concavity, after they have paſſed thro' the glaſs, they will ceaſe to converge and become parallel, for R and D being equal, R — D is o ; conſe-quently, F is infinite ; that is, the point to which the rays converge, is at an infinite diſtance, and the rays of courſe muſt be parallel.

L E C T U R E   XXII.

Of Vision.

M Y deſign in this lecture, is to explain the manner of Vision with the naked eye ; and likewiſe to ſhew you, what aſſiſtances the ſight re-ceives from glaſſes ; and in order thereto, I ſhall give you a ſhort deſcription of the eye.

If a ſmall portion be cut off of a globe, and in the room thereof a portion of a ſmaller globe, but of an equal circular baſe, be ſubſtituted, the com-pound will exhibit the true figure of the eye ; for it is of a globular form, but more *convex* before than in any other part. It conſiſts of ſeveral mem-branes which lie contiguous one to another, of which the outermoſt is called the *tunica adnata* or *conjunc-tiva* ;· it has its riſe from that membrane which in-veſts the ſkull, and it covers the whole ball of the eye, except the foremoſt tranſparent part ; that portion of it which is viſible, is called the *white of the eye*. Beſides, this membrane, which is not reckoned among the proper coats of the eye, there are three others, which conſtitute the proper coats ; the firſt of which is called the *ſclerotica*, it is a tough

membrane

membrane derived from the *dura mater*, which paſſes to the eye from the brain along with the *optick nerve*, and is thence propagated over the whole globe of the eye; on the fore part it becomes tranſparent like thin poliſhed horn, which has given anatomiſts occaſion to make two membranes of it, and to call the tranſparent part *cornea*; this part is repreſented by A B F.

Pl. 9.
Fig. 12.

The ſecond membrane, called *tunica choroides*, is derived from the *pia mater*, and tranſmitted likewiſe from the brain along with the *optick nerve*; this is much thinner and tenderer than the former, and tinged on the hinder part with a black liquor. The fore part is called the *uvea*, and ſometimes the *iris*, from its variety of colours. In its middle is a ſmall hole called the ſight or pupil; the *iris* conſiſts of ſeveral circular concentrick muſcular fibres, which are cut acroſs at right angles by other ſtrait fibres in the manner of ſo many *radii*; by the contraction of the former the pupil is leſſened, and is inlarged by the contraction of the latter.

The third coat is uſually called the *retina*, and ſometimes the *nervous coat*, being nothing elſe but the *optick nerve*, which ſpreads itſelf in the form of a membrane over the bottom of the eye, overagainſt the ſight. Theſe coats lying contiguous, form a *capſula* or bag, wherein are contained the three humors of the eye, called the *aqueous*, the *chryſtalline*, and the *vitreous*.

At a little diſtance behind the pupil is placed the chryſtalline humor, which is *convex* on both ſides, but ſomewhat flatter before than behind; it is ſupported by ſmall muſcular fibres, called the *ciliary ligaments*, which are inſerted into the edges of the chryſtalline humor at one end, and at the other, into the *tunica choroides*, and being cloſely united, form a kind of membrane, whereby the cavity of the eye is divided into two parts; in the foremoſt of which is lodged the *aqueous humor*, ſo called,

becauſe

becaufe in confiftence and colour it fomewhat refembles water, being almoft equally limpid and
tranfparent. In the hindmoft is lodged the *vitreous
humor*, which has its name from the refemblance it
is fuppofed to bear to melted glafs.

It has been generally thought by anatomifts, that
the humors of the eye are of different denfities,
and that the chryftalline is much more denfe than
either of the other two ; but Doctor ROBINSON has
informed us in his lecture on the eye, that upon
weighing thefe humours in an hydroftatical balance, he found the *aqueous* and *vitreous* to be very
nearly of the fame fpecifick gravity ; and that the
fpecifick gravity of the chryftaline, did not exceed the fpecifick gravity of the others, in a greater proportion than that of eleven to ten ; whence
it follows, that the chryftalline is not of fuch great
ufe in bringing the rays together, and thereby
forming on the *retina* the pictures of outward objects, as it has been commonly thought to be by
optical writers ; for tho' in fhape it refembles a double *convex lens*, and on that account is fitted to make
the rays converge, yet forafmuch as it is fituated
between two humors, which are nearly of the
fame denfity with itfelf, it can have but little force
on the particles of light ; for they are found by experience, to be refracted very little in paffing out
of one *medium* into another, when the difference in
the denfities of the *mediums* is but fmall.

Behind all the coats and humors is fituated the
optick nerve, which paffes out of the fkull thro' a
fmall hole in the bottom of the *orbit* which contains
the eye. O reprefents the *optick nerve*, SS the *fclerotica* or outermoft coat, whofe foremoft tranfparent part A B F, is the *cornea*, C C is the *choroides*,
the fore part whereof A P, and F P conftitutes the
uvea or *iris*, with the pupil P P in the middle ; R R
is the *retina*, A D and F E the *ciliary ligaments*, D E

Pl. 9.
Fig. 1.

the

the *chryſtalline humor*, VV the *vitreous humor*, and WW the *watry humor*.

Underneath the white of the eye are inſerted into the *ſclerotica* ſix muſcles, which take their riſe from different parts of the orbit, and are diſtinguiſhed by different names, taken from the different motions which they give the eye ; their tendons ſpread themſelves over the *ſclerotica*, ſo as to terminate in the confines of the *cornea* ; by which means, when the ſix muſcles act together, they preſs the ſides of the eye towards each other, whereby the eye is lengthened, and at the ſame time the convexity of the *cornea* is increaſed ; both which effects are in ſome caſes abſolutely neceſſary in order to diſtinct viſion, as will appear preſently.

Having given this ſhort account of the conſtituent parts of the eye, I now proceed to lay before you, the *manner of viſion*. If an object as A B, be placed at a convenient diſtance before the eye, the rays which flow from the ſeveral points of the object, and falling on the *cornea* paſs thro' the pupil, will be brought together by the refractive power of the eye on ſo many correſponding points of the *retina*, and there paint the image or repreſentation of the object, in the ſame manner as the images of objects placed before a *convex lens* are exhibited on white paper, placed at a proper diſtance behind.

Pl. 9.
Fig. 13. Thus the rays which flow from the point A, are united on the *retina* at C, and thoſe which iſſue from B, are collected at D ; and in like manner, the rays which proceed from the intermediate points of the object, are again united at ſo many intermediate points on the *retina*. On this union of the rays at the bottom of the eye, depends diſtinct

Pl. 9.
Fig. 14.
Pl. 10.
Fig. 1. viſion, for ſhould they be united before they arrive at the *retina*, or ſhould the point of their union lie beyond the *retina*, it is evident, that the rays from each point muſt take up ſome ſpace on the *retina*, and

and of confequence, thofe which flow from contiguous points of the objeƈt will be mixed and blended together on the fund of the eye, fo as to exhibit a confufed reprefentation of the objeƈt.

Now forafmuch as the rays which fall upon the eye from radiating points, whofe diftances from the eye are different, have different degrees of divergence, the divergency of the rays increafing as the diftance of the radiating point leffens, and leffening as that increafes; and whereas thofe rays which have greater degrees of divergence, require a ftronger refraƈtive power to bring them together at a given diftance, than what is requifite to make thofe meet which diverge lefs, it is manifeft, that in order to fee objeƈts diftinƈtly at different diftances, the eye muft have a power of increafing and leffening its refraƈtive force, and thereby of adapting itfelf to the different diftances of objeƈts; and this it does by means of the fix mufcles which are inferted into the *fclerotica*; for when a radiating point is placed fo near, as that the rays which iffue from it fall upon the eye with a confiderable degree of divergence, the mufcles aƈt ftrongly on the eye, whereby the *cornea* is rendered more convex, and of confequence refraƈts the rays with greater force; befides by the lengthening of the eye from the joint aƈtion of the mufcles, the *retina* is removed to a greater diftance from the *cornea*, by which contrivance, the rays are made to convene at the *retina*, notwithftanding the great degree of divergence wherewith they enter the eye. As the radiating point recedes from the eye, and the divergency of the rays of courfe grow lefs, the mufcles relax themfelves in order to leffen the convexity of the *cornea*, and to fhorten the eye, a lefs convexity of the *cornea*, as alfo a lefs diftance between the *cornea* and *retina*, being requifite to diftinƈt vifion in greater diftances of the objeƈt than in fmaller.

Tho'

Tho' moſt mens eyes are ſo framed as to be able
to ſee diſtinctly at different diſtances, yet ſome there
are which are defective in this point, as being unable
to ſee any thing diſtinctly but when placed very
near; and this is the caſe of their eyes who are
called *myopes*, purblind, or ſhort-ſighted; in ſuch
the *cornea* is too convex in proportion to the length
of the eye; for which reaſon, all thoſe rays which
iſſue from diſtant points, and of conſequence diverge
but little when they enter the eye, are made to convene before they reach the *retina*. As theſe men
advance in years, their eyes like thoſe of other old
men, for want of a due ſupply of humors, abate
of their convexity and grow flatter; upon which
account they begin to ſee objects diſtinctly at a diftance, without the help of ſpectacles, and are for
that reaſon deemed to have the moſt laſting eyes.

By the help of concave glaſſes, purblind perſons
may ſee diſtant objects diſtinctly; for as it is the
property of ſuch glaſſes to make the rays diverge,
if the rays which flow from a diſtant point, and fall
upon the eye with a ſmall degree of divergence, be
made to paſs thro' a concave *lens* of a proper concavity, they will thereby be made to diverge ſo
much, as that the eye, notwithſtanding the great
convexity of the *cornea*, ſhall not be able to bring
them together till they arrive at the *retina*.

Pl. 10.
Fig. 2.

If CD be a concave *lens*, and if B be the *focus* of
the rays which flow from the point A; that is, if
the rays which diverge from A, paſs thro' the glaſs,
and by the refraction which they ſuffer in their paſſage, proceed in ſuch a manner as if they had diverged from B; and if the diſtance at which a
purblind perſon ſees diſtinctly with his naked eye,
be equal to the diſtance of B from the glaſs, ſuch
a perſon will by the help of the glaſs CD, be able to
ſee the point A diſtinctly; becauſe the rays which
flow from A, after they paſs thro' the glaſs, fall

upon

upon his eye with the same degree of divergence, as if they had issued from B, the point of distinct vision. Hence it follows, that if in the *Theorem* laid down in my last lecture, for finding the *focus* of double concaves exposed to diverging rays, namely $F = \dfrac{RD}{R + D}$, wherein F denotes the *focus*, D the distance of the point of divergence, and R the *radius* of the concavity, we suppose F to denote the distance at which the purblind person sees distinctly without a glass, and D the distance at which he sees distinctly by the help of the glass, by clearing R we shall have $R = \dfrac{FD}{D - F}$; that is to say, the *radius* of the concavity of a double concave of equal concavities, which enables a purblind person to see an object distinctly, when placed beyond the reach of his naked eye, must be equal to a rectangle, under the distance at which he sees distinctly with his naked eye, and the distance at which it is required he should see distinctly by the help of the glass, divided by the difference of those distances. For instance, if a person with his naked eye can read at the distance of three inches only, and it be required to find the *radius* of such a glass as shall enable him to read at the usual distance of eighteen inches; in this case, F being equal to three inches, and D to eighteen, their product is 54; which being divided by their difference, which is 15, gives three and $\frac{3}{5}$ in the quotient, which shews, that the *radius* of the glass must be three inches and $\frac{3}{5}$ths nearly.

Where the distance at which it is required the purblind person shall see distinctly is infinite, or in other words, where it is so great, as that the distance to which the power of his naked eye reaches, bears no sensible proportion to it, there D — F becomes equal to D, and of course, R becomes equal to F; so

Y   that

that in order to fee fuch objects as are very remote,
purblind perfons muſt make uſe of concave glaſſes,
whoſe *radii* are equal to the diſtances at which they
fee diſtinctly with their unarmed eyes.

As purblind perſons cannot fee remote objects
diſtinctly, fo on the other hand, thoſe who are old
cannot, generally ſpeaking, fee fuch as are nigh;
the reaſon of which is, that in old men the *cornea*,
for want of a due ſupply of humor to plump out
the eye, has not a degree of convexity ſufficient to
bring the rays together on the *retina*, when they
fall upon the eye with a conſiderable degree of divergence; as is the caſe of all thoſe rays which flow
from points ſituated near the eye. The proper remedy for this defect is a *convex lens*, becauſe it leſſens the divergency of the rays, and brings them
nearer to a parallelifm. If with reſpect to the
convex lens CD, A be the *focus* of the rays which
diverge from B; that is to fay, if the rays which
flow from B and paſs thro' the *lens*, do afterwards
proceed in fuch a manner as if they had diverged
from A, and if the diſtance at which an old man
can fee diſtinctly with his naked eye, be equal to
the diſtance of A from the glaſs, he will be able by
the affiſtance of the glaſs, to fee the nearer point B
diſtinctly; becauſe the rays which iſſue from that
point in paſſing thro' the glaſs acquire the fame degree of divergence, with thoſe which flow from A,
the point of diſtinct viſion, and of conſequence,
may as eaſily be brought together on the *retina*, by
the refractive power of the eye; hence, if we take
the *Theorem* laid down in my laſt lecture for finding the *foci* of double *convexes* of equal convexities,
and fit it to the caſe before us, where the *focus* is
imaginary, by making $F = \dfrac{RD}{R-D}$, if then we
ſuppoſe F to denote the diſtance at which an old eye
fees diſtinctly, and D the nearer diſtance at which it

Pl 10.
Fig. 3.

2   is

is required to make it fee diſtinctly with the aſſiſtance of a glaſs, by clearing R, we ſhall find it
equal to $\frac{FD}{F-D}$. So that the *radius* of ſuch a double convex of equal convexities as enables an old
man to fee a nigh object diſtinctly, muſt be equal
to a rectangle under the diſtance at which he fees
diſtinctly with his naked eye, and the diſtance at
which he is to fee by the help of the glaſs, divided
by the difference of thoſe diſtances. To illuſtrate
this by an example; ſuppoſe an old man cannot
with his naked eye read at a leſs diſtance than of
four feet, and it is required to aſſign the *radius* of
ſpectacles which ſhall enable him to read at the diſtance of a foot and an half; in this caſe, F is four
feet, and D is one and an half, and their product is
fix, which when divided by their difference, to wit,
two and an half, gives $2\frac{4}{10}$ in the quotient; which
ſhews, that the ſpectacles muſt be ground to a *radius* of two feet and four tenths.

If F be infinite, which is the caſe where the eye
can fee nothing but what is extremely remote, then
F—D is equal to F, and of conſequence, R is equal
to D; ſo that where an old man can fee no objects
diſtinctly but ſuch as are very far off, in order to fee
diſtinctly at nearer diſtances, he muſt for each
diſtance uſe ſuch ſpectacle glaſſes as have their *radii*
equal to the diſtance.

If D be given, then R becomes equal to $\frac{F}{F-1}$;
and foraſmuch as the proportion of F to F—1 increaſe as F leſſens, R muſt do ſo too, which ſhews,
that where the diſtances at which two old eyes when
unarmed can fee diſtinctly are different, in order to
make them fee diſtinctly at any leſſer given diſtance,
the eye which can fee at the ſmaller diſtance muſt
be furniſhed with a glaſs of a greater *radius* than
the other. And herein lies the whole ſecret of
younger and older ſpectacles, thoſe being deemed

the

the youngeſt, which are ground to the largeſt *radius.*

Having ſhewn you of what uſe both convex and concave glaſſes are in aſſiſting defective eyes, I ſhall now lay before you the alterations which they produce in the appearances of objects; and Firſt, as to *convexes*; if an object be viewed thro' a *convex lens*, at a leſs diſtance than the *focus*, it appears more remote and bigger than it does to the naked eye. That it muſt appear more remote, will be evident, if we conſider what has been already proved in a former lecture, namely, that where rays fall upon a *convex lens*, from a point leſs diſtant than the *focus*, after they have paſſed the glaſs, they proceed in ſuch a manner as if they had iſſued from a more diſtant point; and ſince this is the caſe of the rays which flow from each point in the object, the object muſt of conſequence ſeem to be more diſtant than it is;

Pl. 10.
Fig. 4.
and it muſt likewiſe appear greater; for if AB be an object expoſed to a naked eye at O, its extream points A and B will be perceived by the eye by means of the rays AO and BO, which flow directly from thoſe points to the eye, but if a *convex lens* as C, be interpoſed, the eye will no longer perceive the extremities by means of the rays AO and BO, becauſe as they are refracted by the *lens*, they are made to concur before they can reach the eye; the eye therefore muſt now perceive thoſe points by means of ſome other rays as A E and B D, which falling upon the glaſs at a greater diſtance from each other, are by the refractive power of the glaſs thrown into the directions EO and DO, and made to concur at O; ſo that continuing thoſe lines directly backward as far as the object, to wit, to I and H, the eye at O will perceive the extream points of the object as ſituated at I and H; that is, it will perceive the object magnified. And if the eye be farther removed from the glaſs ſuppoſe to P, the object will appear ſtill greater, its extremities in
that

that cafe appearing at L and K in the lines PG and
PF produced. And on the other hand, if the eye
continuing in its place, the object be farther removed from the *lens*, it will appear larger; for whereas
at the nearer diftance the eye perceives the extream
points of the object by means of the rays AE and
BD, which fall upon the *lens* at E and D, and are
thence refracted to O; when the object is at the
greater diftance, its extremities cannot be feen by
means of the rays incident on the glafs at E and D;
for fince the interval between the extremities continues the fame, the rays which flow from them and
fall upon the *lens* at E and D, will diverge lefs at a
greater diftance of the object than at a fmaller;
confequently, they will concur before they reach
the eye; and therefore in this cafe the extream
points of the object muft be conveyed to the eye by
fome rays as aG and bF, which diverging more
than the former, fall without them at G and F,
whence they are refracted to the eye at O, in the
lines GO and FO, which being continued backward as far as the nearer diftance of the object, to
wit, to L and K, fhew that the object which at the
nearer diftance appeared to extend itfelf only from
I to H, does at the greater diftance feem to reach
from L to K, and of confequence, appears more
magnified.

If the object be removed beyond the *focus*, it
will appear ftill greater; but whereas before it paffes
the *focus* it appears diftinct, as alfo more and more
diftant the farther it is removed from the glafs,
when it gets beyond the *focus* it appears confufed,
and the farther it is removed from the glafs, the
more confufed it appears, and the nearer it feems to
approach the eye, provided its diftance from the
glafs be not fo great as to make it project its image
between the eye and the glafs.

This feeming approach of the object at a time
when it really recedes, and in a cafe where, accord-

Y 3   ing

ing to the received principles of *Dioptricks*, it ought
to appear at a diftance, if poffible more than infinite, has very much puzzled the writers of *Opticks*,
and was looked upon as an infuperable difficulty,
till Doctor BERKELEY took it into confideration
in his *Effay upon Vifion*, wherein, among other difficulties which he has cleared up relating to vifion,
he has given us a natural and fatisfactory account
of this. The fubftance of what he has there delivered concerning this matter is, that by cuftom and experience we are taught to judge thofe objects near
which appear confufed, becaufe, according to the
ordinary courfe of nature, thofe objects, and thofe
only, appear confufed which are brought very near
the eye, and therefore if an object fhall at any
time appear confufed, tho' from another caufe, the
mind will immediately connect nearnefs of diftance
in the object, with that confufion in the appearance,
as having always experienced them to go together;
and the greater the confufion is, the nearer it will
judge the object to be, becaufe it has always obferved the neareft diftances to be attended with the
greateft confufions; now if in the cafe before us,

Pl. 10.
Fig. 6.
we fuppofe A to be an object placed before the
convex lens BC, at a greater diftance than the *focus*,
the rays after they have paffed thro' the glafs will
converge towards fome point as D; if then an eye
be placed at a little diftance behind the glafs, fuppofe at E, it will perceive the object confufed, becaufe as the rays fall upon it converging, they will
be made to meet before they arrive at the fund of
the eye, and confequently, will be fcattered on the
retina, and thereby render the appearance confufed;
if the eye be moved gradually backward to F, G,
and D, or which is the fame thing, if by carrying
the object forward, the rays be made to fall upon
the eye at lefs and lefs diftances from the *focus*, they
will be fcattered more and more upon the *retina*, becaufe the convergency wherewith they fall upon the
eye

eye is by so much the greater, by how much the
nearer the eye is placed to the *focus* or the point D;
consequently, the object as it is more and more removed from the glass, will appear more and more
confused; for which reason, the mind which has
been used to connect nearer distances with greater
degrees of confusion, will in this case judge the object to approach, tho' in reality it recedes; and what
fully confirms this is, that if by placing a concave
glass at a proper distance between the eye and the
convex, the convergency of the rays be taken off,
and the appearance thereby rendered distinct, the
object will then appear at its due distance.

If an eye be removed from a *convex lens* beyond
the place where the image is projected, that is, if
the eye be farther from the *lens* than is the point D,
the object will appear in an inverted position,
and seem to be situated between the eye and the
glass; for in this case, the eye sees only the image
or representation of the object, which, as I shewed in a former lecture, is projected at D in an inverted position; upon looking at the image with
both eyes, it appears double, and upon shutting either eye, the image on the contrary side disappears;
the reason of which is this, the eye at O perceives
the image by means of the rays ODC, and therefore sees it on the same side with C, whereas the
eye at P perceives it by means of the rays PDB,
and on that account sees it on the same side with B;
as the head is moved farther back, the distance between the two images must decrease, and at length
vanish; for since the interval between the eyes continues unvaried, the rays which exhibit the image
to each eye, will diverge less and less as the head
is more and more removed from D, as is evident
from the bare inspection of the scheme; consequently, the distance between the two images must
continually decrease, and at last become so small as
to be insensible.

Pl. 10.
Fig. 6.

Y 4   As

As to concave glaffes, fince it is their property
to make the rays which flow from any point
to diverge, in fuch a manner as if they had iffued
from a point lefs diftant, it is evident, that an object feen thro' a *concave lens* muft appear nearer than
it really is, and it muft likewife appear diminifhed;
for the extream points of the object A B, are feen
by the naked eye by means of the rays AO and
BO, which when the *concave lens* CD is interpofed,
are made to diverge, fo as not to meet at O, confequently, upon the interpofition of the glafs, the
eye will not perceive the extremities of the object by
thofe rays, but by fome others as A K and BL,
which falling within the former, are by the refractive power of the glafs made to proceed in the
lines KO and LO, fo as to meet at O ; wherefore
continuing OK and OL backward to the object,
the extremities of the object will be feen at E and
F, that is, the object will appear to be lefs than it
really is ; and by the fpreading of the rays in their
paffage thro' the glafs, fome of them are made to
efcape the eye, which if the glafs were removed,
would fall upon the pupil; for which reafon, the
object muft appear lefs luminous ; fo that the property of concave glaffes is to make objects appear
fmaller, nearer, and more faint and obfcure, than
they do to the naked eye.

.Pl. 10.
Fig. 7.

LECTURE XXIII,

Of Catoptricks.

Lect.
XXIII. IN this lecture, wherewith I fhall clofe this courfe,
I fhall explain to you the *Doctrine of* Catop-
tricks, or that part of *Opticks* which treats of the
reflexion

reflexion of light; in doing of which, I shall first Lect. say something concerning the cause of that reflexi- XXIII. on; Secondly, I shall lay down two principles, which are the chief foundation of *Catoptricks*; and lastly, I shall lay before you the most remarkable properties of plain and spherical mirrors.

As to the first, before Sir Isaac Newton pub-lished those wonderful and surprising discoveries which he made, concerning the nature and proper-ties of light, it was an opinion generally received by the writers of *Opticks*, that the rays of light were reflected in the manner of other bodies, by striking on the solid and impervious parts of bodies; but that great Philosopher has fully proved this opi-nion to be erroneous; and has shewn, that the par-ticles of light are turned back before they touch the reflecting body, by some power of the body which is equally diffused all over its surface; what he has delivered concerning this matter, is to be met with in the eighth *Proposition* of the *second Book of his Opticks*, wherein, after he has offered several reasons to prove, that light is not reflected by striking a-gainst bodies, he at last expresses himself in the fol-lowing manner; " Were the rays of light reflected " by impinging on the solid parts of bodies, their re-" flexions from polished bodies could not be so re-" gular as they are; for in polishing glass with sand, " putty, or tripoly, it is not to be imagined, that " those substances can, by grating and fretting the " glass, bring all its least particles to an accurate po-" lish, so that all their surfaces shall be truly plain, " or truly spherical, and look all the same way, so " as together to compose one even surface. The " smaller the particles of those substances are, the " smaller will be the scratches by which they con-" tinually fret and wear away the glass until it be " polished; but be they never so small, they can wear " away the glass no otherwise than by grating and

2   " scratching

" scratching it, and breaking the protuberances,
" and therefore polish it no otherwise than by bring-
" ing its roughness to a very fine grain; so that the
" scratches and frettings of the surface become too
" small to be visible. And therefore, if light were
" reflected by impinging on the solid parts of the
" glass, it would be scattered as much by the most
" polished glass, as by the roughest. So then it re-
" mains a *Problem*, how glass polished by fretting
" substances can reflect light so regularly as it does;
" and this *Problem* is scarce otherwise to be solved,
" than by saying, that the reflexion of a ray is effect-
" ed, not by a single point of the reflecting body,
" but by some power of the body, which is evenly
" diffused all over its surface, and by which it acts
" upon the ray without immediate contact."

Now taking it for granted, that this repelling
power is the true cause of reflexion, if it be supposed to act upon the rays of light in lines perpendicular to the surface of the reflecting body; it
will thence follow, that the angle of incidence, or
the angle contained between the incident ray, and a
line drawn perpendicular to the reflecting surface at
the point of incidence, is equal to the angle of reflexion, or the angle contained between the same

Pl. 10.
Fig. 8.

perpendicular and the reflected ray. For if we suppose a ray of light to move in the direction A C,
towards the reflecting surface B C D; and if we suppose that motion to be resolved into two, one in the
direction A E, parallel to B D, and the other in the
direction A B, perpendicular to B D, it is manifest,
that of those two motions, the latter only is opposed to the repelling force; and of consequence, the
ray after reflexion, will go on in the parallel direction, with the same velocity it did before; and forasmuch as the repelling force which opposes the perpendicular motion, acts incessantly, it no sooner de-
stroys the motion of the ray towards the body, but

it

it gives it an equal degree of motion the contrary way; that is, it throws it back with the same perpendicular velocity wherewith it approached. If therefore EG be taken equal to AE, and from G be let fall GD equal and parallel to AB, EG will exprefs the parallel motion of the ray after reflexion, and DG its perpendicular motion; and the diagonal line CG, will be actually deſcribed by the ray, by virtue of its compound motion; and from the nature of ſimilar triangles, the angle of incidence ACE, muſt be equal to ECG, the angle of reflexion; and this is the firſt of thoſe principles whereon the doctrine of *Catoptricks* is founded. The ſecond is, that every radiant point when ſeen by reflexion, appears in that place where the reflected ray meets the perpendicular, drawn from the radiant point to the reflecting ſurface; for inſtance, if from a radiant point as R, placed before the plain *ſpeculum* AB, be let fall the line REM, perpendicular to the plane of the *ſpeculum*; and if RC and CD be ſo drawn, as that the former may denote the incident ray, and the latter the reflected; and if DC be continued on, till it meets the perpendicular REM; an eye at D will perceive the radiant point, as placed at M, the point of interſection of the reflected ray, and the perpendicular; and thus it is in all caſes of reflexion, except two, wherein this principle ſeems to fail; one whereof relates to plain glaſs *ſpeculums*, and the other to concave ſpherical mirrors; the latter has been obſerved by TAQUET, Doctor BARROW, and others; but the former has not been mentioned by any one of the *optick* writers that I know of; I ſhall take notice of each in its proper place, and proceed now to conſider the chief properties of mirrors, and firſt, of ſuch as are plain.

Pl. 10.
Fig. 8.

When an object is ſeen by reflexion from a plain *ſpeculum*, its image appears as far behind the *ſpeculum*, as the object is before; for the proof of which,
let

let R be an object placed before the plain *speculum*
AB, and let it be feen by reflexion from the point
C, by an eye fituated fomewhere in the line CD,
then producing CD, till it meets the perpendicular
REM, the image will, by the fecond principle,
appear at M; now the angles of incidence and reflexion being equal, their complements are fo too,
that is to fay, the angle RCE is equal to DCB or
MCE; fo that in the two right-angled triangles,
the angles at C being equal, and the fide EC
common to both, the triangles muft be equal,
and the fide ME, that is, the diftance of the
image behind the *speculum* muft be equal to RE,
the diftance of the object before the *speculum*; and
the fame thing is in like manner demonftrable,
tho' the point of reflexion be taken different from
C; for the reflected ray will conftantly meet the
perpendicular in the point M; whence it follows,
that however the fituation of the eye with refpect
to the mirror may be changed, yet if the object
and mirror remain unmoved, the image will always appear in the fame place; it likewife follows, that there cannot appear more than one
image of one and the fame object; but then this
is to be underftood with refpect to fuch mirrors, as
being opaque, have but one reflecting furface; for
in looking-glaffes, which by reafon of their tranf-
parency, have a double reflexion in fome certain
pofitions of the eye and object, feveral images
may be feen. Thus if AB be a looking-glafs, R
the flame of a candle, placed at a fmall diftance
before AH, the plane of the glafs produced, an
eye being placed at Q, fhall fee feveral images
ftanding at fmall diftances one beyond another, in
the fame pofition with the letters, C, D, E, F,
whereof the firft and fecond appear bright and luminous, and the reft but faint and obfcure; for the
feveral images taken in their order from the fecond,
grow more and more dark and obfcure, till at length
they

Pl. 10.
Fig. 10.
Exp. 1.

they become too weak and feeble to affect the fight, and of confequence vanifh.

Pl. 10.
Fig. 11.

In order to account for this multiplicity of images, let A B C D be a looking-glafs, whofe near-eft furface, or that which lies next the eye is A B, and its farther or filvered furface is D C, R the place of the candle, and Q the place of the eye, R S a line drawn from the candle perpendicular to A O and D Y the two furfaces of the glafs pro-duced; the angle REA being made equal to QEB, and the line Q E being produced till it cuts the per-pendicular R S in T, the eye fhall fee the firft image at T, by means of the reflexion from the outward furface A B, the ray R E being reflected to the eye from the point E. Let a fecond ray as R G, pafs into the glafs at G, and being refracted to the point H of the farther furface, let it thence be reflected to K, and there paffing out of the glafs, let it by refraction be carried to the eye; let then Q K be pro-duced, and the eye fhall fee a fecond image fituated in that line, and that at a little diftance beyond the perpendicular R S; for if the rays fuffered no re-fraction in paffing in and out of the glafs, the fe-cond image would not be feen by means of the ray R G, but by means of the ray R H, which paffing directly from R to H, is thence reflected directly to Q, and being produced till it cuts the perpen-dicular in X, would exhibit the fecond image at X; but forafmuch as the place of the image is changed by the refraction, and brought nearer to the glafs, if we suppofe the line Q X to be moved upward about the point Q, till it coincides with the line Q V, in which the fecond image really appears, the point X muft neceffarily fall beyond the perpen-dicular, and fo of confequence, muft the place of the image. Let now a third ray as R F, pafs into the glafs at F, and be refracted to L, and from thence let it be reflected to E, and from E to M, and from

M to

M to N, where let it go out, and be refracted to the eye at Q; then producing QN to W, a third image will appear in that line fomewhere beyond the perpendicular; for were there no refraction, the ray which after three reflexions exhibits the third image, would when produced cut the perpendicular in the point S; and therefore, fince the line QS is raifed up by the refraction, and made to coincide with the line QW, the point S, that is, the place of the third image, muft fall beyond the perpendicular.

As a third image is feen by means of three reflexions, fo is a fourth by five reflexions, a fifth by feven, a fixth by nine, and fo on, according to the progrefs of the odd numbers, every fucceeding image being feen by two reflexions more than the preceding; and this is the true reafon why, fetting afide the firft and fecond, which being feen each by one fingle reflexion, appear almoft equally bright, every fucceeding image appears more dim and faint than the foregoing, the rays of light being rendered more weak and feeble by reflexion.

K-

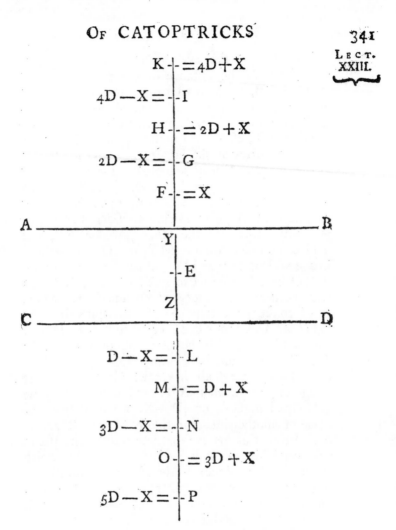

If two plane *speculums* as AB and CD, be let parallel to one another, and an object be placed any
where between them as at E, the rays of light
which issue from the object and fall upon each *speculum*, will be reflected backward and forward from
one to the other a great number of times; by which
means, there will appear in each *speculum* a great
number of images situated one behind another, in
a right line perpendicular to the *speculums*, and
passing thro' the object at E; in order to determine
the

the diftances of the feveral images from the *fpecu-
lums, let E Y, the diftance of the object from the
fpeculum A B, be denoted by X, and let Z Y, the
interval of the glaffes, be denoted by D; and let
us firft confider the reflexion which begins from the
fpeculum A B; if Y F be taken equal to X, then F
will be the place of the firft image; and forafmuch
as the image at F may be looked upon as an object
placed before the *fpeculum* C D, if Z M be taken
equal to F Z, that is, to D + X, there will appear
an image at M; which being confidered as an object with refpect to the other *fpeculum* A B, and
Y H being taken equal to M Y, or 2 D + X, another
image will be feen at H; and for the fame reafon,
if Z O be taken equal to H Z, or 3 D + X, there
will another image appear at O, and fo on; again,
if we confider the reflexion which begins from the
fpeculum C D, by taking Z L equal to E Z, or D—X,
we fhall have L for the place of an image in the
fpeculum C D; and by making Y G equal to L Y,
or 2 D—X, we fhall have the place of another
image in the *fpeculum* A B; and again, by taking
Z N equal to G Z, or 3 D—X, we fhall have the
place of another image in the *fpeculum* C D, and fo
on. From this manner of determining the places
of feveral images, it is evident, that if D, which
ftands for the diftance of the glaffes, be multiplied
into each of the even numbers taken in their order,
and if X, which denotes the diftance of the object
from A B, be fubducted from each product, and
likewife added to each, the differences, and the
fums taken in their order, will exprefs the diftances
of the feveral images from the *fpeculum* A B, the
diftance of the firft being X; that is to fay, the
diftance of the fecond image will be 2 D—X, of
the third 2 D + X, of the fourth 4 D—X, of the
fifth 4 D + X, and fo on, according to the firft
Table.

Table

TABLE I.

Distances of the several images from the speculum
AB.

$$1 = X$$
$$2 = 2D - X$$
$$3 = 2D + X$$
$$4 = 4D - X$$
$$5 = 4D + X$$
$$6 = 6D - X$$
$$7 = 6D + X$$
$$8 = 8D - X$$

&c.

If D be multiplied into each of the odd numbers taken in their order, and if X be deducted from each product, and likewise added to each as before, the differences and the sums taken in their order, will express the distances of the several images from the *speculum* CD, as in the second Table.

TABLE II.

The distances of the several images from the speculum
CD.

$$1 = D - X$$
$$2 = D + X$$
$$3 = 3D - X$$
$$4 = 3D + X$$
$$5 = 5D - X$$
$$6 = 5D + X$$
$$7 = 7D - X$$
$$8 = 7D + X$$

&c.

If by moving the object nearer to AB, X becomes less, then all those images whose distances are expressed by those symbols wherein X is affirmative, will come nearer to the *speculums*, whilst those whose distances are expressed by symbols

Z   wherein

wherein X is negative, move farther off; thus in
the *speculum* AB, the firſt, third, fifth, ſeventh,
and ſo on, will approach, and the ſecond, fourth,
ſixth, eighth, and ſo on, will recede; ſo that the ſeveral images, beginning from the firſt, will approach and recede alternately; and on the other
hand, thoſe in the *speculum* CD will recede and approach alternately, beginning from the firſt; hence,
if a man puts his hand between the two *speculums*,
and moves his palm towards one of them, a perſon looking into the other, ſhall ſee ſeveral pairs of
hands, palm to palm, approaching each other.

Pl. 10.
Fig. 12.
Exp. 3.   If two plane *speculums* as AC and BC, be inclined to one another, ſo as to meet in an acute
angle at C; and if an objeċt be placed any where
between them, ſuppoſe at F, an eye looking into
either, ſhall ſee ſeveral images ſtanding in the circumference of a circle, whoſe center is at C the concourſe of the *speculums*, and its *radius* equal to CF
the diſtance of the objeċt from the concourſe; for
if from F be drawn FD perpendicular to the *speculum* CA, and KD be made equal to FK, D will
be the place of an image in the *speculum* CA; and
if from D be drawn DE perpendicular to the *speculum* CB, and produced till HE is equal to DH,
E will be the place of an image in the *speculum* CB,
and thus by drawing perpendiculars continually from
the place laſt found to the oppoſite *speculum* may the
places of all the images be found which are ſeen by
means of thoſe reflexions, the firſt whereof is made
from the *speculum* CA; and in the ſame manner,
by drawing the perpendiculars FG, GL, and ſo on,
may the places of all thoſe images be found which
are ſeen by means of the reflexions whereof the firſt
is made from the *speculum* CB. Now that the
points D and E are in the circumference of the
circle whoſe *radius* is CF, I thus prove in the triangles CFK and CDK, the ſides FK and DK are
equal by the conſtruċtion, and CK is common to
both,

both, and the angles at K are right ones, wherefore the two triangles are equal, and of confequence CD is equal to CF; again, the triangle CDH is equal to the triangle CEH, the fides DH and EH being by conftruction equal, as are alfo the angles at H, wherefore CE is equal to CD, which is equal to CF, confequently, a circle defcribed on the center C, with the *radius* CF, will pafs thro' the points D and E; and by the fame way of reafoning it will be found to pafs thro' G and L, and thro' the extremities of all the other perpendiculars; and therefore the feveral images muft of neceffity appear in the circumference of a circle whofe center is at the concourfe of the *fpeculums*, and whofe *radius* is equal to the diftance of the object from that concourfe. From what has been faid it follows, that if the diftance of the object from the concourfe of the *fpeculums* be given, the images will ftill appear in the circumference of the fame circle, notwithftanding any alteration that may be made in the angle whereat the *fpeculums* meet; if that be inlarged, the images will be fewer in number, and at greater diftances from one another; and on the other hand, if it be made lefs, the images will be more in number, and ftand clofer together, but the circle in whofe circumference they appear, will be the fame in both cafes; for that is not to be leffened or inlarged otherwife, than by leffening or inlarging the diftance of the object from the concourfe of the *fpeculums*.

Having laid before you the chief properties of plain *fpeculums*, I come now to confider fuch *fpeculums* as are fpherical; and they are of two forts, *concave* and *convex*; concerning which it muft be obferved, that as all the rays which fall upon them from a radiating point, are reflected in fuch manner as to meet the perpendicular very nearly in one and the fame point; in order to find out the *focus*,

Z 2

or

Pl. 10.
Fig. 13.

or the place where the reflected rays crofs one another, nothing more is neceffary, but to determine
the point wherein any one reflected ray meets the
perpendicular; which may be done in the following
manner; let A be a radiating point, expofed directly before the concave glafs F G, whofe center is
C, A B a perpendicular from the radiating point to
the *fpeculum*, which likewife denotes the diftance of
the radiating point from the *fpeculum*, A D a ray
falling on the *fpeculum* at D, whofe diftance from
B is indefinitely fmall, D E the reflected ray meeting the perpendicular in E, C D a *radius* drawn to
the point of incidence, and of confequence bifect-
ing the angle A D E in the triangle A D E; fince the
angle at D is bifected by the line D C, which cuts
the oppofite fide, A D is to D E, as A C to C E;
but forafmuch as the points D and B are fuppofed
to be indefinitely near, A D is equal to A B, and
E D is equal to E B, wherefore A B is to E B, as A C
is to C E; that is, the diftance of the radiating
point from the *fpeculum*, is to the diftance of the
point E, where the reflected ray cuts the perpendicular, commonly called the *point of interfection*, as
the diftance of the radiating point, leffened by the
radius, is to the *radius*, leffened by the diftance of
the point of interfection; that is, putting D for
the diftance of the radiating point, F for the diftance of the point of interfection, and R for the
radius, D : F :: D—R : R—F; confequently,
reducing this analogy into an equation, and clearing F, F will be found equal to $\dfrac{DR}{2D-R}$; that is,
the diftance of the point of interfection from the
fpeculum, and confequently, the diftance of an image
formed by reflexion from a concave *fpeculum*, is
equal to a rectangle under the diftance of the object
from the *fpeculum*, and the *radius*, divided by twice
the diftance of the object leffened by the *radius*.

Hence

Hence it follows, that if an object be placed before a concave *speculum*, at an infinite distance, that is, if the distance be so great as that the *radius* of the *speculum* bears no sensible proportion to it, the image will appear on the same side of the *speculum* with the object, at the distance of one half the *radius* from the *speculum*; for in this case, D being infinite, 2D—R becomes equal to 2D, and of consequence, F is equal to R divided by 2; so that one half the *radius* is the least distance at which an image can be projected from a concave *speculum* on the same side with the object; and forasmuch as the sun's image, which, by reason of the immense distance of his body, is formed at the distance of half the *radius* from the *speculum*, is there apt to burn, that place is usually called the *focus* or *burning point*.

As the object approaches the *speculum*, the image recedes; for as in one and the same *speculum*, the *radius* is a standing quantity, it is manifest, that as D lessens, the proportion of DR to 2D—DR must increase, consequently, F, or the distance of the image from the *speculum* must do so too; and when the object has approached so near the *speculum* as to be at the center, the image will have receded so far as to be there likewise; for in this case, D being equal to R, 2D—R is equal to R, and of consequence, F is equal to D; so that the object and its image meet at the center of the *speculum*; upon the object's passing from the center towards the glass, the image is projected beyond the center, and when the object has approached so near the *speculum*, as to be distant from it but half the *radius*, the image is at an infinite distance; for in this case, D being equal to half the *radius*, 2D—R is nothing, consequently, F, that is the distance of the image, is infinite; or to speak more properly, the rays after reflexion proceed parallel; for which reason, if the flame of a candle be placed directly before a con-

cave

cave *fpeculum*, at the diftance of half the *radius*, the *fpeculum* will feem to be in flames, and the re- flected light will be fo intenfe, as that by the help of it one may be able to read at a very confiderable diftance from the *fpeculum*. Taquet afferts, that he has read at the diftance of no lefs than 400 feet; and to fay the truth, the diftance would be without li- mits, were it not for the atmofphere, whofe par- ticles continually intercept the rays, and by fo do- ing, at length totally extinguifh the light. It fome- times happens, that when the flame of a candle is placed in the *focus* of a concave *fpeculum*, its image is projected on a diftant wall, which feems to in- validate the truth of what I juft now proved con- cerning the parallelifm of the rays after reflexion, but this is occafioned by the flames being too large to be contained totally within the *focus*, for were it fo fmall as to lie wholly within the *focus*, it would not project an image, but the rays after reflexion, would form a cylindrical body of light, which when projected on a diftant wall, would have a cir- cular figure, of an equal circumference with the *fpeculum*.

When the diftance of the object from the *fpecu- lum* is lefs than half the *radius*, the image appears behind the *fpeculum*; for in this cafe, 2D—R is a negative quantity, and of confequence, fo is F, which fhews, that the diftance of the image which is denoted by F, muft be taken on the other fide of the *fpeculum*, with refpect to the object; as the ob- ject moves nearer to the *fpeculum* before, fo likewife does the image behind; and when the object is fo near as to touch the *fpeculum*, the image does the fame; for in this cafe, D being nothing, F, that is, the diftance of the image from the *fpeculum*, is like- wife nothing.

As to the pofition of the images which are feen by reflexion from a concave *fpeculum*, thofe which

appear

appear on the fame fide of the *fpeculum* with the
object muft be inverted, and thofe which appear
behind the *fpeculum* muft be erect. For the proof Pl. 10.
of which, let A B be an object placed before the Fig. 14.
concave *fpeculum* F G, at any diftance beyond the
center C, in which cafe the image will be feen between the center and the *fpeculum*, fuppofe at D E ;
from A and B, the extream points of the object,
let the lines A H and B I be drawn perpendicular to
the *fpeculum*, and of confequence croffing one another at the center ; this being done, fince the image
is fuppofed to be at D E, and fince every point of
an image is feen in the perpendicular drawn from
the correfponding point in the object, it is manifeft,
that D, the loweft point of the image, will correfpond to A, the higheft point of the object, and E,
the higheft point of the image, will correfpond to
B, the loweft point of the object, that is, the
image will appear inverted. And by the fame
way of reafoning, if D E be the object, fituated
at fuch a diftance between the *fpeculum* and the
center, as to have its image projected beyond
the center at A B, the image muft appear inverted.
On the other hand, where an object as D E, is
placed between the *fpeculum* and the center, and
confequently, projects an image behind the *fpeculum* ; for it muft be obferved, that the fame object
D E, which when fituated between the center and
the *fpeculum*, at a lefs diftance from the center than
half the *radius*, projects an image as A B beyond
the center, does likewife project another image as
H I, behind the *fpeculum* ; and as the former image
is vifible to an eye placed beyond it, fo the latter
image is vifible to an eye placed between the
object and the *fpeculum*, and it muft appear erect,
inafmuch as the perpendicular C H, which paffes
thro' D the higheft point of the object, does likewife
pafs thro' H, the higheft point of the image.

As

LECT.
XXIII.

As to the magnitudes of an object and its image, they are to one another in the same proportion with the squares of their distances from the *speculum*; for if the line L C M be drawn thro' the center C, perpendicular to B A, D E, and H I, and consequently, bisecting the angle at C, if B A be the length or breadth of an object, and DE the length or breadth of its image projected on this side the *speculum*, then L A and O D will be half the length or breadth of the object and its image, and the triangle C L A being similar to C O D, L A is to O D, and consequently B A to E D, as L C to O C, that is, the length or breadth of the object, is to the length or breadth of its image, as the distance of the object from the center of the *speculum*, to the distance of

Pl. 10.
Fig. 13.

its image from the same center; but it has been proved, that as A C, the distance of the object from the center, is to E C, the distance of the image from the center, so is A B, the distance of the object from the *speculum*, to E B, the distance of the image from the *speculum*; consequently, the length or breadth of an object, is to the length or breadth of its image, as the distance of the object from the *speculum*, to the distance of the image from the *speculum*; and forasmuch as similar surfaces are to one another, as the squares of their homologous sides, the magnitude of the object, is to the magnitude of the image, as the square of the object's

Pl. 10.
Fig. 14.

distance from the *speculum*, to the square of the image's distance. And by the same method of arguing, if D E be an object whose image behind the *speculum* is H I, the magnitude of the former will be found to be to the magnitude of the latter, as the square of K O, to the square of K M. Hence it follows, that the object during its continuance beyond the center, must appear larger than its image, as being more distant from the *speculum*, and when it is in the center, where it meets the image, it

must

muſt appear equal to it, but being on the ſame ſide of the center with the *ſpeculum,* it muſt be leſs than its image, which in that caſe lies beyond the center, and conſequently, is at a greater diſtance from the *ſpeculum.*

It likewiſe follows, that the image which appears behind the *ſpeculum* is ever larger than the object; for ſince MK, the diſtance of the image behind the *ſpeculum,* is to OK, the diſtance of the object before the *ſpeculum,* as MC, the diſtance of the image from the center, to OC, the diſtance of the object from the center; and ſince in this caſe, the object is always leſs diſtant from the center than its image, during the appearance of the image behind the *ſpeculum,* it is evident, that the image muſt appear larger than the object; but then this is to be underſtood with reſpect to ſuch images only, as are projected by objects leſs diſtant than the center; for if an object be beyond the center, an eye being cloſe to the *ſpeculum,* ſhall ſee the image at the ſame diſtance, and of an equal magnitude with the object; and in this caſe, the ſeveral parts of the image do not appear in thoſe points where the perpendiculars from the correſponding points of the object meet with the reflected rays; the reaſon of all which ſeems to be this, the portion of the *ſpeculum* which the eye makes uſe of in this caſe is ſo exceedingly ſmall, that notwithſtanding the ſpherical figure of the *ſpeculum,* it may be looked upon as plane, and conſequently, the appearances muſt be the ſame as in other plain *ſpeculums;* that is, the image muſt appear as far behind the *ſpeculum* as the object is before it, and of the ſame magnitude with the object.

If an image formed on this ſide a concave *ſpeculum* be looked at with both eyes, it will appear double, provided the diſtance of the eyes from the image be but ſmall, and upon ſhutting either eye, the

Pl. 10. Fig. 14.

the contrary image will difappear; for fince the reflected rays which form the feveral points of an
image meet and crofs one another at the image,
thofe which enter the right eye muft be reflected
from the left fide of the *fpeculum*, and thofe which
fall upon the left eye, muft be reflected from the
right fide of the *fpeculum*, and of confequence, one
and the fame point of the image muft appear to the
right eye, as fituated before the left fide of the *fpe-*
culum, and to the left eye, as fituated before the
right fide of the *fpeculum*; that is, it muft appear
double, and the right or left image muft vanifh upon clofing the contrary eye.   Thus, if the point
C of the image A B, be looked at with both eyes,
one whereof is at O, and the other at Q, the eye
at O fhall fee it by means of the rays O N, which
are reflected from N, and of confequence, fhall fee
it as placed before N, but the eye at Q feeing it by
means of the rays Q M, which proceed from M,
fhall fee it as fituated before M, for which reafon,
the point C will appear double; and what has been
thus fhewn with refpect to the point C, may in the
fame manner be fhewn, with regard to all the other
points in the image, and therefore the whole image
muft appear double; as the eyes are more and more
removed from the images, they approach nearer
together, and at length coincide; the reafon of
which is plain, from the bare infpection of the figure; for fince the interval of the eyes continues
the fame, it is evident, that when they are farther
removed from the image, the rays whereby they fee
the point C muft be reflected from parts of the
fpeculum lefs diftant from one another than M and
N, and the diftance of the parts of the *fpeculum*
which reflect the rays to each eye, muft continually leffen as the eyes are more and more removed
from the image, and at certain diftances of the
eyes, muft become fo fmall as not to be fenfible.

Pl. 10.
Fig. 15.

5

And

Pl. 10.
Fig. 16.

And thus much concerning such spherical *speculums* as are concave; as to *convex speculums*, in order to determine the places of images formed by reflexion from them, let A be a radiating point, exposed directly before the *convex speculum* HK, whose center is C, AB a perpendicular from the radiating point to the *speculum*, which likewise denotes the distance of the radiating point from the *speculum*, AD a ray falling on the *speculum* at D, whose distance from B is indefinitely small, DE the reflected ray meeting the perpendicular in E, CD a *radius* drawn to the point of incidence, and of consequence bisecting the angle FDE; let the angle FCD be made equal to ECD, and let CF be continued till it meets AD produced; this being done, it is evident, that the angle at C in the triangle ACF, is bisected by the line CD, which cuts the opposite side, consequently, AC is to FC, as AD to DF; but forasmuch as D and B are supposed to be indefinitely near, AD is equal to AB, and DE to BE; and because the triangles CFD and CED are equal, DF is equal to DE, and FC is equal to CE; wherefore, AB is to BE, as AC to CE; that is, the distance of the radiating point from the *speculum*, is to the distance of the point E where the reflected ray cuts the perpendicular, which is called the *point of intersection*, as the sum of the distance of the radiating point and the *radius*, to the *radius* lessened by the distance of the point of intersection; that is, putting D for the distance of the radiating point, F for the distance of the point of intersection, and R for the *radius* as before, D : F :: D+R : R—F; consequently, reducing this analogy into an equation, and clearing F, F will be found equal to $\frac{DR}{2D+R}$; that is, the distance of the point of intersection behind the *speculum*, and consequently the distance of an image behind

behind the *speculum*, is equal to a rectangle under the distance of the object from the *speculum* and the *radius*, divided by the sum of twice the distance of the object added to the *radius*. Hence it follows, that if an object be placed so near a *convex speculum* as to touch it, its image will do so too; for in this case, D being nothing, F is likewise nothing; as the object recedes from the *speculum*, the image goes off behind; and when the object is removed to an infinite distance, the image appears behind in the midway between the *speculum* and its center; for in this case, D being infinite, 2D + R becomes 2D, and of consequence, F is equal to $\frac{R}{2}$, so that objects seen by reflexion from convex spherical *speculums*, appear constantly behind the *speculum*, within the limits of half the *radius*; and forasmuch as the images constantly appear on the same side of the center with the objects, they must be less than the objects; for if we suppose H I to be an object placed before the *convex speculum* F G, and projecting its
Pl. 10.
Fig. 14.
image at D E, it is manifest, that the image subtends the same angle at a smaller distance, than the object does at a larger distance, and consequently, must be less; and the disproportion between the object and its image, must increase as the object recedes, and decrease, as it approaches, because, as the object recedes from the center, the image approaches, and as that approaches, the image recedes; but as the image can never be more distant from the center than the object, it can in no case appear larger. The proportions which the magnitudes of the object and its image bear one to another, is the same with the squares of their distances from the *speculum*, as in the case of concave *speculums*; the proof of which being exactly the same with that made use of in the case of concaves, I shall not here repeat it.

As

As to the pofition of fuch images as are feen by reflexion from convex fpherical *fpeculums*, they muft always appear erect; for as they ever appear on the fame fide of the center with the objects, the perpendiculars which are drawn from the upper-moft parts of the objects, muft pafs thro' the up-permoft parts of the images; and thofe from the lower parts of the objects, muft likewife pafs thro' the lower parts of the images; thus, the perpendi-cular HC, which comes from H, the higheft point in the object, paffes thro' D, the higheft point of the image, and IC, which comes from I, the loweft point of the object, paffes thro' E, the loweft point of the image; and fo it is with regard to the perpendiculars which come from the intermediate points; fo that the feveral parts of the image have the fame fituation with the correfponding parts of the object, and of confequence the image appears erect.

Pl. 10.
Fig. 14.

APPENDIX.

OF THE COLLISION OF NON-ELASTICK AND ELASTICK BODIES

PROBLEM I.

IF *two bodies be either entirely void of elasticity or perfectly elastick, and one strike the other directly; if* A *and* B *denote the quantities of matter or weights of the two bodies,* a *and* b *their velocities before the stroke; and if* A *be the swifter body when the bodies move the same way, the body which has the greater motion when they move contrary ways, and the moving body when one of them is at rest before the stroke; to determine the* ratio *of the bodies when their velocities before the stroke are given, or the* ratio *of their velocities before the stroke when the bodies are given; that is, to determine* $\frac{A}{B}$ *when* a *and* b *are given, or* $\frac{a}{b}$ *when* A *and* B *are given; so as that the motion of* A *before the stroke, shall be to its motion after the stroke, in the given* ratio *of* m *to* 1.

To give a solution of this *Problem,* it is necessary to know the motions of A before and after the stroke, both when the bodies are entirely void of elasticity, and when they are perfectly elastick; and likewise to know the motion of A after the stroke, when the bodies move the same way, when they move contrary ways, and when B is at rest before the stroke. The motion of A before the stroke, is Aa in all cases. And from what has been delivered by

our

our Author, when the bodies are entirely void of elaſticity, the motion of A after the ſtroke, is $\frac{AAa+ABb}{A+B}$ when before the ſtroke the bodies move the ſame way, $\frac{AAa-ABb}{A+B}$ when they move different ways before the ſtroke, and $\frac{AAa}{A+B}$ when before the ſtroke B is quieſcent. And when the bodies are perfectly elaſtick, the motions of A after the ſtroke, when before the ſtroke the bodies move the ſame way, contrary ways, and B is quieſcent, are $\frac{2ABb+AAa-ABa}{A+B}$, $\frac{AAa-ABa-2ABb}{A+B}$, and $\frac{AAa-ABa}{A+B}$. Hence, this *Problem* contains ſix *Caſes*, three when the bodies are entirely void of elaſticity, and three when they are perfectly elaſtick; which *Caſes* are thus ſolved.

When the bodies are entirely void of elaſticity.

Case I. If the bodies move the ſame way, Aa will be to $\frac{AAa+ABb}{A+B}$, as m to 1; whence we have $\frac{A}{B} = \frac{a-mb}{ma-a}$, and $\frac{a}{b} = \frac{mB}{A+B-mA}$.

Case II. If the bodies move contrary ways, Aa will be to $\frac{AAa-ABb}{A+B}$, as m to 1; whence we have $\frac{A}{B} = \frac{mb+a}{ma-a}$, and $\frac{a}{b} = \frac{mB}{mA-A-B}$.

Case III. If B be at reſt before the ſtroke, then will Aa be to $\frac{AAa}{A+B}$, as m to 1; whence we

have

have $\frac{A}{B} = \frac{1}{m-1}$. In this cafe b is nothing, and confequently $\frac{a}{b}$ is infinite.

When the bodies are perfectly elaſtick.

CASE IV. If the bodies move the fame way, Aa will be to $\frac{2ABb + AAa - ABa}{A + B}$, as m to 1; whence we have $\frac{A}{B} = \frac{ma + a - 2mb}{ma - a}$, and $\frac{a}{b} = \frac{2mB}{mB + A + B - mA}$.

CASE V. If the bodies move contrary ways, Aa will be to $\frac{AAa - ABa - 2ABb}{A + B}$, as m to 1; whence we have $\frac{A}{B} = \frac{2mb + ma + a}{A + B}$, and $\frac{a}{b} = \frac{2mB}{mA - A - B - mB}$.

CASE VI. If B be at reſt before the ſtroke, Aa will be to $\frac{AAa - ABa}{A + B}$, as m to 1; whence we have $\frac{A}{B} = \frac{m+1}{m-1}$. In this cafe b is nothing, and confequently $\frac{a}{b}$ is infinite.

EXAMP. I. If the bodies be entirely void of elaſticity, and move the fame way, A with a velocity of 7, and B with a velocity of 3; and A loſe half its motion by the ſtroke, or, which amounts to the fame, if the motion of A before the ſtroke be to its motion after, as 2 to 1. In this cafe, a, b, m, b, m, are 7, 3, 2; and $\frac{A}{B}$, which is equal to $\frac{a-mb}{ma-a}$, by *Caſe* 1, will be equal to $\frac{1}{7}$; ſo that A and B will be as 1 and 7. Here Aa, the motion of A before the ſtroke, is 7, and $\frac{AAa+ABb}{A+B}$, its motion after the ſtroke is $3\frac{1}{2}$; but 7 is to $3\frac{1}{2}$, as 2 to 1.

EXAMP. II. If the bodies be entirely void of elaſticity, and move the ſame way, if A and B be as 1 and 4, and the motion of A before the ſtroke be to its motion after, as 3 to 1, in which caſe m will be 3; then $\frac{a}{b}$, which is equal to $\frac{mB}{A+B-mA}$ by *Caſe* 1, will be equal to $\frac{6}{1}$; ſo that a and b will be as 6 and 1. Here Aa, the motion of A before the ſtroke, is 6, and $\frac{AAa+ABb}{A+B}$, its motion after the ſtroke by *Caſe* 1, is 2; but 6 is to 2, as 3 to 1.

EXAMP. III. If B be at reſt before the ſtroke, and the motion of A before the ſtroke, be to its motion after, as 10 to 1, in which caſe m will be 10; then will $\frac{A}{B}$ be $\frac{1}{9}$, or A and B will be as 1 and 9. If the velocity of A before the ſtroke be expreſſed by 1, that is, if a be 1, then will Aa be 1, and $\frac{AAa}{A+B}$ be $\frac{1}{10}$; but 1 is to $\frac{1}{10}$, as 10 to 1.

It is to be obſerved, that A can never communicate all its motion to B, except when it is infinitely greater than B, in which caſe B will become nothing. For if A communicate all its motion to B, m will be 1; and $\frac{A}{B}$, which is as $\frac{1}{m-1}$, will be as $\frac{1}{0}$; but $\frac{1}{0}$ is infinite; and therefore A muſt be in-

A a   finitely

finitely greater than B, to lofe all its motion by the ftroke.

Examp. IV. If the bodies be perfectly elaftick, and move the fame way with velocities which are as 3 and 2 ; and if the motion of A before the ftroke be to its motion after, as 2 to 1 ; then will a, b, m, be 3, 2, 2 ; and $\frac{A}{B}$, which is as $\frac{ma + a - 2mb}{ma - a}$ by *Cafe* 4, will be $\frac{1}{3}$; fo that A and B will be as 1 and 3. Here Aa, the motion of A before the ftroke, is 3 ; and $\frac{2ABb + AAa - ABa}{A + B}$, its motion after the ftroke, is $\frac{3}{2}$; but 3 is to $\frac{3}{2}$, as 2 to 1.

Examp. V. If the bodies be perfectly elaftick, and move the fame way, if A and B be as 4 and 5, and the motion of A before the ftroke be to its motion after, as 3 to 1 ; then will A, B, m, be 4, 5, 3 ; and $\frac{a}{b}$, which is as $\frac{2mB}{mB + A + B - mA}$ by *Cafe* 4, will be $\frac{5}{2}$, fo that a and b will be as 5 and 2. Here, Aa, the motion of A before the ftroke, is 20 ; and $\frac{2ABb + AAa - ABa}{A + B}$, the motion of A after, is $\frac{60}{9}$; but 20 is to $\frac{60}{9}$, as 3 to 1.

Examp. VI. If A and B be perfectly elaftick, and B be at reft before the ftroke, if A move with a velocity of 4, and its motion before the ftroke be to its motion after, as 3 to 1 ; then will a, b, m, be 4, 0, 3 ; and $\frac{A}{B}$, which is as $\frac{m + 1}{m - 1}$ by *Cafe* 6, will $\frac{4}{2} = \frac{2}{1}$; fo that A and B will be 2 and 1. Here, Aa, the motion of A before the ftroke, is 8 ; and $\frac{AAa - ABa}{A + B}$, its motion after the ftroke, is $\frac{8}{3}$; but 8 is to $\frac{8}{3}$, as 3 to 1.

SCHOLIUM.

S C H O L I U M.

If it be required to know the motion of B after the ſtroke in the ſix *Caſes* before mentioned, that motion may be had, from what our Author has delivered, when the weights of the bodies, and their velocities before the ſtroke, are given.

If the bodies be intirely void of elaſticity; the motion of B after the ſtroke, when before the ſtroke, the bodies move the ſame way, when they move contrary ways, or when B is quieſcent, is

$$\frac{BAa+BBb}{A+B}, \quad \frac{BAa-BBb}{A+B}, \quad \text{or} \quad \frac{BAa}{A+B}.$$

And if the bodies be perfectly elaſtick; the motions of B after the ſtroke, when before the ſtroke the bodies move the ſame way, when they move contrary ways, or when B is quieſcent, is

$$\frac{2BAa+BBb-BAb}{A+B}, \quad \frac{2BAa-BBb+BAb}{A+B}, \quad \text{or} \quad \frac{2BAa}{A+B}.$$

PROB. II. *If two bodies* A *and* B *be given, and be perfectly elaſtick, if* A *be the leſſer body, and* B *be at reſt before the ſtroke; it is required to find an intermediate body of ſuch a weight or quantity of matter, which I ſhall denote by* x, *as that* A *ſtriking* x *at reſt, and* x *with the motion acquired by the ſtroke ſtriking* B *at reſt, the motion produced in* B *ſhall be greater than can be produced by an intermediate body of any other weight, or, in other words, that the motion in* B *ſhall be a* maximum.

The motion of x after it is ſtruck by A, is $\frac{2Aax}{A+x}$, and the motion of B after it is ſtruck by x, is $\frac{4ABax}{AB+Ax+Bx+xx}$, by *Schol. Prob.* 1.

But

But by fuppofition the motion of B is a *maximum*, and confequently its fluxion is nothing. The fluxion therefore of $\dfrac{4ABax}{AB+Ax+Bx+xx}$ is nothing ; that is, $\dfrac{4A^2B^2a\dot{x}-4ABxa\dot{x}^2}{\overline{AB+Ax+Bx+xa}^2} = 0.$ Confequently, $4A^2B^2a\dot{x}-4ABxa\dot{x}^2 = 0$; and, by dividing by $4ABa\dot{x}$, $AP - x^2 = 0$; and $AB = x^2$; whence x is a mean proportional between A and B.

Our Author has given a clear folution of this *Problem*, but in a different manner.

Cor. I. If a number of bodies be in a continual geometrical progreffion, if the leaft of the bodies be A, the *ratio* of the increafe be e, and the number of bodies n ; and if A ftrike the fecond body at reft; and the fecond with the motion acquired ftrike the third body at reft, and fo on to the laft ; the bodies, their velocities and motions, will be thus expreffed.

Bodies - - A, $\quad eA, \quad\quad e^2A, \quad\quad e^3A$ &c. $\overline{e}^{n-1}A.$

Velocities - - a, $\dfrac{2a}{1+e}, \dfrac{4a}{\overline{1+e}^2}, \dfrac{8a}{\overline{1+e}^3},$ &c. $\overline{\dfrac{2}{1+e}}^{n-1}a.$

Motions - - Aa, $\dfrac{2Aae}{1+e}, \dfrac{4Aae^2}{\overline{1+e}^2}, \dfrac{8Aae^3}{\overline{1+e}^3},$ &c. $\overline{\dfrac{2e}{1+e}}^{n-1}Aa.$

Examp. I. If the number of bodies increafing in geometrick proportion be 20, and the common *ratio* of the terms be 2, n will be 20, and e be 2. The laft body will be 524288 times greater than the firft ; the velocity of the laft will be to the velocity of the firft, as 1 to $2216\frac{4}{5}$; and the motion of the laft will be about $236\frac{1}{2}$ times greater than the motion of the firft.

Examp. II. If the number of bodies be 100, and the common *ratio* of the progreſſion be 2 ; then will n be 100, and e will be 2. In this caſe, the laſt body will be above 6338253000000000000000000000000 times greater than the firſt, its velocity will be to the velocity of the firſt, as 1 to 271022000000000000 nearly ; and the motion of the laſt will be to the motion of the firſt, nearly as 2338480000000 to 1.

Cor. II. If the motion of the firſt body be to the motion of the laſt, as 1 to D, that is, if Aa

be to $\overline{\dfrac{2e}{1+e}}\Big|^{n-1}$ Aa, as 1 to D, then will e be equal

to $\dfrac{D^{\frac{1}{n-1}}}{2 - D^{\frac{1}{n-1}}}$.

For example, if the number of bodies be 20, and the motion of the laſt be 100000 times greater than the motion of the firſt, n will be 20, D will be 100000, and e will be 10.9746 nearly ; ſo that each preceding body in the 20 bodies muſt be 10.9746 times greater than the body lying next behind it.

Cor. III. If the motion of the firſt body be to the motion of the laſt, as 1 to D, that is, if Aa be to $\overline{\dfrac{2e}{1+e}}\Big|^{n-1}$ Aa, as 1 to D, D will be equal to $\overline{\dfrac{2e}{1+e}}\Big|^{n-1}$, and putting R for $\dfrac{2e}{1+e}$, and L for lo-

garithm, we ſhall have $D = R^{n-1}$, and L, $D = \overline{n-1} \times$ L, R.

For example, if e be 4, and n be 25, $\dfrac{2e}{1+e}$ will

be

be $\frac{4}{3}$, the logarithm of which number is 0.2041299 $= L, R$; and $\overline{n-1} \times L, R = 4.8988795 = L, D$. The natural number of this logarithm is 79228 nearly; so that in this case the motion of the last body will be nearly 79228 times greater than the motion of the first.

COR. IV. If D and R be given, n may be found by being equal to $\dfrac{L, D + L, R}{L, R}$; for by the last *Corollary* L, D $= \overline{n-1} \times$ L, R, and consequently n $= \dfrac{L, D + L, R}{L, R}$.

For example, if D be 100000, and e be 2, in which case R will be $\frac{4}{3}$; then will L, D, be 5.000000 and L, R 0.1249387; and $\dfrac{L, D + L, R}{L, R}$, will be 41. 02 $= n$; so that more than 41 bodies will be necessary to make the motion of the last 100000 times greater than the motion of the first.

Of the Motion *of a* Globe *in a* Fluid Medium.

PROB. III. *If the diameter and density of a Globe moving in a fluid medium, if the density of the medium, the velocity with which the globe sets out, and the time of the motion, be all given; to determine the part of the velocity which is destroy'd by the resistance of the medium, the remaining part of the velocity, and the space described by the globe in the given time.*

Let D denote the diameter of the globe, d its density, $\mathcal{D}$ the density of the fluid *medium*, V the velocity with which the globe sets out, t the time of the motion expressed in seconds, m the part of a diameter or number of diameters of the globe which it would describe with the velocity V in the time t,

and

and T the time in which the globe with the velocity V would *in vacuo* defcribe a fpace which is to $\frac{8D}{3}$ as d to δ; and then the part of the velocity deftroyed by the refiftance of the medium, will be $\frac{m\delta V}{\frac{8}{3}d + m\delta}$; the remaining part of the velocity will be $\frac{\frac{8}{3}dV}{\frac{8}{3}d + m\delta}$; and the fpace defcribed in the *medium* in the time t, will be $\frac{8Dd}{3\delta} \times \text{Log.} \, \overline{1 + \frac{m\delta}{\frac{8}{3}d}} \times 2.302585093.$

For Sir ISAAC NEWTON has proved, that the part of the velocity which is deftroy'd by the refiftance of the medium in the time t, is $\frac{Vt}{T+t}$; that the remaining part, is $\frac{VT}{T+t}$; and that the fpace defcribed in the time t, is $TV \times \text{Log.} \, \overline{\frac{T+t}{T}} \times 2.302585093.$ But by conftruction, T is as $\frac{8Dd}{3\delta V}$, and V is as $\frac{mD}{t}$. And therefore, by fubftituting $\frac{8Dd}{3\delta V}$ and $\frac{mD}{t}$ inftead of T and V in the foregoing expreffions, the part of the velocity deftroyed by the refiftance of the medium in the time t will be $\frac{m\delta V}{\frac{8}{3}d + m\delta}$, the remaining part of the velocity will be $\frac{\frac{8}{3}dV}{\frac{8}{3}d + m\delta}$, and the fpace defcribed in the time t will be $\frac{8dD}{3\delta} \times \text{Log.} \, \overline{1 + \frac{m\delta}{\frac{8}{3}d}} \times 2.302585093.$

COR. I. If the denfity of the globe be equal to the denfity of the *medium*, that is, if d be equal to δ, the velocity deftroy'd by the refiftance of the *medium* in the time t, will be $\dfrac{mV}{\frac{8}{3}+m}$.

This *Corollary* will obtain, if the globe and the *medium* be perfectly denfe or void of pores; for by being entirely void of pores, they will have equal denfities. And fuch a globe moving in fuch a *me-dium* the length of 3 times its diameter, will lofe above half its velocity; for if m be 3, $\dfrac{mV}{\frac{8}{3}+m}$ will be $\dfrac{9V}{17}$. And this will always be the velocity loft in moving three times the length of the diameter, when the globe and the *medium* have equal denfities.

COR. II. If a globe in moving through m times its diameter in a fluid *medium*, lofe the n part of its velocity; then will $n = \dfrac{m\delta}{\frac{8}{3}d + m\delta}$, $d =$ $\dfrac{\delta \times \overline{m - nm}}{\frac{8}{3}n}$, $\delta = \dfrac{\frac{8}{3}dn}{m - nm}$, and $m = \dfrac{8nd}{3\delta - 3n\delta}$. For $\dfrac{m\delta V}{\frac{8}{3}d + m\delta} = nV$; whence $n = \dfrac{m\delta}{\frac{8}{3}d + m\delta}$, $d = \dfrac{\delta \times \overline{m - nm}}{\frac{8}{3}n}$ $\delta = \dfrac{\frac{8}{3}dn}{m - nm}$, and $m = \dfrac{8nd}{3\delta - 3n\delta}$.

EXAMP. I. If a globe lofe $\frac{3}{4}$ of its velocity in moving the length of 10 times its diameter in wa-ter, in which cafe n will be $\frac{3}{4}$, m will be 10, and δ will be 1; then d will be $\frac{5}{4}$, that is, the globe will be denfer than water in the proportion of 5 to 4.

EXAMP. II. If a globe 10 times as denfe as water, lofe $\frac{3}{4}$ths of its velocity in moving 10 times its diameter in a fluid; the denfity of that fluid will

will be 8 times as great as the denſity of water. In this caſe d is 10, m is 10, and n is $\frac{3}{4}$; and d, which is equal to $\frac{\frac{8}{3}dn}{m-nm}$, is 8.

EXAMP. III. If a globe twice as denſe as water, loſe $\frac{3}{4}$ths of its motion by moving in a ſluid 14 times as denſe as water; it will ſuffer this loſs of velocity in moving the length of $1\frac{1}{7}$ D. For in this caſe d, d, n, are 2, 14, $\frac{3}{4}$; and m, which is equal to $\frac{8nd}{3d-3nd}$, will be $\frac{8}{7}=1\frac{1}{7}$.

EXAMP. IV. If a perfectly ſolid globe move 24 times the length of its diameter in a perfectly ſolid *medium*, it will loſe 9 parts in 10 of the velocity it had at the beginning of the motion. For in this caſe d is equal to d, and m is 24; and n, which is equal to $\frac{m}{\frac{8}{3}+m}$, will be equal to $\frac{72}{80}=\frac{9}{10}$.

EXAMP. V. If a globe of equal denſity with water, move half the length of its diameter in air, it will loſe the $\frac{1}{4587\frac{2}{3}}$ part of its velocity, on ſuppoſition that the denſity of water is to the denſity of air, as 860 to 1. For in this caſe, d, d, m, are 860, 1, $\frac{1}{2}$; and n, which is equal to $\frac{md}{\frac{8}{3}d+md}$, will be $\frac{1}{4587\frac{2}{3}}$.

EXAMP. VI. If the earth moved round the ſun in a ſluid *medium* of equal denſity with the air at the ſurface of the earth, it would by the reſiſtance of the *medium* loſe almoſt all its motion in 10000 years, on ſuppoſition that the denſities of the earth, of water, and of the *medium*, are 5, 1, $\frac{1}{860}$, or in

decimals

decimals 0.0011628. For the earth moves in its orbit with a velocity that carries it at the rate of 4893938782791 miles, or 617142343 times the length of its own diameter in 10000 years, on suppofition that the fun's horizontal parallax is $10\frac{1}{4}$ feconds. In this cafe therefore d, ♂, m, are 5, 0.0011628, and 617142343; and confequently n, which is equal to $\frac{m\text{♂}}{\frac{8}{3}d + m\text{♂}}$, will be a $\frac{7177131}{7117727}$th part, which is nearly the whole, of its prefent velocity.

By the *French* meafures, a degree of a great circle of the earth contains 342366 *Paris* feet, or 365403.3158 *Englifh* feet, on fuppofition that a *Paris* foot is to an *Englifh* foot, as 1142 to 1076: And confequently the diameter of the earth, fuppofing the earth to be fpherical, will be 41870881 *Englifh* feet, or 7930 miles. The mean diftance of the fun from the earth, reckoning the parallax at $10\frac{1}{4}$ feconds, is about 19644.2675 femidiameters of the earth, or 77889520.6375 miles; confequently the circumference of the earth's orbit is 489393878.2791 miles, which the earth defcribes in one year, or 29558161.6 feconds of time.

EXAMP. VII. If the earth move in an *Æther* 700000 rarer than the air at the furface of the earth, it will lofe about $\frac{1}{14}$th part of its prefent velocity in 10000 years; for in this cafe d, ♂ and m, are 5, 0.0000000166, and 617142343; and confequently n, which is equal to $\frac{m\text{♂}}{\frac{8}{3}d + m\text{♂}}$ will be equal to a $\frac{1.02445628938}{14.35778962271}$th part, that is $\frac{1}{14}$th part of the prefent velocity very nearly.

And if the earth moves 100000 years in this *Æther*, it will lofe almoft half of its prefent motion in that time.

EXAMP.

EXAMP. VIII. If we fuppofe the earth to lofe the $\frac{1}{100}$th part of its prefent velocity by moving in an *ætherial medium* for 400000 years, in which time it will have defcribed 24685693680 times its diameter, the denfity of the *medium* will be above 200 millions of times lefs than the denfity of the air at the furface of the earth. For in this cafe d, n, m, are 5, 0.01, 24685693680, and confequently d, which is equal to $\frac{\frac{8}{3}dn}{m-nm}$, will be

$$\frac{0.1333333}{24438836743.2000000} = \frac{1}{183366000000}.$$ But the denfity of water being 1, the denfity of air is $\frac{1}{850}$; and confequently, the denfity of the air at the furface of the earth will be to the denfity of this *medium*, as above 213200000 to 1.

Of the Motion *of* Wheels *over* Obftacles.

PROB. IV. *If a wheel moving on an horizontal plane, meet with an immoveable obftacle in its way, over which it is to be drawn by a force fixed to its center; if the weight and diameter of the wheel, the height of the obftacle, and the direction of the force drawing the wheel, be all known; thence to determine the force that is fufficient to draw the wheel over the obftacle.*

Let GPME be the wheel, ND the horizontal plane on which it moves from N towards D, EF the obftacle over which it is to be drawn; let the wheel arrive at the obftacle, and touch its top E; and there let it be fuppofed to ftand preffing the horizontal plane at G with its whole weight. Draw OEK a tangent to the wheel in the point E, draw the diameter ACG perpendicular to the horizontal plane, and produce it till it meet the tangent in O; from

Pl. 11. Fig. 1.

from E draw the *radius* EC; draw EH perpendicular to AG; and mr, MC, perpendicular to EC, and consequently parallel to the tangent OK; and laftly, draw the *radius* Cm; if the whole weight of the wheel be expreffed by CO, in the direction of which line that weight acts when the wheel is wholly fupported by the horizontal plane at G, that weight may be refolved into two others CE and OE, acting according to the directions of thofe lines, the weight CE preffing againft the top of the immoveable obftacle, and being wholly fuftained by it, and the weight OE drawing the wheel down in a direction parallel to the tangent OEK. Let W denote the whole weight of the wheel, r its *radius*, h the height of the obftacle, and x the part of the whole weight which draws the wheel down in a direction parallel to OEK; and then we fhall have this analogy; as x is to W, fo is OE to CO, or HE to CE, from the fimilarity of the triangles CEO, and CEH; whence $x = \dfrac{W \times HE}{r}$; but HE from the nature of the circle, is equal to $\sqrt{AH \times HG}$, or to $\sqrt{AH \times EF}$, that is, in fymbols, to $\sqrt{2rh - hh}$; and therefore $x = \dfrac{W \times \sqrt{2rh - hh}}{r}$. A force juft equal to this weight, and acting in direct oppofition to it, that is, drawing the wheel upward in the direction CM parallel to OK, will juft be able to make the wheel reft on E the top of the obftacle, without fuffering any part of its weight to reft on the horizontal plane at G. This force muft be increafed to produce the fame effect, if it act in any other direction than that of CM. For let it draw the wheel in the direction Cm, m lying between E and M, and then the force acting in this direction may be refolved into two forces, which will be as Cr and rm, whereof Cr draws the wheel directly againft E the

top

top of the obftacle, and fo is loft, and mr draws it up in a direction parallel to OK. But mr is lefs than Cm or CM, and to become equal to it, and confequently, fufficient to fupport the wheel againft the top of the obftacle without fuffering any part of its weight to reft on the horizontal plane, it muft be increafed in the *ratio* of Cm or CM to rm, that is, putting s for the fine of the angle which the direction of the force makes with CE, in the *ratio* of r to s ; but the force rm cannot be increafed, but the whole force CM muft be increafed in the fame proportion. And therefore the force $\dfrac{W \times \sqrt{2rh - hh}}{r}$ muft be increafed in the proportion of r to s, and then, putting F for the force, acting in the direction Cm, which is juft fufficient to fupport the wheel on the obftacle without fuffering it to prefs on the plane ND, $F = \dfrac{W \times \sqrt{2rh - hh}}{s}$; and the fmalleft addition to this force will make it draw the wheel over the obftacle.

Since the refiftance given by the obftacle, is equal to the force that is juft fufficient to make the wheel reft on the obftacle without fuffering any part of its weight to prefs on the plane of the horizon, that is, putting R for the refiftance given by the obftacle, fince R is equal to F ; R will be equal to $\dfrac{W \times \sqrt{2rh - hh}}{s}$.

It is to be obferved, that the direction of the force muft lie between CE and CA ; for if the force draw the wheel in the direction CE it will be wholly fpent upon the obftacle, and not in the leaft contribute to draw the wheel over it ; and if it draw the wheel directly upwards from C to A, it will not make it to prefs againft the obftacle, and confe-

confequently, however great we may fuppofe it to be, can never draw it over it.

Cor. I. If the direction of the moving force change continually, paffing from C E to C M, and thence to C P, the fine of the angle which the line of direction makes with C E, will increafe in the paffage of that line from C E to C M, and decreafe in its paffage from C M to C P; but as s increafes or leffens, $\dfrac{W \times \sqrt{2rh - hh}}{s}$ will leffen or increafe; and confequently the force F will leffen in the paffage of the line of its direction from C E to C M, and thence increafe in the paffage of that line to C A. So that the force will be leaft when it acts in the direction C M, in which cafe the whole force will be employed in drawing the wheel over the obftacle; whereas in all other directions, part of the force will be loft by drawing directly againft the top of the obftacle. Hence the moft advantageous direction of the force, will be that which makes a right angle with C E, in which cafe s will be equal to r, and $F = \dfrac{W \times \sqrt{2rh - hh}}{r}$

Cor. II. If the height of the obftacle be given, in which cafe h will be as 1, and the force draw the wheel in the direction C M parallel to O K; then F will be as $\dfrac{W \times \sqrt{2r - 1}}{r}$.

If the *radii* of four wheels be 1, 2, 3, 4, then will $\dfrac{\sqrt{2r - 1}}{r}$, be 1, $\dfrac{\sqrt{3}}{2}$, $\dfrac{\sqrt{5}}{3}$, $\dfrac{\sqrt{7}}{4}$, that is, as the numbers 1000, 866, 745, 661; and the forces requifite to fupport thefe wheels on the point E, fo as not to fuffer any part of their weight to reft on the horizontal plane, will be as their weights multiplied

plied into theſe numbers reſpectively. The orce requiſite to ſupport the firſt wheel, will be as its weight multiplied into 1000, the force requiſite to ſupport the ſecond wheel as its weight multiplied into 866; and ſo of the reſt. And if the weights of all the wheels be equal, the forces neceſſary to ſupport them, and conſequently the reſiſtances given by the obſtacle to which theſe forces are equal, will be as the numbers 1000, 866, 745, 661. So that in wheels of a given weight, the leſſer the wheel is, the greater will be the reſiſtance which is given to it by an obſtacle of a given height.

COR. III. If the height of the obſtacle be indefinitely ſmall and given, in which caſe the tangent O K will coincide with the horizontal plane N D, and the point E coincide with the point G; and if the force draw the wheel in a direction parallel to O K or N D; then will F be as $W \times \frac{\sqrt{2r}}{r}$, or, becauſe 2 is a given quantity, as $\frac{W}{rr}$; and if the weight of the wheel be given, F will be as $\frac{1}{\sqrt{r}}$.

If the *radii* of four wheels of equal weights be 1, 2, 3, 4, and the wheels be drawn on a ſmooth plane parallel to the horizon; the forces neceſſary to put them in motion, when they draw in directions parallel to that plane, will be as $1, \frac{1}{\sqrt{2}}, \frac{1}{\sqrt{3}}, \frac{1}{\sqrt{4}}$; that is, as the numbers 1000, 707, 577, 500. And therefore, of wheels drawn on the plane of the horizon by forces acting in directions parallel to that plane, leſſer wheels will require a greater force to put them in motion than greater.

COR.

Cor. IV. If the height of the obstacle be proportional to the *radius* of the wheel, and if the force draw the wheel in a direction parallel to OK; that is, if h be as r, and F be as $\dfrac{W \times \sqrt{2rh - hh}}{r}$; then will the force, and consequently the resistance given by the obstacle, be as the weight of the wheel; for $\dfrac{\sqrt{2rh - hh}}{r}$ will be as $\dfrac{\sqrt{2rr - rr}}{r}$, that is, as 1; and therefore F will be as W.

Cor. V. If the direction of the force drawing the wheel be parallel to the horizontal plane, that is, if mC be parallel to ND; then will the force that is requisite to sustain the wheel on the point E, be $\dfrac{W \times \sqrt{2rh - hh}}{r - h}$. For in this case the angle mCE is equal to the angle CEH, and consequently, their sines are equal, that is, s is equal to CH, which in symbols is r — h. And therefore F, which universally is as $\dfrac{W \times \sqrt{2rh - hh}}{s}$, is in this case as $\dfrac{W \times \sqrt{2rh - hh}}{r - h}$.

If the height of the obstacle be given, in which case h will be as 1, then will F be as $\dfrac{W \times \sqrt{2r - 1}}{r - 1}$.

If the *radii* of four wheels of equal weight, be 1, 2, 3, 4; then will F with respect to these four wheels, be as $\dfrac{1}{0}$, $\dfrac{\sqrt{3}}{1}$, $\dfrac{\sqrt{5}}{2}$, $\dfrac{\sqrt{7}}{3}$, that is, as infinite, 1732, 1128, 882. The height of the obstacle is equal to the *radius* of the first wheel, inasmuch as I have supposed them both to be as 1; and consequently the force must be infinite to make the wheel rest against E, and hinder any part of

its

its weight from prefling on the horizontal plane at G.

Cor. VI. The force, is to the weight of the wheel, as the fine of the angle ECH, is to the fine of the angle which the line of direction of the force makes with EC; that is, $\dfrac{F}{W} = \dfrac{\sqrt{2rh - hh}}{r}$.

If the force be one half of the weight of the wheel, that is, if F be one half of W, $\sqrt{2rh - hh}$ will be one half of s; if P be equal to W, $\sqrt{2rh - hh}$ will be equal to s; and if F be as W, $\sqrt{2rh - hh}$ will be as s.

Of the Motion *of* Water *through* Orifices *and* Pipes.

Prob. V. *To determine the motion of water running out of a hole made in the bottom of a veffel.*

Sir Isaac Newton has given a general folution of this *Problem* in the following paragraph, which is contained in *prop.* 36. *prob.* 8. *lib.* 2.

" Sit ACDB vas cylindricum, AB ejus orificium
" fuperius, CD fundum horizonti parallelum, EF
" foramen circulare in medio fundi, G centrum fo-
" raminis, et GH axis cylindri horizonti perpendi-
" cularis. Et finge cylindrum glaciei APQB ejuf-
" dem effe latitudinis cum cavitate vafis, et axem
" eundem habere, et uniformi cum motu perpetuo
" defcendere, et partes ejus quam primum attingunt
" fuperficiem AB liquefcere, et in aquam converfas
" gravitate fuâ defluere in vas, et cataractam vel
" columnum aquæ ABNFEM cadendo formare,
" et per foramen EF tranfire, idemque adæquate
" implere. Ea vero fit uniformis velocitas glaciei
" defcendentis ut et aquæ contiguæ in circulo AB,

Pl. 11. Fig. 2.

B b " quam

" quam aqua cadendo et cafu fuo defcribendo alti-
" tudinem I H acquirere poteft ; et jaceant I H et
" HG in directum, et per punctum I ducatur recta
" KL horizonti parallela et lateribus glaciei occur-
" rens in K et L. Et velocitas aquæ effluentis
" per foramen EF ea erit quam aqua cadendo ab I
" et cafu fuo defcribendo altitudinem IG acquirere
" poteft. Ideoque per theoremata GALILÆI erit
" IG ad IH in duplicata ratione velocitatis aquæ
" per foramen effluentis ad velocitatem aquæ in
" circulo AB, hoc eft, in duplicata ratione circuli
" AB ad circulum EF ; nam hi circuli funt reci-
" proce ut velocitates aquarum quæ per ipfos eo-
" dem tempore et æquali quantitate, adæquate
" tranfeunt. De velocitate aquæ horizontem verfus
" hic agitur. Et motus horizonti parallelus quo
" partes aquæ cadentis ad invicem accedunt, cum
" non oriatur a gravitate, nec motum horozonti
" perpendicularem a gravitate oriundum mutet,
" hic non confideratur. Supponimus quidem quod
" partes aquæ aliquantulum cohærent, et per co-
" hæfionem fuam inter cadendum accedant ad in-
" vicem per motus horizonti parallelos, ut unicam
" tantum efforment cataractum et non in plures
" cataractas dividantur : fed motum horizonti pa-
" rallelum, a cohæfione illâ oriundum, hic non
" confideramus."

This *Theory* Sir ISAAC corrected by experiments,
proved it in fix different cafes, and drew feveral
corollaries from it. The reafon why a correction
was neceffary will be fhewn in the *Scholium.* And
the truth of his and other corollaries flowing from
this theory, will more eafily appear by expreffing
the foregoing proportions of the velocities in fymbols; to do which let A denote the *area* of the
circle AB, a the *area* of the hole EF, H the line
HG, which is the perpendicular height of the water
in the veffel above the hole, x the height IH, from
which water or any other body muft fall by the

2   force

force of gravity from a ſtate of reſt, to acquire the velocity of the water in AB, V the velocity of water in its paſſage through the hole EF, and v its velocity in the ſurface AB; and then the proportions will be thus expreſſed, $H + x . x :: V^2 . v^2 :: A^2 . a^2$; whence, $\sqrt{H + x} . \sqrt{x} :: V . v :: A . a$.

COR. I. The height from which a body muſt fall to acquire a velocity equal to the velocity of the water in the ſurface AB, is equal to $\frac{v^2 H}{V^2 - v^2}$, or $\frac{a^2 H}{A^2 - a^2}$. For by inverſion and diviſion of proportion, $x . H :: v^2 . V^2 - v^2 :: a^2 . A^2 - a^2$; whence $x = \frac{v^2 H}{V^2 - x^2} = \frac{a^2 H}{A^2 - a^2}$. But x denotes IH. And therefore $IH = \frac{v^2 H}{V^2 - v^2} = \frac{a^2 H}{A^2 - a^2}$.

COR. II. The perpendicular height of the water in the veſſel, denoted by H, is equal to $\frac{IH \times \overline{V^2 - v^2}}{v^2}$, or $\frac{IH \times \overline{A^2 - a^2}}{a^2}$, by *Cor.* 1.

COR. III. The height from which a body muſt fall, to acquire a velocity equal to that with which the water flows through the hole, is equal to $\frac{V^2 H}{V^2 - v^2}$, or $\frac{A^2 H}{A^2 - a^2}$. For by diviſion of proportion, $H + x = IG . H :: V^2 . V^2 :: A^2 . A^2 - a^2$, whence $IG = \frac{V^2 H}{V^2 - v^2} = \frac{A^2 H}{A^2 - a^2}$.

COR. IV. The perpendicular height of the water in the veſſel, denoted by H, is equal to $\frac{IG \times \overline{V^2 - v^2}}{V^2}$, or to $\frac{IG \times \overline{A^2 - a^2}}{A^2}$, by *Cor.* 3.

COR.

Cor. V. If the *area* of the furface be equal to the *area* of the hole, H will be nothing in comparifon of IH and IG which will be equal. For if A be equal to a, H will be nothing, by *Cor.* 2. and IH and IG will be equal and infinite, by *Cor.* 1, and *Cor.* 3.

The truth of this *Corollary* may likewife appear from the nature of gravity. For if A be equal to a, V muft be equal to v. But V can never be equal to v while there is any acceleration of the motion of the water in its defcent thro' the veffel, as there will always be till H becomes nothing in comparifon of the equal lines IH and IG, which in this cafe muft be confidered as infinite.

Cor. VI. If a be greater than A, in which cafe $A^2 - a^2$ will be negative, H will be negative, by *Cor.* 4; and IG, and confequently V, will be affirmative, by *Cor.* 3. But a negative perpendicular height of the water in the veffel, and an affirmative velocity of the water flowing through the hole, require an inverfion of the veffel or a turning of its bottom upwards; by which inverfion the hole will become the upper orifice, and the upper orifice the hole; a will become A, and A become a; and the velocity will be affirmative, that is, the water will move downwards, as it ought to do from the nature of gravity. Farther, when a is greater than A, the veffel will be conical with its wider end downwards; but from the nature of gravity, water poured in at the top or narrower end of fuch a veffel, will defcend in a cylindrical column, which will not fill the bafe, as the foregoing account of this motion requires; and therefore, to give this cafe the conditions required, there muft be an inverfion of the veffel.

Cor. VII. If the hole be fmall, and the furface of the water infinitely large, both a and v may

be,

be confidered as o with refpect to A and V; confequently IH will be o, by *Cor.* 1. and IG will be
equal to H, by *Cor.* 3.

In this cafe, and this only, the fuperficial parts
of the water have no velocity at the very beginning
of the motion, but begin to defcend from a ftate
of reft, as quiefcent bodies do when the fupport is
taken away. In all other cafes, in which a and v
have fome magnitudes when compared with A and
V, the fuperficial parts of the water fet out with
fome velocity, and do not begin to defcend, on
the water's beginning to flow through the hole, as
heavy bodies near the furface of the earth begin to
defcend from a ftate of reft.

Cor. VIII. If the *ratio* of the furface to the
hole be given, as it will be when each of them continues the fame, or when both of them change in
the fame proportion; the velocity in the furface
will be proportional to the velocity through the
hole, and both will be proportional to the velocity
which would be acquired by a body in falling
through a height equal to the perpendicular height
of the water in the veffel. If $\frac{A}{a}$ be given, $\frac{V}{v}$ will
be given; and confequently v will be as V. And
fince $\frac{A}{a}$ is given, $\frac{a^2}{A-a^2}$, and $\frac{A^2}{A^2-a^2}$, will both
be given; and confequently both IH and IG will
be as H, by *Cor.* 1. and *Cor.* 3. But v and V are
as $\sqrt{IH}$ and $\sqrt{IG}$. And therefore, both v and V
will be as $\sqrt{H}$.

By this *Corollary*, when A and a continue invariable, and the heights of the water in the veffel
are 1, 4, and 16 feet; the velocities in AB and
EF will be as 1, 2, and 4. But bodies placed at
fmall diftances from the furface of the earth, do all
begin to defcend with the fame velocity very nearly,

as

as has been proved by experiments. And there-fore the fuperficial parts of the water, in this cafe, begin to defcend in a very different manner, or with very different velocities from that with which a heavy body placed at thofe heights, begins to defcend from a ftate of reft. The velocity in AB is regulated by the velocity in EF, and the velocity in EF is always meafured by $\sqrt{H}$, when $\frac{A}{a}$ is given.

COR. IX. The velocity of the water in the fur-face AB is always the $\frac{a}{A}$ part of the velocity thro' the hole, that is, v is the $\frac{a}{A}$ part of V, or in other words, $v = \frac{aV}{A}$. When a is nothing in proportion to A, as we may fuppofe it to be, when a is very fmall, and A exceedingly great, then will v be no fenfible part of V, that is, it will be nothing; and confequently, the fuperficial parts of the water will in this cafe begin their motion, as heavy bodies do, from a ftate of reft.

COR. X. The whole motion of the defcending column AMEFNB, is equal to the motion of a cylinder of water, whofe bafe is a, whofe altitude is H, and whofe velocity is V, that is, to the mo-tion aH × V. For Va is equal to vA, that is, the motion of the water in EF is equal to its motion in AB; and from the nature of the defcending co-lumn, each of them is equal to the motion in any fection of the column parallel to EF or AB; and confequently, the motion in all the fections, fuppof-ing them to be indefinitely many, that is, the whole motion of the defcending column, will be equal to the motion in the hole multiplied into the number

of

of sections, that is, to Va × H, or aH × V. This property has been proved by Dr. JURIN.

COR. XI. The force which can generate the whole motion of the water running out of the hole, is equal to the weight of a cylinder of water whose base is a, and altitude is 2IG, by *Cor.* 3 ; that is, equal to the weight of a cylinder of water, whose magnitude is 2aH × $\frac{A^2}{A^2 - a^2}$. For in the same time, in which the water running out is equal to this cylinder, this cylinder, by falling from the height IG by the force of its gravity, will acquire a velocity equal to that with which the water runs out. But when the quantities of matter and velocities of two bodies are equal, their motions, and consequently the forces which can generate those motions in equal times, will likewise be equal. And therefore the force which can generate the whole motion of the water running out of the hole, is equal to the weight of a cylinder of water, whose magnitude is 2aH × $\frac{A^2}{A^2 - a^2}$.

COR. XII. The weight of the descending column AMEFNB is equal to the weight of a cylinder of water, whose base is a, and whose height is $\frac{2HA}{A + a}$, that is, whose magnitude is 2aH × $\frac{A}{A + a}$. For let IO be a mean proportional between IH and IG, and then $\sqrt{IH} . \sqrt{IG} :: IH . IO :: IO . IG :: a . A$; and, by division of proportion, HO . IH :: OG . IO ; and by alternation and composition, HO + OG . 2HO :: IH + IO . 2IH :: a + A . 2a. But, by *Cor.* 11. in the time a drop of water falls by its own gravity from I to G, the quantity of water discharged by the hole will be equal to a × 2IG, or A × 2IO ; and in the time the drop

descends

defcends from I to H, the quantity of water paffing through the furface AB, and difcharged by the hole, will be equal to A × 2IH, and the difference of thefe quantities, namely A × 2HO, will be the quantity difcharged in the time the falling drop defcends from H to G, which quantity is the column AMEFNB; for in the time the drop defcends from H to G, the fuperficial parts of the water, fetting out with the velocity of the drop at H, and defcending freely and without refiftance, will reach the hole. And therefore, all the water in the veffel will be to the water in the column AMEFNB, as A × H is to A × 2HO, or as H = HO + OG to 2HO; or as $a + A$ to $2a$; whence, putting Q for the quantity of water in the defcending column, $A × H . Q :: A + a . 2a$; and confequently,
$$Q = 2AH × \frac{A}{A + a}.$$

This *Corollary* may be proved in another manner, thus. The cataract is the difference of the two hyperboloids KAMEFBL and KABL, fuppofing the affymptote KL to be infinitely extended both ways, and the *area* AB to be infinite; but by fluxions, as Dr. JURIN has fhewn, the hyperboloid KAMEFNBL is equal to $2a × \overline{H + x}$, or to $\frac{2A^2x}{a}$, becaufe H is equal to $\frac{A^2x - a^2x}{a^2}$ by *Cor.* 2; and the hyperboloid KABL, is equal to $2Ax$, and the difference of the two is $\frac{2A^2x}{a} - 2Ax = \frac{2A^2x - 2Aax}{a}$. All the water in the veffel is AH or, by fubftituting $\frac{A^2x - a^2x}{a^2}$ in the room of H, $\frac{A^3x - Aa^2x}{a^2}$; and confequently, the water in the veffel is to the water in the cataract, as A^3x

$\dfrac{A^2x - Aa^2x}{a^2}$ is to $\dfrac{2A^2x - 2Aax}{a}$, that is, after due reduction, as $A + a$ is to $2a$. Therefore $AH . Q$

$:: A + a . 2a :$ whence, $Q = 2aH \times \dfrac{A}{A + a}$.

Cor. XIII. The weight of all the water in the veffel, is to the weight of that part of it which is fuftained by the bottom, as the fum of the circles AB and EF is to their difference. For, fince $A \times H . Q :: A + a . 2a$, by *Cor.* 12. $A \times H$. $A \times H - Q :: A + a . A + a - 2a = A - a$ by divifion of proportion.

Cor. XIV. The weight of the water which the bottom fuftains is to the weight of the cataract, as the difference of the circles AB and EF, to twice the leffer circle EF. For $A \times H . Q :: A + a . 2a$, by *Cor.* 12. And by divifion of proportion, $A \times - Q . Q :: A + a - 2a = A - a . 2a$.

Cor. XV. The weight of water which the bottom fuftains, is to the weight of water perpendicularly incumbent thereon, as the circle AB, is to the fum of the circles AB and EF. For the weight of water which the bottom fuftains is $A \times H - Q$ $= AH - \dfrac{2aHA}{A + a}$, by *Cor.* 12. $= \dfrac{A^2H - aAH}{A + a}$; and the weight perpendicularly incumbent on the bottom is $\overline{A - a} \times H = AH - aH$. But $\dfrac{A^2H - aAH}{A + a} . AH - aH :: A^2 - aA . A^2 - a^2$ $:: A . A + a$ by dividing by $A - a$.

Cor. XVI. The quantity of water in the defcending column is to the quantity perpendicularly incumbent on the hole, as twice the circle AB, is to the fum of the circles AB and EF. For the

5 quantity.

quantity of water in the descending column is $2aH \times \frac{A}{A+a}$. But $2aH \times \frac{A}{A+a} \cdot aH :: \frac{2A}{A+a}$. $1 :: 2A \cdot A + a$.

Hence, when a is nothing, as we may suppose it to be when A is infinitely great, the descending column will be equal in magnitude to $2aH$, as Dr. JURIN has shewn it to, be by determining its magnitude by fluxions.

COR. XVII. The weight of the descending column, is to the weight of water which can generate the whole motion of the water running out of the hole, as the difference of the circles AB and EF, is to the greater circle AB. For, putting F for the force or weight which can generate the whole motion of the water running out of the hole, and supposing Q to denote the weight of the descending column, we shall have F equal to the weight of a quantity of water whose magnitude is $2aH \times \frac{A^2}{A^2-a^2}$, by *Cor.* 11. and Q equal to the weight of a quantity, whose magnitude is $2aH \times \frac{A}{A+a}$, by *Cor.* 12. And therefore, $Q \cdot F :: 2aH \times \frac{A}{A+a} \cdot 2aH \times \frac{A^2}{A^2-a^2} :: 1 \cdot \frac{A}{A-a} :: A-a \cdot A$.

Hence $Q = \frac{F \times A - a}{A}$, and $F = \frac{QA}{A-a}$; and consequently, the force which can generate the whole motion of the water running out of the hole, will always exceed the weight of the descending column, except when a becomes o, as we may suppose it to do, when it is very small, and A exceedingly great.

COR. XVIII. The force which can generate the whole motion of the water running out of the hole,

hole, is to the weight of water perpendicularly incumbent on the hole, as twice the square of the greater circle AB, to the difference of the squares of the circles AB and EF. For the force which can generate the whole motion of the water runing out of the hole, is the weight of $2aH \times \dfrac{A^2}{A^2 - a^2}$ quantity of water, by *Cor.* 11. and the weight of water perpendicularly incumbent on the hole, is the weight of the cylinder aH. But $2aH \times \dfrac{A^2}{A^2 - a^2}$. aH :: $2A^2 . A^2 - a^2$. In the same *ratio* is the whole motion of the effluent water to the motion of the water in the cataract.

Cor. XIX. If in the middle of the hole be placed a little circle PQ parallel to the horizon, whose center is G, and if the *area* of this circle be called o; the weight of water which it sustains during the efflux of the water through the ring surrounding it, is to the weight of half the cylinder oH, as a to $a - \frac{1}{2}o$; if R denote the weight sustained, R is to $\dfrac{oH}{2}$, as a to $a - \frac{1}{2}o$, and R is equal to $\dfrac{aoH}{2a - o}$. For if we suppose A to be contracted till it becomes equal to a, in which case IH will be infinite, by *Cor.* 1. the water, notwithstanding this, will descend about the column PQH which the little circle sustains with velocities, which are every where in the subduplicate *ratio* of the distance from KL, and likewise in the reciprocal *ratio* of the several sections through which it passes; consequently, the cataract AEPHQFB, is equal to the difference of the two hyperboloids PEAKLBFQH and AKLB. But the hyperboloid PEAKLBFQH $= 2a - 2o \times \overline{H + x} = 2aH - 2oH + 2ax - 2ox$; and the hyperboloid AKLB is 2ax; and the differ-

Pl. 11. Fig. 3.

rence

rence of the two is $2aH - 20H - 20x$, which is the cataract AEPHQFB. The *ratio* of all the water in the veſſel to this annular cataract, is $\dfrac{aH}{2aH - 20H - 20x}$. But from the nature of the motion of the deſcending water, a is to $a - o$, as $\sqrt{H + x} \cdot \sqrt{x}$, whence $H = \dfrac{2aox - oox}{a^2 - 2ao + oo}$. The foregoing *ratio*, when this value of H is ſubſtituted in its room, will, after due reduction, become $\dfrac{2a - o}{2a - 2o}$. Therefore aH, the whole quantity of water in the veſſel, is to the annular cataract, as $2a - o$ to $2a - 2o$; whence the annular cataract is $\dfrac{2a^2H - 2aoH}{2a - o}$, which being ſubducted from aH, leaves $\dfrac{aoH}{2a - o}$ for the quantity ſuſtained by the little circle o. Conſequently, $R = \dfrac{aoH}{2a - o}$; and $R \cdot \dfrac{oH}{2} :: a \cdot a - \frac{1}{2}o$.

S C H O L I U M.

Upon examining this motion by experiments, Sir ISAAC NEWTON found the velocity of the water in its paſſage through the hole to be leſs than it ought to be, if the water in the veſſel deſcended from the ſurface to the hole freely and without reſiſtance, in the proportion of 1 to $\sqrt{2}$. For he obſerved the vein of the effluent water, and found it to contract and grow narrower, to the diſtance of about a diameter of the hole below it, at which place he meaſured the diameter of the vein, and found it to be leſs than the diameter of the hole in the proportion of 21 to 25, and conſequently, the *area* of a ſection of the vein at that place to be leſs than the *area* of the hole, in the proportion of 441

to

to 625, that is, of 1 to $\sqrt{2}$. But as the vein contracts the velocity increases. And therefore, at the distance of a diameter of the hole below it, the velocity will be greater than in the hole in the proportion of $\sqrt{2}$ to 1. If IG be four feet or 48 inches, and the diameter of the hole be 1 inch, 1 added to 48 will make the height from the place where the velocity is greatest to be 49 inches; and if the velocities of the descending column in the hole and that place, were truly measured by the subduplicate *ratios* of those heights, as they would be if the water descended freely and without resistance, they would be nearly equal, being as the numbers 69 and 70. And therefore, the velocity of the water in the hole is less than it would be if it was proportional to $\sqrt{IG}$, in the *ratio* of 1 to $\sqrt{2}$. This diminution of velocity can be owing to nothing but the lateral motion of the descending water, retarding its perpendicular motion downwards, and making it less than it otherwise would be, in the said *ratio* of 1 to $\sqrt{2}$. Hence, the velocity with which the water flows through the hole is very nearly equal to the velocity which a body, by falling freely and without resistance from a state of rest at I, would acquire in descending through $\frac{1}{4}$IG. For the velocity acquired in falling through $\frac{1}{4}$IG, is to the velocity acquired in falling through IG, as 1 to $\sqrt{2}$.

According to Sir Isaac Newton, a body falling *in vacuo* from a small height above the surface of the earth, will describe $193\frac{1}{3}$ inches, or $16\frac{1}{9}$ feet in one second minute of time, and will have acquired a velocity at the end of the fall, which being continued uniform, would carry it through twice that space, that is, $386\frac{2}{3}$ inches or $32\frac{2}{9}$ feet, in an equal time. But uniform velocities are as the spaces described by them in the same time, and the velocities acquired by a body falling *in vacuo* through the

Pl. 11.
Fig. 2.

the fpaces $16\frac{1}{3}$, and GI or $\frac{A^2H}{A^2-a^2}$, are in the
fubduplicate *ratios* of thofe fpaces; and therefore
$32\frac{2}{3} . V :: \sqrt{16\frac{1}{3}} . \sqrt{\frac{A^2H}{A^2-a^2}}$. Whence, $V = 8.02773$
$\sqrt{\frac{A^2H}{A^2-a^2}}$ feet, $= 96.33276 \sqrt{\frac{A^2H}{A^2-a^2}}$ inches. And
leſſening thefe meafures of the velocity of the water
flowing through the hole, the *ratio* of 1 to $\sqrt{2}$, that is,
dividing each by 1.414, we ſhall have $V = 5.6773196$
$\sqrt{\frac{A^2H}{A^2-a^2}}$ feet, $= 68.1278352 \sqrt{\frac{A^2H}{A^2-a^2}}$ inches.
Thefe are the true meafures of the velocity of the
water in its paſſage through the whole; which velocity is therefore fuch as carries it at the rate
of $5.6773196 \sqrt{\frac{A^2H}{A^2-a^2}}$ feet, or 68.1278352
$\sqrt{\frac{A^2H}{A^2-a^2}}$ inches, in a fecond minute of time.
Thefe expreſſions may be ſhortened, if A be con-
fiderably greater than a, for in all fuch cafes $\frac{A^2H}{A^2-a^2}$
will be fo nearly equal to H, that $\frac{A^2}{A^2-a^2}$ may be
fafely rejected; and then the foregoing meafures of
the velocity will become $5.6773196 \sqrt{H}$ feet, or
$68.1278352 \sqrt{H}$ inches. To ſhew the truth of this
by an example, let A be 100 fquare inches, and
a 1 fquare inch, and then $\frac{A^2H}{A^2-a^2}$ will be $\frac{10000H}{9999}$;
if H be four feet or 48 inches, $\frac{10000H}{9999}$ will be
48.0048 inches, which is only greater than 48 by
48 parts of an inch divided into 10000. The excefs is fo fmall, that it may be fafely rejected.

Another true meafure of the velocity of the water
flowing through the hole, will be had by dividing
the quantity of water difcharged, by the *area* of the
                                                    hole

hole and time of the difcharge, taken together; the quantity of water difcharged being expreffed in cubick inches, the *area* of the hole in fquare inches or parts of a fquare inch, and the time of the difcharge in feconds. Let Q denote the quantity difcharged, d the diameter of the hole, and t the time of the difcharge, and then V will be meafured by $\frac{Q}{at} = \frac{Q}{0.78539816d^2t}$ inches, which will be the fpace defcribed in one fecond of time.

This meafure is equal to the former, that is, $\frac{Q}{0.78539816d^2t} = 68.1278352\sqrt{H}$; and confequently, $Q = 53.5074764d^2t\sqrt{H}$ cubick inches; or $13555.227d^2t\sqrt{H}$ grains; becaufe a cubick inch of water weighs $253\frac{1}{3}$ grains. If W denote the weight of water difcharged, then will $W = 13555.32d^2t\sqrt{H}$ grains.

In order to know, whether the velocities of water flowing through circular holes of different diameters, when placed at the fame perpendicular diftance from the furface of the water, be all equal; what relation the velocity of water flowing through a hole, bears to the velocity of water flowing through an horizontal pipe of an equal diameter, inferted into the fide of a veffel at an equal perpendicular diftance from the furface of the water; and under what circumftances the meafure of the velocity laid down in my *Animal Œconomy* obtains; I fay, in order to know thefe things, I caufed a proper *apparatus* to be made, and from the experiments made with it, I compofed the following Tables.

TABLE

TABLE I.

t	H	d	W	w	$\frac{w}{W}$
10	4	$\frac{1}{10}$	2711	2944	1086
		$\frac{4}{10}$	43377	47040	1084
		$\frac{5}{10}$	67776	72960	1076
		$\frac{8}{10}$	173507	178560	1029
	2	$\frac{1}{10}$	1917	2087	1088
		$\frac{4}{10}$	30672	33600	1095
		$\frac{5}{10}$	47925	51840	1082
		$\frac{8}{10}$	122688	128400 ·	1046

TABLE II.

d	l	w	d	l	w	d	l	w
$\frac{2}{10}$	0	12736	$\frac{4}{10}$	0	47040	$\frac{8}{10}$	0	178560
	d	14385		d	54720		d	204720
	2d	14400		2d	56160		2d	224640
	3d	13792		3d	52800		3d	217440
	4d	13728		4d	52220		4d	212160
	5d	13663		5d	51600		5d	203520
	10d	12683		10d	47040		16d	188160
							23d	178560

The firſt Table contains, in the firſt column, under t, the time of the diſcharge in ſeconds; in the ſecond column, under H, the perpendicular heights of the water above the hole in *London* feet; in the third, the diameters of the hole in parts of an inch; in the fourth, under W, the weights of water in grains, which ought to have been diſcharged by the theory or foregoing rule; in the fifth, under w, the weights of water in grains which were diſcharged by experiment, each weight being a mean taken from five or ſix experiments; and in the ſixth column, under $\frac{w}{W}$, the *ratio* of the weight diſcharged by experiment,

periment,

periment to the weight which ought to have been difcharged by the theory.

The fecond Table confifts of three parts, and each part of three columns. The firft column of each part, contains the diameter of the pipe in parts of an inch; the fecond contains the lengths of the pipe in the terms of the diameter, beginning with the hole, which may be confidered as a pipe of an infinitely fmall length expreffed by 0; and the third column contains the weights in grains difcharged in ten feconds, each weight being a mean taken from particular experiments. The holes and pipes were all at the perpendicular diftance of four feet from the furface of the water, fo that here t was 10 feconds, and H four feet.

Table III.

d	H	l	W	w	$\frac{w}{W}$	H	W	w	$\frac{w}{W}$
$\frac{1}{10}$	2	1	2180	2180	1000	$\frac{1}{2}$	1090	982	901
		2	1541	2080	1349		770	922	1196
		3	1258	2057	1634		629	877	1393
		4	1090	1874	1719		545	762	1398
		5	980	1759	1804		490	720	1469
		6	890	1690	1899		445	665	1494
		7	824	1564	1898		412	620	1505
		8	770	1520	1972		385	585	1519
		9	727	1440	1982		363	553	1522
		10	689	1410	2045		344	525	1523
		12	629	1320	2098		314	470	1493
		14	582	1225	2102		291	430	1476
		16	545	1163	2134		272	383	1405
		18	514	1086	2113		257	350	1362
		20	487	1030	2113		243	320	1313
		24	445	866	1946		222	260	1168
		25	436	860	1972		218	253	1160
		28	412	844	2048		206	230	1116
		32	385	758	1967		192	202	1048
		36	363	659	1814		181	185	1018
		48	314	509	1618				
		60	281	421	1496				
		72	257	345	1342				

TABLE

TABLE IV.

d	H	l	W	w	$\frac{w}{W}$	H	W	w	$\frac{w}{W}$
$\frac{2}{10}$	2	1	12332	10040	814	$\frac{1}{2}$	6166	5018	814
		2	8720	9270	1063		4360	4630	1062
		3	7120	8820	1238		3560	4400	1235
		4	6166	8570	1389		3083	4270	1385
		5	5515	8240	1494		2758	4040	1465
		6	5034	7840	1557		2517	3880	1541
		7	4661	7580	1626		2330	3766	1616
		8	4360	7360	1688		2180	3668	1682
		9	4111	7150	1739		2055	3570	1737
		10	3900	6950	1782		1950	3414	1751
		16	3083	5776	1873		1541	2955	1918
		25	2466	4785	1940		1233	2460	1995
		36	2055	4048	1970		1027	2120	2064
		49	1762	3480	1975		881	1730	1963
		64	1542	3250	2108		771	1326	1720
		81	1370	3062	2235		685	1120	1635
		97	1252	2700	2156		626	940	1502

The third and fourth Tables confift each of two parts correfponding to different perpendicular heights of the water in the veffel, and different diameters of the pipes. In both Tables, H denotes the perpendicular height of the water in the veffel above the pipe in feet ; l the length of the pipe in inches ; W the weight in grains which ought to be difcharged by the firft *Propofition* of my *Animal Œconomy* ; w the weight in grains which was difcharged by experiment ; and $\frac{w}{W}$ the *ratio* of the weight difcharged by experiment to the weight which ought to have been difcharged by that *Propofition*. The diameter of all the pipes in the third Table was $\frac{1}{10}$ of an inch, and of all the pipes in the fourth Table $\frac{2}{10}$

of

of an inch. And the time of the difcharge was
10 feconds in all the experiments of both Tables.

The quantity or weight of water which ought to
be difcharged by the firft *Propofition* of the *Animal
Œconomy*, may be thus found. I there proved,
that the velocity of water flowing thro' a pipe, is
as $\sqrt{\frac{F}{dl}}$. But if the force which can generate the
motion of water flowing through a pipe lying parallel to the horizon, be equal to the force which
can generate the motion of water flowing through
a hole of an equal diameter with the pipe, when
placed at an equal perpendicular diftance from the
furface of the water; F, by *Cor.* 11. of this *Problem*,
will be as $2d^2H$, on fuppofition that the *area* of the
hole is extreamly fmall in comparifon of the *area*
of the furface of the water. And therefore the velocity of water flowing through a pipe lying parallel
to the horizon, is as $\sqrt{\frac{2dH}{l}}$. The weight of water difcharged, is as the orifice of the pipe, the time
of the difcharge, and velocity, taken together;
that is, as $d^2t\sqrt{\frac{2dH}{l}}$. And therefore, W is as
$d^2t\sqrt{\frac{2dH}{l}}$.

A pipe of $\frac{1}{10}$ of an inch in diameter, and 1 inch
in length, difcharged 2180 grains of water in 10
feconds, when it was inferted into the fide of the
veffel at the perpendicular diftance of two feet from
the furface. In this cafe therefore, d, t, H, l, were
0.1, 10, 2, 1; and $d^2t\frac{2dH}{l}$ was equal to 0.06326.
Hence we may find W in other cafes by this analogy; $2180 : 0.06325 :: W : d^2t\sqrt{\frac{2dH}{l}}$; whence
$W = 48746.3d^2t\sqrt{\frac{dH}{l}}$ grains.

In

In the firſt part of the third Table, W is $\frac{2180}{\sqrt{l}}$,

and in the firſt part of the fourth Table, $\frac{12332}{\sqrt{l}}$;

and W in the ſecond part of each Table is one half
of W in the firſt part.

OBSERVATIONS *on the* TABLES.

OBS. I. By the firſt Table the diſcharges by experiment are nearly proportional to the diſcharges
by the theory, that is, w is nearly proportional to
W, or $\frac{w}{W}$ is nearly the ſame, whatever be the diameter of the hole, provided the time of the diſcharge, and the perpendicular height of the water
in the veſſel abóve the hole, be given. The diſcharges by experiment were all ſomething larger
than the diſcharges by the theory, which might be
partly owing to the pouring in of the water at the
top of the veſſel, in order to keep the veſſel conſtantly full during the time of the diſcharge; for
the pouring, tho' it was done gently, might a little
increaſe the velocity wherewith the water ran out
of the hole.

OBS. II. By the ſecond Table, the weight of
water diſcharged, and conſequently the velocity, increaſes from the hole till the length of the pipe becomes equal to about twice its diameter, that is,
till l becomes equal to about 2d, and is greater there
than at any other length of the pipe. The greateſt
velocities in theſe pipes in proportion to the velocities through their reſpective holes, are as the numbers 1130, 1258 to 1000.

OBS. III. From the length of twice the diameter, that is from the length 2d, the velocity leſſens
continually on increaſing the length of the pipe, and
becomes equal to the velocity through the hole
when the length of the pipe becomes equal to about
22.3657d√d inches. For, by the ſecond Table,

the

the velocities of the water flowing through the
pipes, were nearly equal to the velocities through
their refpective holes, when the lengths of the pipes
were 10d, 16d and 23d, that is 2 inches, 6.4 inches,
and 18.4 inches. But 2, 6.4, and 18.4, are nearly as
1, 2.8, and 8, the fefquiplicate *ratios* of 1, 2 and 4,
and 1, 2 and 4, are as the diameters $\frac{2}{10}$, $\frac{4}{10}$ and $\frac{8}{10}$.
And therefore the velocities of the water flowing
through the pipes, were nearly equal to the velocities through their refpective holes, when the lengths
of the pipes were in the fefquiplicate *ratios* of their
diameters. The diameter of the fmalleft pipe be-
ing $\frac{2}{10}$ of an inch, d$\sqrt{}$d is 0.0894; and if d be of
any other magnitude, and l be the length of a pipe
of that diameter through which the water flows
with a velocity equal to that with which it flows
through its correfponding hole, we fhall have this
proportion; as 2 is to 0.0894, fo is l to d$\sqrt{}$d,
whence l $=$ 22.3657d$\sqrt{}$d.

Obs. IV. By the third and fourth Tables, the
quantity of water difcharged by experiment in proportion to the quantity which ought to have been
difcharged by the theory, that is $\frac{w}{W}$, increafes gradually till the pipe comes to be of a certain length,
and after that it decreafes gradually on increafing
the length of the pipe. In the two parts of the
third Table this *ratio* was greateft, when the lengths
of the pipes in inches were about 20 and 10, and it
was greateft in the two parts of the fourth Table,
when the lengths of the pipes were 81 and 36. But
from the courfe of the numbers exprefling $\frac{w}{W}$ in
the fecond part of the fourth Table, I think this
ratio would have been greater in a pipe of 40 inches
in length, than in the one I ufed of 36, and therefore fhall fuppofe that it would have been greateft
at the lengths of 81 and 40. Confequently, putting x for the length of the pipe in inches, at which

this

this *ratio* is greateſt, x will be as $\sqrt{H}$ when d is given, and as d^2 when H is given, and when nei-ther d nor H is given, as $d^2\sqrt{H}$. Hence we may form a rule for finding the length of the pipe, at which this *ratio* ſhall be a *maximum* ; for it was a *maximum* in a pipe of $\frac{1}{10}$th of an inch in diameter, when its length was 20 inches, and the perpendicular height of the water in the veſſel two feet. In this caſe therefore x, d and H, are 20, 0.1, and 2, and $d^2\sqrt{H}$ is 0.01414; and in other caſes, x may be found by this analogy ; as 20 is to 0.01414, ſo is x to $d^2\sqrt{H}$; whence x is equal to $1414d^2\sqrt{H}$. To ſee whether this rule be univerſal, and obtain in pipes of greater diameters, and at greater diſtances from the ſurface of the water, I ſhall ſuppoſe d and H to be 0.5 and 3, as in our Author's Table p. 227, and then $1414d^2\sqrt{H}$ will be about 600 inches or 50 feet, which length is twice as great as it was in reality ; for the *ratio* was a *maximum* by that Table, when the length of the pipe was 25 feet ; ſo that the value of x here determined ſeems to obtain only in pipes of ſmall diame-ters.

Obs. V. By the third and fourth Tables, the quantity diſcharged by experiment in proportion to the quantity which ought to have been diſ-charged by the theory, that is $\frac{w}{W}$, does not differ much in pipes whoſe lengths are within certain li-mits. $\frac{w}{W}$, in the pipes, whoſe lengths were 6 and 32 in the firſt part of the third Table, is leſs than in the pipe where this *ratio* is a *maximum*, in the proportions of 100 to 112 and 110, and the difference of $\frac{w}{W}$ and the *maximum* is ſtill leſs in pipes of all other lengths between 6 and 32 ; ſo that in this part of the Table, 6 and 32 are the limits, at and within which there is a near agreement be-

tween

tween theory and experiment. $\frac{w}{W}$ in the pipes whofe lengths are 4 and 16 in the fecond part of this Table, is lefs than in the pipe where this *ratio* is a *maximum*, in the proportion of about 100 to 109 and 108, and it is ftill lefs in pipes of all other lengths within thefe limits. And $\frac{w}{W}$ in the pipes whofe lengths are 9 and 64 in the fecond part of the fourth Table, for the pipes were not carried to fuch lengths as were neceffary to fettle the limits in the firft part, is lefs than the *maximum* in the proportion of 100 to 118 and 120; and it is ftill lefs in pipes of all other lengths within thefe limits.

OBS. VI. By the third and fourth Tables, the quantity of water difcharged by experiment always exceeds the quantity which ought to be difcharged by the theory; it was near double within the limits of the firft part of the third Table and fecond part of the fourth, and greater in the fecond part of the third Table in the proportion of 5713 to 3858. If we fuppofe it to be double within the limits, in pipes of all lengths, then will w be equal to 2W, or to $97492.6 d^2 t \sqrt{\frac{dH}{l}}$ grains, W being equal to $48746.3 d^2 t \sqrt{\frac{dH}{l}}$ grains, as was fhewn above.

Of the Foci *of* Optick Glaffes.

PROB. VI. *If the diftance of an object from a double convex lens whofe furfaces are fpherical; if the radii of both the fpherical furfaces, the thicknefs of the lens, and the fines of incidence and refraction, be all given; thence to determine the diftance behind the lens of the principal focus or concourfe of the rays iffuing from the object and falling perpendicularly, or very nearly fo, on that furface of the lens which is turned towards the object.*

Let MN be a *lens*, E and e the centers of its Pl. 11 fpherical furfaces MCN and MDN, Q an object Fig. 5 placed

placed directly before the *lens*, Qq a line drawn
from the object perpendicular to the ſurfaces of the
lens, and conſequently paſſing through the centers
e and E ; let the point A be indefinitely near to C,
in which caſe QA and QC may be looked upon as
equal ; let q be the *focus* or concourſe of the rays
QA and QC after the firſt refraction by the ſurface
MCN, and z their *focus* or concourſe after the ſecond refraction by the ſurface MDN. Put D for
QC the diſtance of the object from the *lens*, r for
the *radius* CE, ρ for the *radius* eD, x for Dq, the
diſtance of the *focus* behind the *lens*, after the firſt
refraction, and z for Dz its diſtance behind the
lens after the ſecond refraction ; and laſtly, let I and
R denote the ſines of incidence and refraction of the
rays paſſing out of air or any other *medium* into the
firſt ſurface MCN, and conſequently R and I the
ſines of incidence and refraction in their paſſage out
of the ſecond ſurface MDN into air or that other
medium.

Pl. 11.
Fig. 4.
To determine z, we muſt firſt determine the
meaſure of x in known terms, to do which draw
AF perpendicular to Qq, EI perpendicular to
QAI the incident ray produced, and ER perpendicular to the refracted ray Aq ; and then, from the
ſimilarity of the two triangles QAF and QIE, and
alſo of the triangles qAF and qER, and from QA
being equal to QC, and qA equal to qC, we ſhall
have AF equal to $\dfrac{QC \times EI}{QE}$, or, in ſymbols, to $\dfrac{DI}{D+r}$,
by the two firſt triangles, and by the two laſt tri-
angles, equal to $\dfrac{Cq \times ER}{Eq}$, or, in ſymbols, to $\dfrac{Rx}{x-r}$.
Conſequently, $\dfrac{DI}{D+r}$, is equal to $\dfrac{Rx}{x-r}$, and x $=$
$\dfrac{DIr}{DI-DR-Rr}$.

Pl. 11.
Fig. 5.
Having found the meaſure of x or Cq in known
terms, z or Cz may be thus determined. For that
meaſure,

meaſure, that is for $\dfrac{DIr}{DI-DR-Rr}$, put A, to za

and qa produced draw the perpendiculars eI and eR, and draw am perpendicular to Qq. And then, from the ſimilarity of the triangles qam and qRe, and alſo of the triangles zam and zIe, and from qa being equal to qD, and za equal to zD a m will be equal to $\dfrac{eR \times qD}{qe}$ from the firſt triangles, and equal to $\dfrac{eI \times zD}{ze}$ from the ſecond. eR is the ſine of inci-dence of the ray Aa falling on the ſecond ſurface MDN, and eI the ſine of its refraction; and there-fore eR will be I, and eI will be R. qD is equal to qC — CD $=$ A — t, putting t for CD the thickneſs of the *lens*; and qe is equal to qC $+$ eC $=$ qC $+$ eD — CD $=$ A $+$ ℮ — t; conſequently $\dfrac{R \times \overline{A-t}}{A+℮-t} = \dfrac{Iz}{z+℮}$; and from this equation, A $=$ $\dfrac{I℮z + Rtz + Rt℮ - Itz}{Rz + R℮ - Iz}$. But A denotes $\dfrac{DIr}{DI - DR - Rr}$.

And therefore $\dfrac{I℮z + Rtz + Rt℮ - Itz}{Rz + R℮ - Iz} = \dfrac{DIr}{DI-DR-Rr}$.

By clearing z in this equation, we ſhall have z $=$

$$\dfrac{DIR r℮ + RR r℮t + DRR℮t - DIR℮t}{DII℮ + 2DIRt - DIIt - DIR℮ - DRRt - IR r℮ - RRrt + IRrt - DIRr + DIIR}.$$

To give this equation a more ſimple form, di-vide both numerator and denominator by I — R; and then the numerator will become $\dfrac{R}{I-R} \times IDr℮$

$+ \dfrac{R}{I-R} \times Rr℮t$ — DR℮t, or by putting B inſtead

of $\dfrac{R}{I-R}$, BIDr℮ $+$ BRr℮t — DR℮t, and the deno-minator will become IDr $+$ ID℮ — IDt $+$ RDt $+$ Rrt — BIr℮; and the equation will be reduced to another form, and ſtand thus;

$$z = \dfrac{BIDr℮ + BRr℮t - DR℮t}{IDr + ID℮ - IDt + RDt + Rrt - BIr℮}.$$

This

This is Dr. HALLEY's univerſal *Theorem* for finding the principal *focus* of rays falling diverging on a double *convex lens*, publiſhed in the *Philoſophical Tranſactions*.

If the rays inſtead of falling diverging, fall parallel on a double convex *lens*, as they will nearly do, when the object is at an immenſe diſtance from the *lens*, D in this caſe may be conſidered as infinite ; and conſequently, all the terms in which D is not found, may be thrown out of the equation, and then

$$z = \frac{BIDr\varrho - DR\varrho t}{IDr + ID\varrho - IDt + RDt} = \frac{BIr\varrho - R\varrho t}{Ir + I\varrho - It + Rt}.$$

And laſtly, if the rays fall converging on a double convex *lens*, the ſigns of all the terms in which D is found muſt be changed ; for when the rays fall converging, the point behind the *lens* to which they tend at their incidence, muſt be conſidered as the place of the object, which, from its being differently ſituated with reſpect to the *lens* from what it is when the rays fall diverging, requires the ſigns of all the terms in which D is found to be changed, which being done, we ſhall have

$$z = \frac{DR\varrho t + BRr\varrho t - BIDr\varrho}{IDt - IDr - ID\varrho - RDt + Rrt - BIr\varrho}.$$

Theſe are the three general *Theorems* for finding the principal *focus* of rays falling, diverging, parallel, or converging on a double convex *lens*.

If the *lens* be made of glaſs, as *lenſes* uſually are, and the object be placed in air, then, ſince the ſine of incidence of a ray paſſing out of air into glaſs, is to the ſine of refraction, as 3 to 2, I, R and B will be 3, 2, and 2 ; and the foregoing general *Theorems* for finding the *foci* of rays falling diverging, parallel, and converging, on a double convex glaſs, will be

$$z = \frac{6Dr\varrho + 4r\varrho t - 2Dr\varrho t}{3Dr + 3D\varrho - 3Dt + 2Dt + 2rt - 6r\varrho},$$

$$z = \frac{6r\varrho - 2\varrho t}{3r + 3\varrho - 3t + 2t}, \text{ and}$$

$$z =$$

$$z = \frac{2Dpt + 4rpt - 6Dr\rho}{3Dt - 3Dr - 3D\rho - 2Dt + 2rt - 6r\rho}.$$

And if the *radii* be equal, and the thickneſs of the glaſs be neglected, or conſidered as 0, then will theſe *Theorems* ſtand thus, $z = \frac{Dr}{D-r}$, $z = r$, and

$$z = \frac{-Dr}{-D-r}.$$

If the *lens* be a double concave glaſs, the *radii* of whoſe two ſpherical ſurfaces are equal, and if the thickneſs of the *lens* be conſidered as 0, the *radii* will lie on different ſides of the *lens* with reſpect to the object from what they did before, and conſequently, the ſigns of the *radii* muſt be changed ; and then the laſt *Theorems*, in which the *radii* were ſuppoſed to be equal, and the thickneſs of the glaſs was neglected or conſidered as 0, will ſtand thus; $z = \frac{-Dr}{D+R}$, $z = -r$, and $z = \frac{Dr}{r-D}$. By theſe *Theorems*, z is always negative when the rays fall upon the double concave, diverging, or parallel, and when they fall converging it is negative when D is greater than r. When z is negative, the *focus* falls on the ſame ſide of the glaſs with the object, contrary to what it does in all caſes of a double convex *lens*, excepting that of diverging rays, when the diſtance of the object is leſs than the *radius*, or D is leſs than r. For in that caſe, z, which is equal to $\frac{Dr}{D-r}$, will be negative,

By this *Problem* we may determine how far a radiating point muſt be diſtant from the eye, to have the principal *focus* of the rays iſſuing from it placed in the *retina*, on ſuppoſition that the coats and humours of the eye are unchangeable as to their figures, magnitudes, and denſities.

Let ABGz repreſent a human eye, in which ABG Pl. 11. is the *cornea*, AMCNGB the cavity containing the Fig. 6. aqueous humour, MCND the cryſtalline humour, and

and AMDNGz the cavity containing the vitreous humour. According to Doctor Jurin, the *radii* of the ſpherical ſurfaces of the *cornea* and of the cryſtalline humour, that is, of the ſpherical ſurfaces ABG, MCN, and MDN, are in 10th parts of an inch, 3.3294, 3.3081, and 3.5056; and the diſtance of the *cornea* from the anterior part of the cryſtalline, the thickneſs of the cryſtalline, the diſtance of the poſterior part of the cryſtalline from the *retina*, and the diſtance of the *cornea* from the *retina*, are in the ſame parts of an inch, 1.0358, 1.8525, 6.2617 and 9.15. Let Q be the radiating point, q the principal *focus* of the rays by the firſt refraction of the aqueous humour, by virtue of which refraction they fall converging on the cryſtalline, and let z be their *focus* after their refractions by the cryſtalline and vitreous humours. By taking the ſpecifick gravities of the humours of the eye, I have found that the ſpecifick gravities of the aqueous and vitreous humours are very nearly equal, and each much the ſame with that of water; and that the ſpecifick gravity of the cryſtalline is greater than the ſpecifick gravity of water, in the proportion of about 11 to 10. For the mean ſpecifick gravities of five cryſtalline humours of oxen's eyes, and of three cryſtalline humours of ſheep's eyes, were 11134 and 11033, the ſpecifick gravity of water being 10000, and the mean of theſe two means, is 11083, which I ſhall ſuppoſe to be the ſpecifick gravity of the cryſtalline humour of a human eye. But the refractive power of the cryſtalline is very nearly proportional to it's denſity, and the ſine of incidence of rays paſſing out of the aqueous humour into the cryſtalline, is to the ſine of refraction, very nearly as 21 to 20, as I ſhall ſhew in the *Scholium*. And conſequently, I will be 21, and R will be 20. From theſe meaſures I now proceed to determine the diſtance of a radiating point from the *cornea*, that is, the diſtance of Q from B, ſo as that the

focus

focus of the rays, issuing from it and falling diverging on the *cornea*, may by the refractive powers of the aqueous, crystalline, and vitreous humours, be placed in the *retina* at z. By the refraction of the aqueous humour, the rays fall on the crystalline with such a degree of convergence as would make them unite at q. In the universal *Theorems* therefore for finding the principal *focus* of rays falling converging on a double convex *lens*, Cq is D, Dz equal to 6.2617 is z, the *radius* of MCN is r, the *radius* of MDN is ϱ, CD the thickness of the crystalline is t, and I and R are 21 and 20. And by clearing D in that *Theorem*, we shall have D $=$

$$\frac{BIr\varrho z + BRr\varrho t - Rrtz}{BIr\varrho + Itz - Irz - I\varrho z - R\varrho t - Rtz} = 10.3102 =$$

Cq. And Cq+BC = 11.346 = Bq.

In the *Theorem* for finding x, Bq is x, QB is D, I is 4, R 3, the sine of incidence of rays passing out of air into water or into the aqueous humour, being to the sine of refraction, as 4 to 3, and the *radius* of the *cornea* is 3.3294 10th parts of an inch; consequently, D is 57.48, that is, about 5 inches and 3 quarters. So that supposing the eye to be unchangeable, a radiating point placed at the distance of $5\frac{3}{4}$ inches from it, will have its image placed in the *retina*.

SCHOLIUM.

" Let AB represent the refracting plane surface Pl. 11.
" of any body, and IC a ray incident obliquely on Fig. 7.
" the body at C, so that the angle ACI may be
" infinitely little, and let CR be the refracted ray.
" From a given point B perpendicular to the re-
" fracting surface erect BR meeting the refracted
" ray CR in R, and if CR represent the motion of the
" refracted ray, and this motion be distinguished in-
" to two motions CB and BR, whereof CB is parallel
" to the refracting plane, and BR perpendicular to
" it : CB shall represent the motion of the incident
" ray

" ray, and BR the motion generated by the refrac-
" tion." NEWT. *Opt. Prop.* 10. p. 245, 246. CBR is
equal to the angle of incidence, and CRB is equal to
the angle of refraction; conſequently, if R be made
the center, and a circle be ſuppoſed to be drawn with
the *radius* CR, CR will be the ſine of the angle of
incidence, and CB the ſine of the angle of refraction; and, putting I and R for thoſe ſines, we ſhall
have this analogy, I . R : : CR . CB. Hence
$\frac{I^2 - R^2}{R^2} = \frac{CR^2 - CB^2}{CB^2}$, or $\frac{I^2}{R^2} - 1 = \frac{BR^2}{CB^2}$. But the
motion of the ray at its incidence repreſented by
CB, is given; and therefore, $\frac{I^2}{R^2} - 1$ is as BR^2.
But by the aforeſaid propoſition BR^2 expreſſes the
refractive force, and is nearly as the denſity of the
body; as Sir I. NEWTON found, by computing BR^2
from the ſines I and R in ſeveral bodies, and then
comparing it with their reſpective denſities. And
conſequently, putting D for the denſity of the
body, $\frac{I^2}{R^2} - 1$ is as D, and $\frac{I}{R}$ as $\sqrt{D + 1}$. In paſſing out of air into water $\frac{I}{R}$ is $\frac{4}{3}$, and, the denſity
of water being 10000, $\sqrt{D + 1}$ is 10004: And in
paſſing out of air into the cryſtalline, whoſe denſity
is to that of water as 11083 to 10000, $\sqrt{D + 1}$ is
10528. Therefore in paſſing out of air into the
cryſtalline $\frac{I}{R}$ will be $\frac{7}{5}$; for 10004 . 10528 : : $\frac{4}{3}$;
$\frac{42112}{30012} = \frac{7}{5}$ very nearly. $\frac{I}{R}$ in paſſing out of the
aqueous humour into the cryſtalline, will be com-
pounded of the *ratio* $\frac{3}{4}$ and $\frac{7}{5}$ by the ſecond *Theo-*
rem of the *Opticks*, p. 113; and therefore $\frac{I}{R}$ will
be equal to $\frac{21}{20}$; or I will be to R, as 21 to 20.

F I N I S.

PLATES

The original plates were inserted within the text pages at the place indicated on each of them.

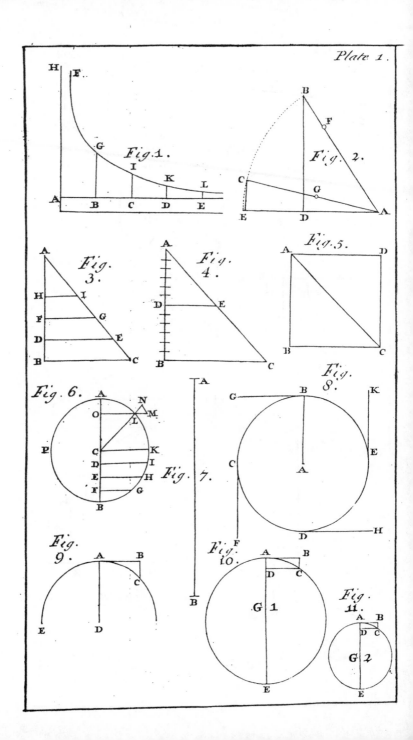

Plate 1.

Plate 2

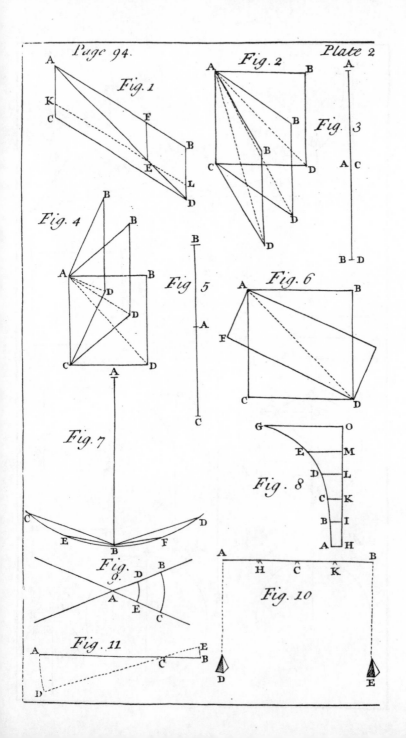

Plate 3.

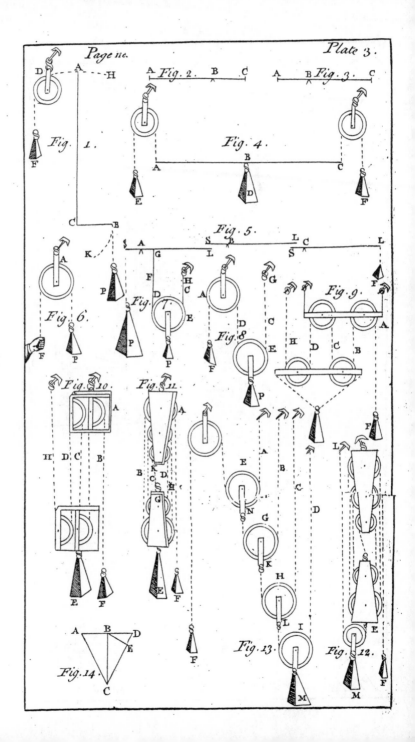

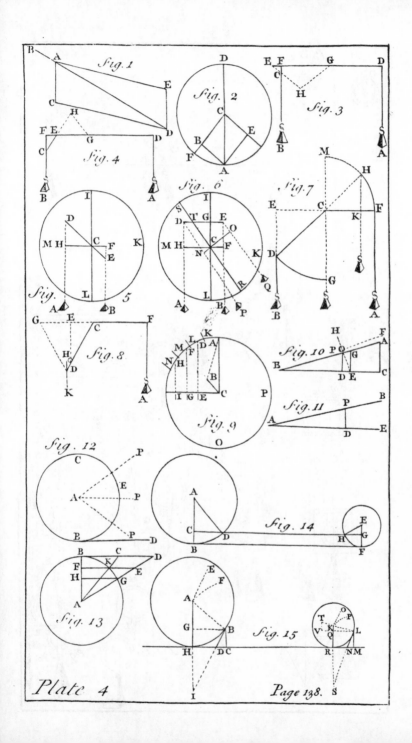

Plate 4

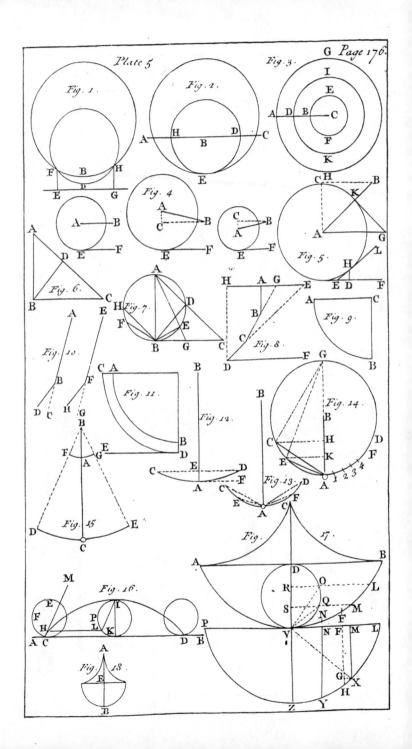

Plate 5

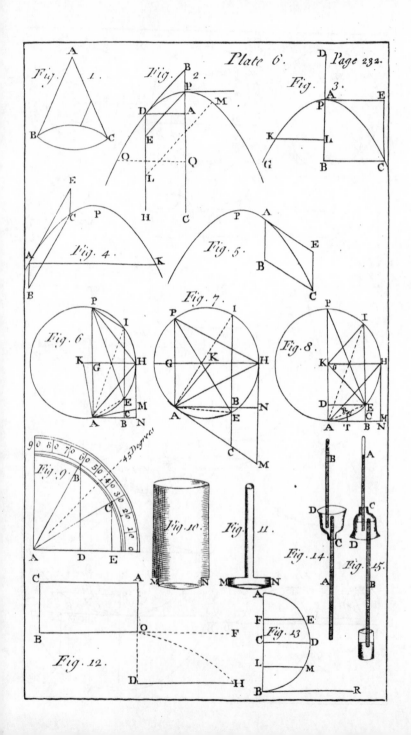

Plate 6. Page 232.

Fig. 1.
Fig. 2.
Fig. 3.
Fig. 4.
Fig. 5.
Fig. 6.
Fig. 7.
Fig. 8.
Fig. 9.
Fig. 10.
Fig. 11.
Fig. 12.
Fig. 13.
Fig. 14.
Fig. 15.

90 80 70 60 50 40 30 20 10 45 Degrees

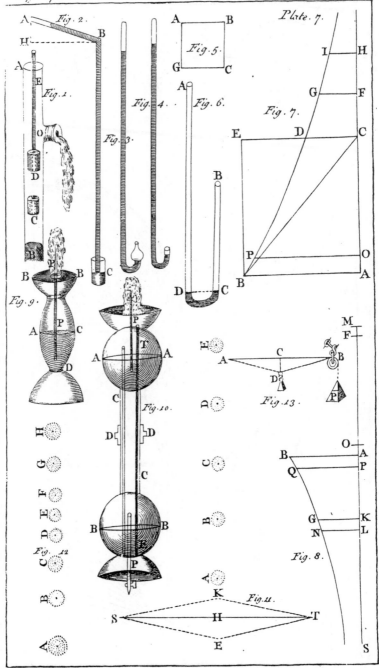

Plate 7.

Fig. 1.
Fig. 2.
Fig. 3.
Fig. 4.
Fig. 5.
Fig. 6.
Fig. 7.
Fig. 8.
Fig. 9.
Fig. 10.
Fig. 11.
Fig. 12.
Fig. 13.

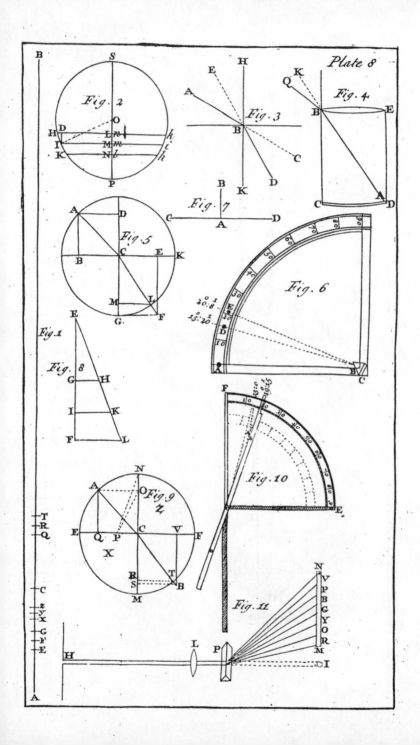

Plate 8

Fig. 1
Fig. 2
Fig. 3
Fig. 4
Fig. 5
Fig. 6
Fig. 7
Fig. 8
Fig. 9
Fig. 10
Fig. 11

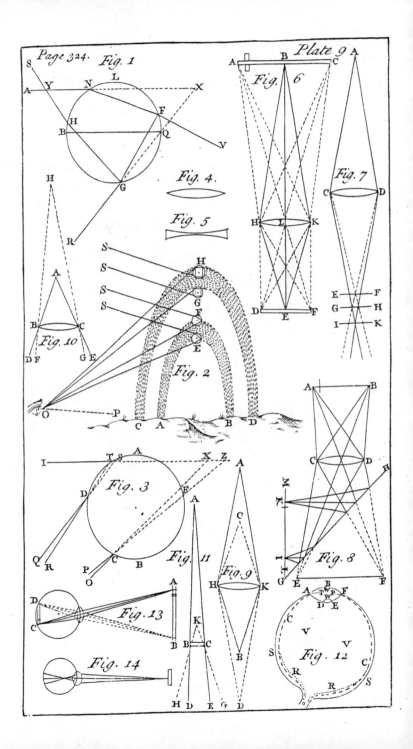

Fig. 1

Fig. 4.

Fig. 5

Fig. 2

Plate 9

Fig. 6

Fig. 7

Fig. 10

Fig. 3

Fig. 11

Fig. 9

Fig. 8

Fig. 13

Fig. 14

Fig. 12

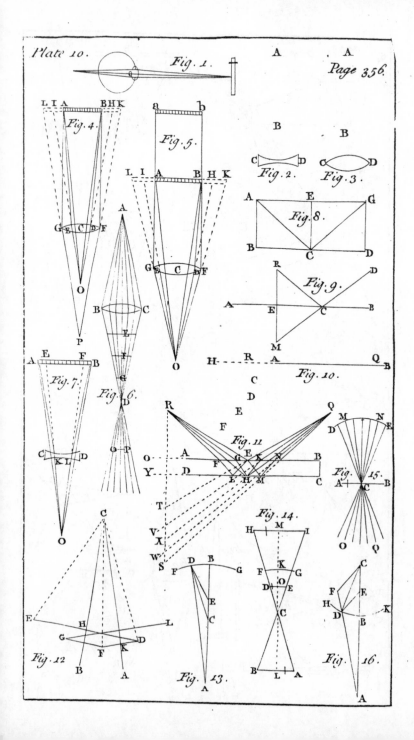

Plate 10.   Fig. 1.   Page 356.

Plate 11.

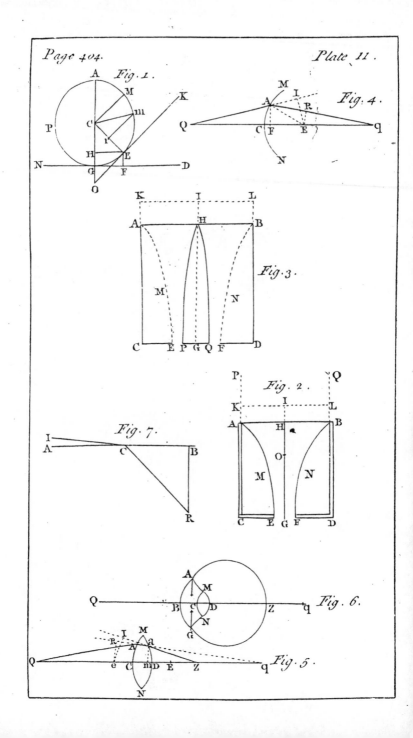

Fig. 1.

Fig. 4.

Fig. 3.

Fig. 2.

Fig. 7.

Fig. 6.

Fig. 5.